Methane and Alkane Conversion Chemistry

Methane and Alkane Conversion Chemistry

Edited by

Madan M. Bhasin

Union Carbide Corporation
South Charleston, West Virginia

and

D. W. Slocum

Western Kentucky University
Bowling Green, Kentucky

Plenum Press • New York and London

Library of Congress Cataloging in Publication Data

Methane and alkane conversion chemistry / edited by Madan M. Bhasin and D. W. Slocum.
    p.   cm.
    "Proceedings of American Chemical Society Symposium on Methane and Alkane Conversion Chemistry, held March 13–17, 1994, in San Diego, California"—T.p. verso.
    Includes bibliographical references and index.
    ISBN 0-306-45212-X
    1. Synthetic fuels—Congresses 2. Methane—Congresses. 3. Alkanes—Congresses. 4. Oxidation—Congresses. I. Bhasin, Madan M. II. Slocum, D. W. (Donald Warren), 1933–   .
III. American Chemical Society Symposium on Methane and Alkane Conversion Chemistry (1994: San Diego, Calif.)
TP360.M465   1996   95-51190
661.8′ 14—dc20   CIP

Proceedings of American Chemical Society Symposium on Methane and Alkane Conversion Chemistry, held March 13–17, 1994, in San Diego, California

ISBN 0-306-45212-X

© 1995 Plenum Press, New York
A Division of Plenum Publishing Corporation
233 Spring Street, New York, N. Y. 10013

Printed in the United States of America

PREFACE

Natural gas, an abundant natural energy and chemical resource, is underutilized. Its inherent high energy content is compromised by its volatility. Storage and transportation problems abound for liquified natural gas. Several of the drawbacks of the utilization of natural gas, particularly its high volatility, could be offset by development of an economical and efficient process for coupling and/or further homologation of its principal component, methane. Alternatively, other conversion strategies such as partial oxidation to methanol and syngas, to oxygenates or conversion to such products via the intermediacy of chlorides should also be considered.

Given the energy-intensive regimes necessary for the likely activation of methane, it was inevitable that researchers would turn to the use of heterogeneous catalysts. Heterogeneous catalysis is now a relatively mature discipline with numerous and diverse reactions being explored alongside informative studies on surface characterization, mechanism, and theory. Relationships to important related areas such as homogeneous catalysis, organometallic chemistry, and inorganic chemistry have become firmly established within this discipline.

The field of methane and alkane activation is now over ten years old. The first decade of investigation produced results plagued by low yields and low–moderate conversions with well-articulated mechanistic limitations. As we begin the second decade of inquiry, novel strategies have brought increasing yields and conversions to such products as ethane, ethylene, methanol, and formaldehyde. These new approaches utilize separation of products via membranes or adsorbents. Moreover, additional mechanistic insight has been forthcoming from theoretical and computational examination as well as experimental investigation.

Catalysis itself is a strategic resource whose importance and remarkable capabilities have been often demonstrated yet whose potential cannot be overestimated. As we progress toward a chemical industry where process improvements for both bulk, commodity, and specialty chemicals will increasingly be in demand, new and less expensive sources of raw material must be sought. Natural gas is an obvious candidate source. Therefore, methods are needed to activate methane for conversion to higher hydrocarbons, particularly to useful olefins, the most important chemical industry building blocks, and to fuels.

The symposium from which this book originates was developed to further these aspirations and concerns. Organization of the symposium was under the auspices of the Catalysis and Surface Science Secretariat of the American Chemical Society, such organization constituting one of the major undertakings of the Secretariat. The symposium took place during the 207th National Meeting of the American Chemical Society which was held in San Diego, March 13–17, 1994. Two ACS Divisions, Petroleum Chemistry and Industrial and Engineering Chemistry, co-sponsored the four-day affair.

Investigations of methane and alkane activation are taking place throughout the world. International representatives from Japan, China, Australia, India, Russia, and several European countries made this symposium even more interesting and valuable to all participants. Contributions by overseas researchers received financial support from the Petroleum Research Fund, Norton Chemical Process Products Corporation, and Union Carbide Corporation. Thus, the observations described in this monograph represent the progress of a broad cross-section of the research going on throughout the world.

We, the editors, are most appreciative of the support that we received. We also would like to express our sincerest thanks to our co-organizers, Professor Jack H. Lunsford (Texas A&M University) and Dr. Jon G. McCarty (SRI International).

M. M. Bhasin
D. W. Slocum

CONTENTS

GAS CONVERSION: METHANE AND ALKANE ACTIVATION CHEMISTRY

MECHANISM AND MODELING OF METHANE-RICH OXIDATION

METHANE TO OXYGENATES AND CHEMICALS

GAS CONVERSION: METHANE AND ALKANE ACTIVATION CHEMISTRY

OXIDATIVE COUPLING OF METHANE — A PROGRESS REPORT

M.M. Bhasin and K.D. Campbell

Union Carbide Corporation
Industrial Chemicals Division, Research and Development
South Charleston, West Virginia 25303 (USA)

INTRODUCTION

A number of different schemes have been proposed to directly convert methane (from abundant natural gas) into useful products, i.e., ethylene, propylene, as the chemical industry building blocks, and liquid fuels for transportation since the publication of early Union Carbide Corporation (UCC) work in 1982 [1]. At the heart of these schemes lies the ability to convert methane in high selectivity to ethylene and ethane. Though homogeneous catalysts, biological as well as biomimetic, and heterogeneous catalyst systems have been extensively studied in recent years, this paper will focus entirely on the heterogeneous catalytic route. Heterogeneous catalysis is

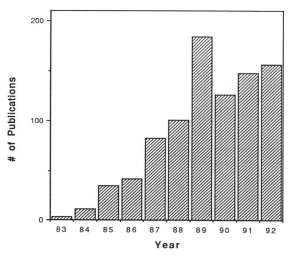

Figure 1. Number of publications/year in the area of heterogeneous oxidative coupling of methane. Number represents those published in English.

considered to be the closest route for potential commerical exploitation [2]. Since 1982, the pace of publications in the heterogeneous oxidative coupling route has accelerated in an exponential manner (Figure 1).

As of the latest count in early 1993, the total number of publications and patents was >950 with over 160 U.S. Patents. In this paper, some of the major advances in the chemistry of methane coupling will be highlighted along with the challenges that lie ahead to make this simple and highly attractive reaction into a commerically viable process.

DISCUSSION

Recent advances in the oxidative coupling of methane will be discussed under the following key subject areas:
1. Homogeneous Reaction
2. Catalysts and Promoters - Cofeed and Redox Mode
3. Mechanism:
   (a) Heterogeneous/Homogeneous versus Heterogeneous;
   (b) Rate Determining Step
   (c) Heterogeneous versus Homogeneous C-H Bond Breakage
4. Economics
5. Future Directions

1. Homogeneous Reaction

Since the initiation step in homogeneous gas phase oxidation of methane is the formation of methyl radicals, the extent and importance of the uncatalyzed homogeneous oxidation of methane during heterogeneous catalytic oxidative coupling studies must be considered. Work at UCC in 1988 examined the extent and selectivity of the homogeneous oxidation of methane in quartz reactors for a temperature range of 650-850 °C and 1-2 atmospheres of pressure. A summary of the results obtained at 850 °C are presented in Table 1.

Table 1. Homogeneous Gas Phase Results at 850 °C

P (psig)	CH_4/O_2 (molar)	R.T.* (sec.)	CH_4 Conv. (%)	O_2 Conv. (%)	C_2 Sel. (%)	C_2 Yield (%)	=/- (molar)
5	2	4.8	35	81	25	8.6	4.1
15	2	14.3	38	94	15	5.7	10
5	15	4.9	3.1	29	59	1.8	1.4
15	14	4.8	4.4	54	53	2.4	2.1

*: R.T. is residence time in heated region.

High C_2 selectivities (51-59%) were obtainable for the homogeneous oxidation; however, the high selectivities occurred only for low methane conversions (<5%). For higher methane conversions (>35%), the best C_2 selectivity observed was only 25%.

High ethylene to ethane ratios were obtained for runs at high temperature (850 °C) and long residence time (14-15 seconds). The highest C_2 yield observed in the experimental set was 8.6%. It is felt that much higher C_2 yields will be difficult, if not impossible, to obtain from the homogeneous oxidation reaction. Literature reports agree well with these results. For example, Wolf and Lane [3] reported that at one atmosphere significant homogeneous gas phase oxidative coupling of methane occurred under certain experimental conditions. For a methane conversion of 2%, a hydrocarbon (C_{2+}) selectivity of 65% was reported. When the methane conversion was increased to 32%, a C_{2+} selectivity of 30% resulted (C_{2+} yield = 9.6%). Yates and Zlotin [4] studied methane coupling over lithium/magnesium oxide (Li/MgO) catalyst at atmospheric pressure and reported the formation of significant quantities of ethane and ethylene in the absence of a catalyst. They concluded that Li/MgO was mainly a combustion catalyst that produced more C_2's than the homogeneous reaction but at the cost of converting carbon monoxide to carbon dioxide.

Two recent articles [5,6] give thorough reviews of the importance of the homogeneous reaction during the partial oxidation of methane.

2. Catalysts and Promoters

A number of periodic table elements have been studied since the first listing disclosed in our 1982 publication [1]. A listing of elements giving $\geq 10\%$ C_{2+} yield is given in reference [2]. Several reviews cover the numerous catalysts of interest as well as new insights obtained from mechanistic and kinetic studies [7-11]. Many of the methane coupling catalysts contain either a co-catalyst and/or a promoter; for example alkali and alkaline earth metal oxides and chlorine or chloride salts. The exact nature or mechanism of such catalytic or promoter action is not known but appears to provide either stabilization of the proper oxidation state and the selective oxygen sites or preventing formation of unselective oxygen species, or by itself participating in the selective oxidation of methane as in the case of chlorine and chloride salt

Since the pioneering work at Union Carbide almost all of the literature studies dealing with the oxidative coupling of methane have utilized the concurrent feed (cofeed mode) of methane and oxygen or air. The main exceptions are the numerous works reported by Atlantic Richfield Company (ARCO)[12] which used a sequential or pulse flow of reactants like that orginally employed by Union Carbide[1]. This mode of operation is commonly referred to as the redox mode. In the redox mode, the catalyst is first oxidized by the passage of air or oxygen over the catalyst. The system is then flushed with an inert gas and methane is passed over the catalysts. The sequence can then be repeated. The redox mode has the advantage that methane and higher hydrocarbon products are not in contact with gaseous oxygen thus homogeneous gas phase oxidation of either is minimal. However, it is a stoichiometric instead of catalytic reaction which requires rapid regeneration of the catalyst in a separate step.

In terms of a common feature for redox mode catalysts (and many cofeed), it appears very important that an element or metal be able to possess more than one oxidation state in such a fashion as to provide negative or near zero free energy for the reaction:

$$CH_4 + \text{Metal Oxide} \longrightarrow 0.5\,C_2H_6 + H_2O + \text{Reduced Metal Oxide or Metal}.$$

Thus, many of the more active (and efficient) metal oxides are indeed of those elements that have favorable free energy for this reaction, for example manganese, lead, bismuth, etc.

An example of a new class of redox catalyst developed at UCC is the double perovskite, $LaCaMnCoO_6$, which possesses unique structure and redox properties [13]. Synthesis of $LaCaMnCoO_6$ was first reported in 1988 [14]. The material is an ordered

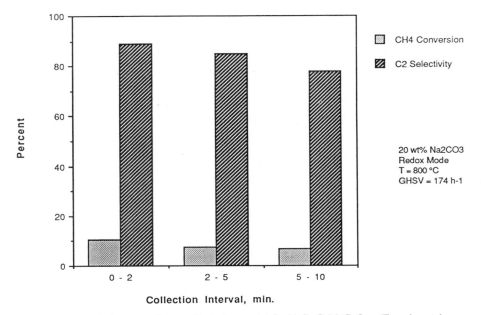

Figure 2. Methane coupling results (redox mode) for Na/LaCaMnCoO$_6$. (Experimental conditions: 800°C; 15 psia; GHSV=174 h^{-1}; collection intervals=times intervals of sample collections)

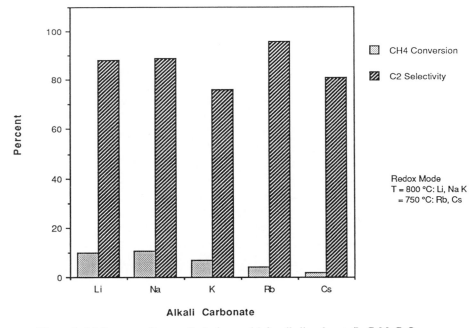

Figure 3. Methane coupling results (redox mode) for alkali carbonate/LaCaMnCoO$_6$. (Experimental conditions: 800°C or 750°C; 15 psia; GHSV=174 h^{-1}; collection intervals= 0-2 minutes)

perovskite showing multiple occupation of both A (La and Ca) and B (Mn and Co) sites in sublattices. X-ray diffraction patterns suggested the crystalline material possessed some ordered domains, while in other areas the cations were distributed at random. For the ordered domains, the most probable structural model was an ABO_3 perovskite-type structure in which Mn^{4+} and Co^{3+} ions occupied B positions in adjacent ABO_3 units while La^{3+} and Ca^{2+} ions alternate in the A positions. The material has a unique redox property in that two ions in the structure undergo reduction.

$$LaCaMn^{4+}Co^{3+}O_6 + H_2 \ \text{------------>} \ LaCaMn^{3+}Co^{2+}O_5 + H_2O$$

This made the material of interest as a methane coupling catalyst for sequential studies.

Redox mode results for $LaCaMnCoO_6$ promoted with Na_2CO_3 are presented in Figure 2.

Sodium addition (Na_2CO_3) to $LaCaMnCoO_6$ drastically changed its catalytic properties for methane coupling. At 850 °C, $LaCaMnCoO_6$ gave very low C_2 selectivities (< 2%) with high activity (29% CH_4 conversion); whereas, $Na_2CO_3/LaCaMnCoO_6$ gave very high C_2 selectivities (> 80%) at lower CH_4 conversions (10%).

Other alkali promoters were added to $LaCaMnCoO_6$ to determine if better methane coupling catalysts would result. Alkali promoters used included K_2CO_3, Li_2CO_3, Rb_2CO_3, and Cs_2CO_3. Representative results obtained in the redox mode with these catalysts are presented in Figure 3.

One trend reported with all of the literature sequential mode catalysts is that the C_{2+} selectivity increases as the methane coupling catalyst is reduced (methane conversion decreases). Thus, initially when the methane conversion is greatest the C_{2+} selectivity is at its minimium. This trend is not observed with the alkali promoted double perovskite $LaCaMnCoO_6$. Both the CH_4 conversions and C_2 selectivities are at maximas during the early stages of the catalyst reduction resulting in greater yields of C_{2+} products (Table 2).

An example of a new class of cofeed catalysts developed at UCC are the layered perovskites [13]. Alkali doped lanthanide oxides were known to be very active catalysts for the oxidative coupling of methane. These materials are prepared using incipient wetness techniques and are operated at high temperatures (usually 800 °C). The high temperatures causes the loss of alkali components during reaction while the preparation methods results in catalysts in which the dispersion and location of the alkali components within the catalyst particle are not readily controlled. In order to alleviate some of these problems, layered perovskites containing layers of alkali ions dispersed and sandwiched between lanthanide perovskite layers were tested as methane coupling catalysts.

Layered perovskites of the form $A_2Ln_2Ti_3O_{10}$ (A=K,Rb; Ln=La,Nd,Sm,Gd,Dy) were first synthesized by J. Gopalakrishnan and V. Bhat [15]. These materials are composed of three octahedra-thick perovskite slabs $[Ln_2Ti_3O_{10}]$ separated by alkali metal ions. In this structure the alkali and lanthanide species are in close contact with the lanthanide species in a perovskite environment. Potassium ions of $K_2Ln_2Ti_3O_{10}$ are easily exchanged with Na^+ or Li^+ by treating with molten alkali-metal nitrates.

Figure 4 presents a summary of results obtained at 800 °C for some of the materials. The results are comparable to many of the reported literature methane coupling catalysts. The highest C_2 selectivities was obtained with $K_2La_2Ti_3O_{10}$ (41%). None of the lanthanide substitutions in the original $K_2La_2Ti_3O_{10}$ catalyst resulted in significantly improved catalytic performance.

Table 2. Comparison of Sequential (Redox) Mode Results Using Different Alkali Carbonate Dopants on $LaCaMnCoO_6$

Results for 20 wt% $Na_2CO_3/LaCaMnCoO_6$

Temp. (°C)	GHSV (Hr-1)	Collection (min.)	CH_4 Conv.%	C_2 Conv.%	C_2 Sel.%	=/- Yield%
Pretreatment: 800°C; Air = 100ccm; overnight						
800	174	0 - 2	10.6	89	9.4	1.6
800	174	2 - 5	7.6	85	6.5	1.6
800	174	5 - 10	6.9	78	5.4	1.7

Results for 24 wt% $K_2CO_3/LaCaMnCoO_6$

Temp. (°C)	GHSV (Hr-1)	Collection (min.)	CH_4 Conv.%	C_2 Conv.%	C_2 Sel.%	=/- Yield%
Pretreatment: 800°C; Air = 50 ccm; overnight						
800	222	0 - 2	6.8	76	5.1	0.61
800	222	2 - 5	5.4	68	3.7	1.0
800	222	5 - 10	3.8	46	1.7	0.89

Results for 15 wt% $Li_2CO_3/LaCaMnCoO_6$

Temp. (°C)	GHSV (Hr-1)	Collection (min.)	CH_4 Conv.%	C_2 Conv.%	C_2 Sel.%	=/- Yield%
Pretreatment: 800°C; Air = 200 ccm; 2 hours						
800	263	0 - 2	9.8	88	8.7	1.4
800	263	2 - 5	5.2	92	4.8	0.89
800	263	5 - 10	1.9	92	1.7	0.56

Results for 48 wt% $Rb_2CO_3/LaCaMnCoO_6$

Temp. (°C)	GHSV (Hr-1)	Collection (min.)	CH_4 Conv.%	C_2 Sel.%	C_2 Yield%	=/-
Regenerated: 750°C; Air = 200 ccm; 4 hours						
750	2000	0 - 1	4.2	96	4.0	0.27
750	2000	1 - 3	2.3	96	2.2	0.21
750	2000	3 - 10	1.3	93	1.2	0.19

Results for 51 wt% $Cs_2CO_3/LaCaMnCoO_6$

Temp. (°C)	GHSV (Hr-1)	Collection (min.)	CH_4 Conv.%	C_2 Sel.%	C_2 Yield%	=/-
Regenerated: 750°C; Air = 200 ccm; 4 hours						
750	1429	0 - 1	1.8	81	1.4	0.18
750	1429	1 - 3	0.74	70	0.52	0.10
750	1429	3 - 10	0.52	62	0.32	0.11

The effects of changing the CH_4/O_2 ratio on the coupling process were determined for $K_2La_2Ti_3O_{10}$ (Figure 5), $Rb_2La_2Ti_3O_{10}$, and $Na_2La_2Ti_3O_{10}$. As expected, higher CH_4/O_2 ratios resulted in higher C_2 selectivities but lower C_2 yields due to the decreased CH_4 conversions. For all three catalysts, the CH_4 conversion, C_2 yield, ethylene/ethane ratio, and CO_2/CO ratio decreased while the O_2 conversion and C_2 selectivity increased with increasing CH_4/O_2 ratio.

The effects of changing the temperature on the coupling process were also determined for $K_2La_2Ti_3O_{10}$ (Figure 6), $Rb_2La_2Ti_3O_{10}$, and $Na_2La_2Ti_3O_{10}$. For each catalyst, increasing temperature results in both increased CH_4 conversion and increased C_2 selectivity. Also, O_2 conversion, ethylene/ethane ratio, and CO_2/CO ratio increased with increasing temperature.

3. Mechanism

It is generally accepted that for high temperature methane coupling, the catalyst surface generates methyl radicals via an active oxygen center which then undergo coupling in the gas phase. Ethane can then undergo dehydrogenation in the gas phase or oxidatively on the surface via ethyl radicals. This mechanistic picture was first proposed by Lunsford's group [16]. A comprehensive picture of this mechanism is presented in reference [2]. Non-selective oxidation is thought to occur via methyl peroxy radicals formed from methyl radicals reacting with O_2 or thru the gas phase (or surface) combustion of C_2 products, especially ethylene. This mechanism has become known as the heterogeneous-homogeneous mechanism. Several studies have provided experimental support for this mechanism. Some studies quantified the methyl radicals with the amount of C_2 hydrocarbon products and showed that the methyl radicals could account for most of the C_2 products [17-19]. Other evidence for the mechanism comes from isotopic labeling studies [20-23] where reaction products monitored from coupling

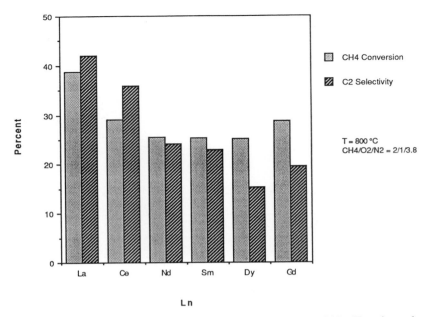

Figure 4. Methane coupling results for $K_2 Ln_2 Ti_3 O_{10}$ [Ln=lanthanide]. (Experimental conditions: 800°C; 15 psia; $CH_4 /O_2 /N_2 =2/1/3.8$)

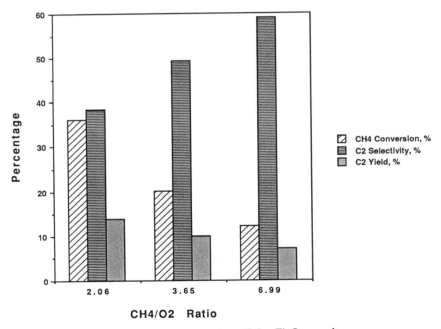

Figure 5. Effect of changing CH_4/O_2 ratio on $K_2Ln_2Ti_3O_{10}$ results. (Experimental conditions: 800°C; 15 psia

of CH$_4$ and CD$_4$ were shown to consist only of CH$_3$-CH$_3$, CD$_3$-CD$_3$ and CD$_3$-CH$_3$. No CD$_2$, CH$_2$, CD or CH units were observed in the ethane product.

Isotopic studies in which the kinetic isotope effect was measured upon substituting CD$_4$ for CH$_4$ have given valuable insight as to the rate determining step in the oxidative coupling of methane [20-23,25]. The kinetic isotope effects indicate that the rate determining step for oxidative methane coupling over a variety of catalysts (Li/MgO, Sm$_2$O$_3$, Na/MnO$_x$/SiO$_2$) is breakage of the C-H bond.

While Lunsford and other researchers have proposed that C-H bond breakage occurs homolytically (resultring in CH$_3\bullet$ and H$\bullet$), some work raises the possibility of heterolytic C-H bond breakage [25,26]. In this model CH$_4$ is adsorbed on the surface and bond breakage results by the abstraction of H$^+$ resulting in the formation of CH$_3^-$

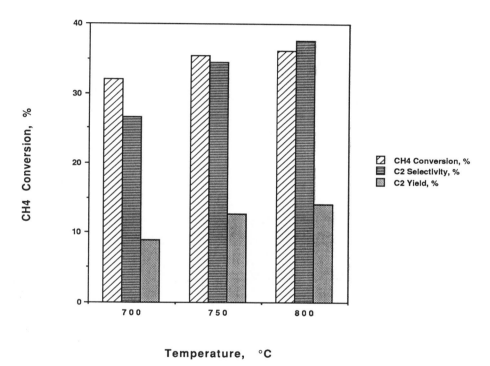

Figure 6. Effect of changing temperature ratio on K$_2$Ln$_2$Ti$_3$O$_{10}$ results. (Experimental conditions: (800°C; 15 psia; CH$_4$/O$_2$/N$_2$=2/1/3.8)

ions adsorbed on surface metal species. A rapid electron transfer to the catalyst surface then give CH$_3\bullet$ radicals. This model is based on the generally accepted conclusion that high basicity is a characteristic of good methane coupling catalysts.

Lapszewicz and Jing [27] present results that contradict the occurrence of heterolytic bond breakage. They studied the abilities of metal oxides of differing basicity to promote oxidative coupling and also the exchange reaction between CH$_4$ and D$_2$. They did not find any correlation between the rates for methane coupling and the isotopic exchange reaction indicating that heterolytic bond breakage could not by itself explain the formation of methyl radicals during methane coupling.

Strong support for a heterogeneous reaction mechanism is provided by the work of Keulks and Yu [28]. These workers have shown that a MgO supported Bi-P-K oxide catalyst operates via a heterogeneous reaction pathway. The experimental evidence is provided thru the use of isotopically labelled switching reaction of CH_4 or CD_4 with oxygen over this catalyst at 650 and 700 °C. Indeed, C_2H_3D production in ethylene is observed in 35% and 17% of the ethylene fraction at 650 and 700 °C, respectively. This represents a very strong indication of surface catalyzed coupling reaction, although the exact mechanism by which this catalysis takes place is not clear. The authors suggest a parallel direct pathway to C_2H_4 via a H-deficient surface intermediate. Another interesting sidenote here is that the C_2H_3D yield is higher at 650 °C than at 700 °C. Again, this is an indication, quite expectedly, that the surface reaction is more dominant at the lower of the two temperatures studied.

Hutchings et al. [29] have also presented results suggesting a heterogeneous reaction mechanism occurs during methane coupling. If the heterogeneous-homogeneous mechanism is the only one occurring, ethane and COx are the only primary products. Hutchings et al. report that direct formation of ethylene from methane is not negligible for all catalysts, and must therefore be considered in the mechanistic scheme.

Russian researchers [30] determined the existence of stationary concentrations of CH_3- and CH_2- surface species bonded with carbon and noncarbon surface sites on Sm_2O_3/MgO coupling catalyst using SERS-in situ method in a 670-970K temperature range. They proposed the fragments could be intermediates in the formation of C_2 products.

Whether heterogeneous only or a heterogeneous-homogeneous mechanism operates would be dependent upon three important variables; these are the catalyst itself, the reaction temperature and partial pressure of reactants. Thus, it is entirely conceivable that a heterogeneous-homogeneous mechanism operates primarily at high

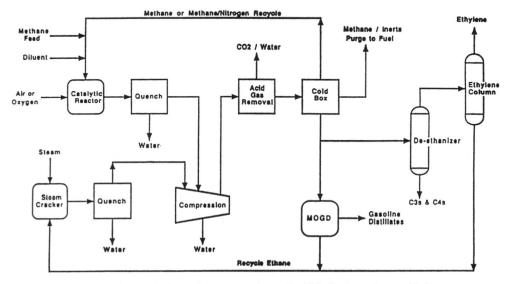

Figure 7. Schematic for methane conversion to liquid fuels via methane oxidative coupling to ethylene.

temperature and a heterogeneous only mechanism could predominate at low temperature. This view is shared by other workers. A large number of workers have convincingly shown that a limited selectivity-conversion as well as C_2 yield limitation of 25% will occur as a consequence of the heterogeneous-homogeneous mechanism[31]. In addition, higher temperatures bring about a large number of undesirable by-products. Such limitations will not necessarily occur with a heterogeneously catalyzed reaction at lower temperatures.

In the context of lower reaction temperatures for methane activation, Kool and Whitesides [32] have observed activation of methane to yield methanol at 60 °C and 2000 psi pressure using a heterogeneous platinum in aqueous $Fe_2(SO_4)_3$ solution alongwith activation of ethane, propane and other hydrocarbons at even lower pressures.

Catalytica researchers [33] recently announced a homogeneous system for the selective, catalytic oxidation of methane to methanol via methyl bisulfate. The reaction is carried out at 180 °C using mercuric ions, Hg(II), as catalyst. The process involves the oxidation of methane by concentrated sulfuric acid resulting in methyl bisulfate, water and sulfur dioxide. These two works are good examples of activating methane at extremely low temperatures.

4. Economics

The following summarizes results of Matherne and Culp (UCC researchers) [34] on the economics of different conceptualized schemes for conversion of methane (natural gas) to liquid hydrocarbons fuels via ethylene and other light olefins. These results were part of a two-year project that Union Carbide conducted under contract for the United States Department of Energy (Contract Number DE-AC-22-87PC79817).

Figure 7 depicts the process arrangement used for the methane coupling studies. It shows not only the integration of the process with the Mobil olefins-to-gasoline and distillates (MOGD) process for producing gasoline and distillate fuels, but also the optional refining system required to produce high purity ethylene.

Results for different methane coupling reactor diluent cases are presented in Table 3 (diluents considered are steam, nitrogen or methane). The maximum amount of oxygen present in the reactor feed was determined by the explosive limits for the mixture and the oxygen conversion was assumed to be 100% for all cases. All cases assumed an ethylene to ethane ratio in the product stream of 3:1 which agrees with experimental results obtained with the UCC $BaCO_3/Al_2O_3/ECl$ catalyst.

Table 4 summarizes the economics of the four methane coupling dilution cases. These results are based on the production of ethylene for sale. The benefits of higher

Table 4 summarizes the economics of the four methane coupling dilution cases. These results are based on the production of ethylene for sale. The benefits of higher methane conversion and yield-to C_2's that characterize the nitrogen and steam diluent cases are more than offset by the costs of their lower selectivity to C_2's. Thus yield-to-C_2's alone is not a good general predictor of economics.

While the methane dilution and redox feed cases were economically equivalent, the methane dilution case was used for futher comparisons due to: (1) it was based on actual UCC laboratory results and (2) redox feed required untested fluidized-bed reactor design.

Table 3. Methane Coupling Case Descriptions Using Different Diluents

Feed System	Diluent			None
	Nitrogen	Methane	Steam	Redox
		Cofeed		
Catalyst	UCC BaCO3/Al2O3/ECl			ARCO Mn/Na/P/SiO2
Reactor				
Type		Fixed-bed		Fluidized-bed
Temperature (°C)		750		850
Pressure (psig)		50		50
GHSV (h^{-1})		2133		1200
CH4 Conv. (%)	45	18	45	24
Selectivity to				
C_2H_4 (%)	39	58	39	48
C_2H_6 (%)	11	19	11	14
C_3H_6 (%)	2.2	4.3	2.2	4.6
C_3H_8 (%)	0.7	1.7	0.7	1.5
C_{4+} (%)	0.2	---	0.2	10
CO (%)	15	3	15	11
CO_2 (%)	32	14	32	11
C_2 Yield (%)	22	14	22	15

Table 4. Methane Coupling Economics[a] for Different Diluents

	Diluent			
	Nitrogen	Methane	Steam	None
Total Fixed Investment	710	530	560	530
Working Capital	65	60	70	60
Total Utilized Investment	775	590	630	590
Operating Costs				
Methane	146	112	132	124
By-Product Credits	(12)	(44)	(22)	(50)
Utilities	24	29	56	14
Variable Cost	159	97	166	90
Fixed Cost	53	40	42	41
Total Cash Cost	212	137	209	130
ROIAT[b] (%)	3	14	6	15

[a] All costs are in millions of dollars per year for a unit producing one billion pounds per year ethylene
[b] ROIAT is based on ethylene sales at $0.32 per pound

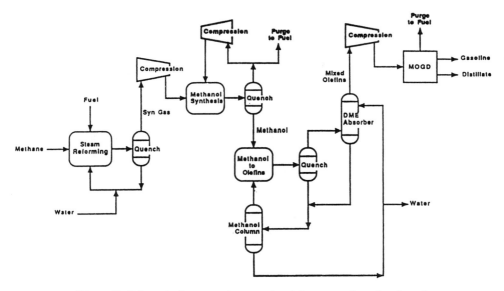

Figure 8. Schematic for comparison case involving conversion of methane to liquid fuels via synthesis gas and methanol.

Table 5. Comparison of Gasoline and Distillate Economics[a]

	Methane Coupling	Syngas and Methanol
OSBL Investment	285	314
Air separation unit	61	---
MOGD	40	35
ISBL Investment	154	141
Total Fixed Investment	540	490
Working Capital	50	46
Total Utilized Investment	590	536
Operating Costs		
Methane	101	60
By-product credits	(26)	(2)
Utilities	23	28
Variable Cost	98	88
Fixed Cost	38	40
Total Cash Cost	136	128
Required Sales Price[b] (per gallon)	$2.09	$1.94

[a] All costs are in millions of dollars per year for a 10,600-bpd unit
[b] Required price to generate a 10% ROIAT

As a comparison case, the existing technology for the conversion of methane to liquid fuels via synthesis gas and methanol as intermediates was considered (Figure 8). The comparison case consist of SRI's process for the steam reforming of methane to synthesis gas, ICI's low pressure synthesis of methanol, UOP's process for the conversion of methanol to olefins and dimethyl ether and Mobil's MOGD process for the conversion of olefins to gasoline and distillate fuels.

A comparison of the economics of making gasoline and distillate fuels via methane coupling versus the synthesis gas and methanol route is presented in Table 5. The two routes are similar in that most of the fixed investment is in the methane conversion portion of the processes. The synthesis gas-methanol route does have an operating cost advantage due to higher selectivity to final products (nearly 100% versus 77% to C_2's for methane coupling). However, the comparison is encouraging for methane coupling considering its early stage of development compared to the many years of development and commerical practice associated with the synthesis gas route.

5. Future Directions

In a previous publication [35], it was indicated that the major challenges (and opportunities) toward developing a commerically feasible process will be in the development of more active and selective catalysts that will permit operation at 400-600°C, as well as resolving the issue of long term catalyst stability. These views, in part, are shared by Hutchings, et al. [29]. Furthermore, catalyst stability becomes less of a problem when coupling catalysts operate at lower temperatures since "catalyst volatility issue" disappears. The authors continue to be optimistic that advances along these directions will be made in the 1990's to make this simple, attractive process a commerical reality in another decade or so.

ACKNOWLEDGEMENTS

The authors thank the Industrial Chemicals Division of Union Carbide Chemicals Corporation for the permission to publish this work. Special recognition is given to Gary Culp and Joe Matherne who are responsible for the economics section.

REFERENCES

1. G.E. Keller and M.M. Bhasin, *J. Catal.* 73, 9 (1982).
2. "Progress in Alkane Activation", *Catalytica Study No. 4188 AA,* 132.
3. G.S. Lane and E.E. Wolf, *J. Catal.* 113, 144 (1988).
4. D.J.C. Yates and N.E. Zlotin, *J. Catal.* 111, 317 (1988).
5. Z. Kalenik and E.E. Wolf, The role of gas-phase reactions during methane oxidative coupling, in "Methane Conversion by Oxidative Processes", Van Nostrand Reinhold (1992).
6. J.C. Mackie, *Catal. Reviews* 33, 169 (1991).
7. M.G. Poirier, A.R. Sanger and K.J. Smith, *Can. J. Chem. Eng.* 69(5), 1027 (1991).
8. A. Holmen, B.F. Magnussen and F. Steineke, "Proc. Europ. Appl. Research Conf. on Natural Gas", June 1-3, 1992.

9. *Catal. Today* 14(3-4), 415 (1992).

10. M. Baerns, Oxidative coupling of methane for the utilization of natural gas, in: "Chemical Reactor Technology for Environmentally Safe Reactors and Products", 283 (1993).

11. Q. Chen, J.H.B.J. Hoebink and G.B. Marin, *Proc. Europ. Appl. ResearchConf. on Natural Gas,* June 1-3, 1992

12. J. A. Sofranco, J. J. Leonard, and C.A. Jones, *J. Catal.* 103, 302 (1987) and *J. Catal.* 103, 311 (1987) and numerous U.S. Patents.

13. K.D. Campbell, U. S. Patents 4,988,660 and 4,982,041 (double perovskites); and 5,026,945 (layered perovskites) to Union Carbide Chemicals and Plastics Co. Inc. (1991)

14. M. Vallet-Regi, E. Garcia and J.M. Gonzalez-Calbet, *J. Chem. Soc. Dalton Trans.* 775 (1988).

15. J. Gopalakrishnan and V. Bhat, *Inorg. Chem.* **26**, 4299 (1987).

16. T. Ito, J.X. Wang, C.H. Lin and J.H. Lunsford, *J. Am. Chem. Soc.* 107, 5062 (1985).

17. K.D. Campbell, E. Morales and J.H. Lunsford, *J. Am. Chem. Soc.* 109, 7900 (1987).

18. Y. Feng and D. Gutman, *J. Phys. Chem.* 95, 6558 (1991).

19. Y. Feng and D. Gutman, *J. Phys. Chem.* 95, 6564 (1991).

20. N.W. Cant, C.A. Lukey, P.F. Nelson and R.J. Tyler, *J. Chem. Soc. Commun.* 766 (1988).

21. P.F. Nelson, C.A. Lukey and N.W. Cant *J. Catal.* 120, 216 (1989).

22. K. Otsuka, M. Inaida, Y. Wada, T. Komatsu and A. Morikawa, *Chem. Lett.* 1531 (1989).

23. C.A. Mims, R.B. Hall, K.D. Rose and G.R. Myers, *Catal. Lett.* 2, 361 (1989).

24. R. Burch, S.C.Tsang, C. Mirodatos and J.G. Sanchez, *Catal. Lett.* 7, 423 (1990).

25. V.D. Sokolovskii, O.V. Buyevskaya, S.M. Aliev and A.A. Davydov, *Proc. Intl. Symp. on "New Dev. in Selective Oxidation"*, Rimini, Italy.

26. V.R. Choudhary and V.H. Rane, *J. Catal.* 130, 411 (1991).

27. J.A. Lapszewicz and X.Z. Jiang, *Catal. Lett.* 13, 103 (1992).

28. G.W. Keulks and M. Yu, *Proc. Intl. Symp. on "New Dev. in Selective Oxidation"*, Rimini, Italy.

29. G.J. Hutchings, M.S. Scurrell and J.R. Woodhouse, *Chem. Soc. Rev.* 18, 251 (1989).

30. A.A. Kadushin, O.V. Krylov, S.E. Plate, Yu.P. Tulenin, V.A. Seleznev, A.V. Bobrov and Ya.M. Kimelfeld, *Proc. Intl. Symp. on "New Dev. in Selective Oxidation"*, Rimini, Italy.

31. (i)J.G. Mccarty, McEwen and M.A, Quinlan, *Proc. Intl. Symp. on "New Dev. in Selective Oxidation"*, Rimini, Italy.

    (ii)J.A. Labinger, *Catal. Lett.* 1, 371(1988)

32. L.K. Kool and G.M. Whitesides, unpublished report.

33. R.A. Periana, D.J. Taube, E.R. Evitt, D.G. Loffler, P.R. Wentrcek, G. Voss and T. Masuda, *Science* 259, 340 (1993).

34. J.L. Matherne and G.L. Culp, *"Direct Conversion of Methane to C_2's and Liquid Fuels: Process Economics"* in "Methane Conversion by Oxidative Processes", Van Nostrand Reinhold (1992).

35. M.M. Bhasin, *Stud. Surf. Sci. Catal.* 36, 343 (1988).

METHANE AND LIGHT ALKANE (C_2-C_4) CONVERSION OVER METAL FLUORIDE-METAL OXIDE CATALYST SYSTEM IN PRESENCE OF OXYGEN

X. P. Zhou, S. Q. Zhou, W. D. Zhang, Z. S. Chao, W. Z. Weng,
R. Q. Long, D. L. Tang, H. Y. Wang, S. J. Wang, J. X. Cai,
H. L. Wan , K. R. Tsai

Department of Chemistry and State Key Laboratory for Physical Chemistry of the Solid Surface, Xiamen University
Xiamen, 361005, CHINA

ABSTRACT

A novel series of metal fluoride-metal oxide catalysts with good to excellent catalytic performance for oxidative coupling of methane and oxidative dehydrogenation of ethane, propane and iso-butane were developed. XRD analysis suggested that, with the addition of metal fluorides to metal oxides, partial substitution of cations and/or anions happened between fluorides and oxides, leading to the formation of new phases (e.g. LaOF) and expansion/contraction of lattices (e.g. in CeO_2-BaF_2). O_2^{n-} ($1 \leq n \leq 2$) species were detected by the Raman spectroscopy over the catalysts after O_2 adsorption. O_2^- species was observed by *in situ* FTIR spectroscopies at 650°C over SrF_2-La_2O_3 catalyst.

KEYWORDS

Methane Oxidative Coupling, Oxidative dehydrogenation of Light Alkane, Fluoride-Oxide Catalyst, Oxygen Species.

INTRODUCTION

The methane oxidative coupling (MOC) to C_2 hydrocarbons and oxidative dehydrogenation of lower alkanes (C_2-C_4) to the corresponding olefins have become of major importance for their potential utilization of the world's abundant natural gas and refinery gas resource, and for their fundamental significance in the catalytic activation and selective conversion of the first and the most inert member of paraffin hydrocarbons, methane, and its correlation with that of the other light alkanes. Recently, much attention has been focused on the development of better catalysts and on mechanistic studies of O_2 activation. Among these reactions, the one that has been mostly studied is MOC, and many catalyst systems including alkaline earth-, rare earth- and reducible transition metal-based oxides or complex oxides have been reported to be effective for the reaction. In some studies,[1,2,3] halides, especially chlorides, have been added to the oxides in order to improve the catalytic activity and selectivity. Since MOC reaction is usually carried out at high temperature (~750°C), the thermal stability of the catalyst is often considered to be an important factor, in consideration of the facts that the fluorides are usually more stable than the corresponding chlorides and bromides, and that F^- and O^{2-} have the similar ionic radii, we have developed a new series of metal fluoride-metal oxide catalysts with good stability and good to excellent catalytic performance for MOC as well as for oxidative dehydrogenation of ethane (EOD), propane (POD) and iso-butane (i-BOD).[4] The details of the catalytic performance for the catalysts are reported in this paper, some new experimental evidences for the catalytic activity and selectivity in relation to catalyst composition, structure and oxygen adspecies are also presented.

EXPERIMENTAL

All of the catalysts were prepared by wet mixing different mole ratios of metal oxides with metal fluorides, followed by drying and calcination (4 ~ 6 h, 850 ~ 900°C).

The catalyst evaluation was carried out in a fixed-bed quartz micro-reactor (3 mm ID) with on line gas chromatography (TCD) for products analysis. 0.20 ml of catalyst was used in each experimental run with stable operation for more than 2 days. Other reaction conditions were as follows : MOC (GHSV = 15000 h^{-1} and $CH_4 : O_2 \approx 3 : 1$); EOD (GHSV = 18000 h^{-1} and $C_2H_6 : O_2 : N_2 = 10 : 5 : 85$); POD (GHSV = 6000 h^{-1} and $C_3H_8 : O_2 : N_2 = 4 : 5 : 11$); i-BOD (GHSV = 6000 h^{-1} and i-$C_4H_{10} : O_2 = 1 : 1$). The conversion of alkane (C_{alkane}) and selectivity (S_i) were calculated from the equations : $C_{alkane} = (\Sigma A_i \times F_i)/ [\Sigma (A_i \times F_i) + A_{alkane} \times F_{alkane}]$ and $S_i = (A_i \times F_i)/\Sigma (A_i \times F_i)$, respectively, where A = peak area and F = correction factor.

The XRD measurements were carried out at room temperature on a Rigaku Rotaflex D/Max-C system with Cu Kα ($\lambda = 1.5406$ Å) radiation.

XPS were recorded on a V. G. ESCALAB MK II instrument with Al Kα as photo source. Raman spectra were taken at room temperature on a Jobin Yvon U-1000 spectrometer with argon laser (5145 Å) as the excitation source.

In XPS and Raman experiments, the catalyst, after calcination in air, was treated in a flow of H_2 at 900°C for 30 minutes, followed by in a flow of He at same temperature for 10 minutes, part of the catalyst was then separated and used as a "blank" sample, the rest of the catalyst was exposed to O_2 at room temperature to prepare the O_2-adsorbed sample.

The *in situ* IR spectra were recorded on a Nicolet 740 FTIR spectroscopy at the resolution of 4 cm^{-1}, the number of the scan was 160. A quartz *in situ* IR cell which can be heated up to 700°C was designed, BaF$_2$ was used as cell windows. The temperature of the cell was measured by a thermal couple mount close to the sample wafer. In the experiments, A self-supported sample wafer was thermally treated at 700°C under H_2 for 20 minute to remove the surface carbonate residue, followed successively by purging with He at 700°C, and treatment with O_2 or CH_4/O_2 (3/1) at 650°C, O_2 used for the adsorption has been purified with solid KOH to removed CO_2 impurity. The O_2-adsorbed sample was then cooling down gradually under O_2 atmosphere from 650°C to 25°C within a period of 3 hr, and the IR spectra were recorded at specified temperature, i.e. 650, 500, 300, 100 and 25°C, respectively during the cooling period. At each specified temperature point, temperature of the IR cell was maintained constant for about 10 min to insure that the IR measurement was conducted under thermostatic condition. The spectra of the CH_4/O_2 (3/1) treated sample was recorded after 45 min on stream at 650°C.

RESULTS AND DISCUSSION

Catalytic Performance Evaluation

Catalytic performance evaluation for MOC showed that, pure CeO_2 was actually a combustion catalyst, while no reaction occurred over pure BaF_2. However, when certain amount of BaF_2 was added to CeO_2, after the activation treatment, CH_4 conversion of 33-34% with C_2 selectivity of 51-55% was obtained (Table 1), indicating that the addition of fluorides played a significant promoting role for MOC reaction. Similar results can also be found over Sr-La catalysts (Table 2), the CH_4 conversion, C_2 selectivity and C_2H_4 to C_2H_6 ratio over SrO/LaF_3 and SrF_2/La_2O_3 were apparently higher than those over SrO/La_2O_3, La_2O_3, SrO and SrF_2. For oxidative dehydrogenation of ethane, propane and iso-butane, this series of fluoride containing catalysts also showed good to excellent catalytic performance (Table 3-5).

Table 1. Catalytic Performance of BaF_2/CeO_2 Catalysts for Methane Oxidative Coupling at 800 °C.

BaF_2/CeO_2 (mole ratio)	CH_4 Conversion (%)	Selectivity (%)				C_2 Yield (%)
		CO	CO_2	C_2H_4	C_2H_6	
1	32.76	0	48.06	31.95	19.99	17.02
2	33.69	0	46.77	32.26	20.97	17.93
3	32.93	0	45.18	33.71	21.11	18.05
4	34.01	1.42	45.12	33.57	19.89	18.18
5	32.75	1.57	46.88	31.09	20.80	16.99

Feed = $CH_4 : O_2 \approx 3 : 1$, GHSV = 15000 h^{-1}.

Table 2. Catalytic Performance of Catalysts for Methane Oxidative Coupling.

Catalyst	Temperature (°C)	CH$_4$ Conversion (%)	Selectivity (%)				C$_2$ Yield (%)
			CO	CO$_2$	C$_2$H$_4$	C$_2$H$_6$	
La$_2$O$_3$	750	27.1	15.9	53.7	19.0	11.4	8.2
	700	28.4	11.7	51.4	21.5	15.4	10.5
	650	29.3	9.0	47.3	24.6	19.2	12.8
	600	29.1	9.2	45.9	24.6	20.2	13.1
	550	27.8	9.7	45.7	23.4	21.2	12.4
LaF$_3$	750	2.1	0	46.2	0	53.8	1.1
SrO	750	0.9	0	48.8	0	51.2	0.5
SrF$_2$*	750	0.7	0	46.0	0	54.0	0.4
20% SrO/La$_2$O$_3$	750	25.4	6.3	57.6	20.8	15.3	9.2
	700	27.0	3.9	50.6	25.4	20.1	12.3
	650	27.3	0	49.2	28.1	22.7	13.9
	600	28.4	5.0	45.2	27.2	22.6	14.1
	550	27.6	6.6	45.4	25.6	22.3	13.2
50% SrO/La$_2$O$_3$	750	29.5	7.1	44.2	29.4	19.3	14.4
	700	30.2	8.2	40.9	29.2	21.7	15.4
20% SrO/LaF$_3$	750	31.8	10.2	35.8	37.8	16.2	17.2
	700	32.0	11.9	31.0	35.7	21.4	18.2
33% SrO/LaF$_3$	700	33.1	11.8	31.4	36.4	20.5	18.8
	650	17.9	17.2	35.0	15.9	31.9	8.5
50% SrO/LaF$_3$	750	33.8	16.0	30.1	40.0	13.9	18.2
	700	33.7	13.6	29.7	36.2	19.4	19.1
20% SrF$_2$/La$_2$O$_3$	750	30.5	4.7	44.3	33.2	17.8	15.6
	700	32.5	4.8	40.8	34.0	20.4	17.7
	650	34.7	5.7	36.8	35.8	21.7	19.9
	600	33.8	7.2	37.4	31.7	23.7	18.7
50% SrF$_2$/La$_2$O$_3$	750	33.9	9.5	33.5	37.9	19.1	19.3
	700	31.9	11.1	31.2	33.4	24.3	18.4
60% SrF$_2$/La$_2$O$_3$	800	34.2	9.5	35.5	39.7	15.4	18.8
	750	34.0	9.7	32.4	38.8	19.1	19.7

Feed = CH$_4$: O$_2$ ≈ 3 : 1, GHSV = 15000 h^{-1}. *GHSV = 20000 h^{-1}.

Table 3. Catalytic Performance of LaF$_3$-BaF$_2$ Catalysts for Oxidative Dehydrogenation of Ethane at 470 °C.

Catalyst	C$_2$H$_6$ Conversion	Content of Product (%)					C$_2$H$_4$ Selectivity	C$_2$H$_4$ Yield
	(%)	CO	CO$_2$	CH$_4$	C$_2$H$_4$	C$_2$H$_6$	(%)	(%)
20% LaF$_3$/BaF$_2$	46.3	0	0.20	0.43	3.93	4.92	92.7	42.9
50% LaF$_3$/BaF$_2$	46.8	0	0.17	0.71	4.01	5.05	90.2	42.2
80% LaF$_3$/BaF$_2$	54.2	0.42	0.25	0.87	4.05	4.08	84.0	45.5

Feed = C$_2$H$_6$: O$_2$: N$_2$ = 10 : 5 : 85, GHSV = 18000 h^{-1}.

Table 4. Catalytic Performance of Catalysts for Oxidative Dehydrogenation of Propane at 500 °C.

Catalyst	C$_3$H$_8$ Conversion (%)	Selectivity (%)						C$_3$H$_6$ Yield (%)
		C$_3$H$_6$	CH$_4$	C$_2$H$_6$	C$_2$H$_4$	CO$_2$	CO	
CeO$_2$	86.7	7.9	39.8	3.1	23.9	17.8	7.0	6.85
CeO$_2$/2CeF$_3$	10.3	72.7	0	0	0	13.2	14.0	7.49
3%Cs$_2$O/2CeO$_2$/CeF$_3$	53.4	67.5	3.5	0	12.2	1.3	3.4	36.0
3%Cs$_2$O/CeO$_2$/CeF$_3$	47.5	74.6	3.4	0	16.5	0.5	5.0	35.4
3%Cs$_2$O/CeO$_2$/2CeF$_3$	41.3	81.1	3.8	0	10.7	0.8	3.6	33.5
3%Cs$_2$O/CeO$_2$/3CeF$_3$	7.7	84.4	0	0	0	0	11.6	6.50

Feed = C$_3$H$_8$: O$_2$: N$_2$ = 4 : 5 : 11, GHSV = 6000 h^{-1}.

Table 5. Catalytic Performance of Catalyst for Oxidative Dehydrogenation of iso-Butane at 500 °C and 520 °C.

Catalyst	Conversion of iso-Butane (%)	Selectivity (%)				Yield of iso-Butene (%)
		i-C$_4$H$_8$	CH$_4$	C$_3$H$_6$	CO+CO$_2$	
Sm$_2$O$_3$/4CeF$_3$	5.61	84.7	2.53	6.60	7.62	3.48
Sm$_2$O$_3$/4CeF$_3$*	7.80	75.0	4.78	10.3	9.93	5.85
Nd$_2$O$_3$/4CeF$_3$	4.31	87.9	1.45	6.27	4.37	3.79
Nd$_2$O$_3$/4CeF$_3$*	4.94	75.7	2.15	17.1	5.15	3.74
Y$_2$O$_3$/4CeF$_3$	6.41	78.6	3.96	14.4	3.04	5.04
Y$_2$O$_3$/4CeF$_3$*	11.9	72.6	5.03	17.8	4.59	8.64

Feed = i-C$_4$H$_{10}$: O$_2$: N$_2$ = 2 : 3 : 5, GHSV = 6000 h^{-1}, * T = 520 °C.

XRD Characterization of Sr/La and BaF$_2$/CeO$_2$ Catalysts

The XRD measurements (Table 6) indicated that new phase such as SrF$_2$, cubic and tetragonal LaOF were formed in the fresh 20-50% SrO/LaF$_3$ catalysts. This result suggested that, during the process of catalyst preparation, part of the F$^-$ (r = 1.33 Å) in LaF$_3$ and O^{2-} (r = 1.35 Å) in SrO were substituted by O^{2-} and F$^-$, respectively, leading to the formation of new phases, such as tetragonal, cubic and rhombohedral LaOF (superstructure of fluorite), which is a solid electrolyte with ionic conductivity,[5] and might be favorable to promoting the transfer of the charge carrier in the catalytic system. In the 20~60 % SrF$_2$/La$_2$O$_3$, only the SrF$_2$, cubic and hexagonal La$_2$O$_3$ Phases, the latter is considered to be consisted of O^{2-} and (LaO)$_2^{2+}$ layers,[6] were found. However, partial ionic exchange between SrF$_2$ and La$_2$O$_3$ might also happen in this catalyst system, in the case when one Sr^{2+} substituted for one La^{3+} in the La$_2$O$_3$ lattice, O$^-$ would be formed, and if the distance between a pair of O$^-$ species in the lattice were close enough, the interaction between O$^-$ ions and the transformation from the O$^-$ to O$_2^{2-}$ would occur.[7] On the other hand, the intrinsic oxygen vacancies in cubic La$_2$O$_3$ would be also favorable to the adsorption and activation of molecular oxygen.

Table 6. The results of XRD analysis of the La-Sr-O-F catalysts.

Catalyst	Composition and Structure*
20% SrO/LaF$_3$	tetragonal LaOF(s) (a=4.091, c=5.837); cubic SrF$_2$(w) (a=5.800); LaF$_3$(w)
33% SrO/LaF$_3$	rhombohedral LaOF(m) (a=7.131, b=32.010); tetragonal LaOF(s); SrF$_2$(m); LaF$_3$(w)
50% SrO/LaF$_3$	SrF$_2$(s); tetragonal LaOF(s); cubic LaOF(m) (a=5.76)
20% SrF$_2$/La$_2$O$_3$	cubic La$_2$O$_3$(s) (a=11.3); SrF$_2$(w); hexagonal La$_2$O$_3$(s) (a=10.30, c=5.81); hexagonal La$_2$O$_3$(s) (a=3.9373, c=6.1299)
50% SrF$_2$/La$_2$O$_3$	hexagonal La$_2$O$_3$(vs) (a=3.9373, c=6.1299); hexagonal La$_2$O$_3$(m) (a=10.30, c=5.81); cubic La$_2$O$_3$(m); SrF$_2$(m)
60% SrF$_2$/La$_2$O$_3$	SrF$_2$(vs); cubic La$_2$O$_3$(w); hexagonal La$_2$O$_3$(m) (a=3.9373, c=6.1299); hexagonal La$_2$O$_3$(m) (a=10.30, c=5.81)

* s = strong; m = medium; w = weak.

In BaF$_2$/CeO$_2$ catalysts, only CeO$_2$ and BaF$_2$ phases, both have the fluorite-type structure, were detected, but the lattice of CeO$_2$ expanded (Table 7), while that of BaF$_2$ contracted (Table 8), implicating that partial isomorphous substitution of the anion (F$^-$ and O^{2-}) and/or cation (Ba^{2+} and Ce^{4+}) happened between the BaF$_2$ and CeO$_2$ lattices. This may lead to the formation of anion vacancies and/or O$^-$ in the case when O^{2-} substituted for F$^-$ or Ba^{2+} for Ce^{4+}, and the formation of Ce^{3+} in the case when F$^-$ substituted for O^{2-} or Ce^{4+} for Ba^{2+}. These defects would be favorable to the activation of molecular oxygen and enhancement of catalytic activity for MOC reaction. In addition, the dispersion of F$^-$ on the surface of above catalysts will also be helpful to the isolation of the surface active center and to the decrease of CO$_2$ inhibition.

Table 7. XRD results of CeO_2 phase in the BaF_2/CeO_2 catalysts.

BaF_2/CeO_2 (mole ratio)		(111)	(200)	(220)	(311)	(222)	(400)	(331)
1	d/Å	3.121	2.707	1.911	1.630	1.563	1.354	1.241
	I/I_0	100	37	82	67	12	10	23
2	d/Å	3.129	2.711	1.916	1.635	1.566	1.354	1.243
	I/I_0	71	26	53	40	7	10	20
3	d/Å	3.142	2.719	1.918	1.635			
	I/I_0	46	14	23	14			
4	d/Å	3.136	2.714	1.916	1.634	1.564		1.243
	I/I_0	18	5	9	6	2		3
5	d/Å	3.140	2.715	1.917				
	I/I_0	17	7	10				
Pure CeO_2	d/Å	3.1234	2.7056	1.9134	1.6318	1.5622	1.3531	1.2415
	I/I_0	100	30	52	42	8	8	

Table 8. XRD results of BaF_2 phase in the BaF_2/CeO_2 catalysts.

BaF_2/CeO_2 (mole ratio)		(111)	(200)	(220)	(311)	(331)	(422)
1	d/Å	3.556	3.074	2.178	1.855	1.411	1.256
	I/I_0	73	22	63	46	18	16
2	d/Å	3.559	3.089	2.177	1.856	1.411	1.257
	I/I_0	100	44	90	60	21	22
3	d/Å	3.582	3.097	2.188	1.863	1.415	1.257
	I/I_0	100	27	52	36	11	9
4	d/Å	3.562	3.079	2.173	1.852	1.407	1.253
	I/I_0	100	29	54	36	10	7
5	d/Å	3.565	3.083	2.170	1.853	1.421	1.253
	I/I_0	100	29	50	37	5	9
Pure BaF_2	d/Å	3.579	3.100	2.193	1.870	1.423	1.266
	I/I_0	100	27	79	51	13	14

XPS and Raman Characterizations of Oxygen Species over the Catalysts

From the XPS of O_2-adsorbed $2BaF_2/CeO_2$ and CeO_2 catalysts, we found that the binding energy of Ce $3d_{5/2}$ in BaF_2/CeO_2 was 2.2 eV lower than that in CeO_2. This result provided evidence for the presence of partially reduced state of Ce^{4+} ion. The O1s spectra showed that, on CeO_2, only one kind of oxygen species with binding energy at 529.1 eV was detected; on $2BaF_2/CeO_2$, four kinds of oxygen species with binding energies at 527.1, 528.9, 530.4 and 531.9 eV were found. The peaks at 527.1, 528.9 and 530.4 eV were assigned to lattice oxygen species (O^{2-}), while that at 531.9 eV was attributable to partially reduced oxygen species.[8] On the O_2-adsorbed CeO_2-$2BaF_2$ sample, Raman bands at 888, 956, 1050, 1062, 1080, 1094 and 1178 cm^{-1} were observed. Among these bands, the peaks at 888 cm^{-1} was tentatively assigned to O_2^{2-} species,[9,10,11] while the rest of the peaks were assigned to O_2^- and some intermediate dioxygen adspecies (O_2^{n-}, $1 < n < 2$).[12,13,14,15,16] No Raman band in this wavenumber region was observed on the "blank" sample of $2BaF_2/CeO_2$.

For the O_2-adsorbed LaF_3/BaF_2 catalysts, XPS analysis showed that the O1s spectra on 20% LaF_3/BaF_2 and 80% LaF_3/BaF_2 can be resolved into three kinds of oxygen species with the binding energy at about 529, 531 and 532 eV, respectively; and the O1s peak on 50% LaF_3/BaF_2 can be resolved into two kinds of oxygen species with the binding energy at 531 and 532.5 eV. The peaks around 531 eV may be assigned to the O_2^{2-} or O^- species; and the peaks at 532 and 532.5 eV may be attributed the CO_3^{2-} or O_2^- species.[17] Since the samples for XPS analysis have been pre-treated with H_2 and He at 900°C for a long time, the CO_3^{2-} could not be the main surface species and the peaks at 532 and 532.5 eV may be due to the O_2^- species. The peak at 529 eV is attributable to the O^{2-} species, which may have resulted from the formation of LaOF or La_2O_3 generated from slow hydrolysis of the LaF_3 in catalyst at calcining temperature of about 900 °C and in atmosphere of low humidity.[18] Two Raman peaks at 920, 1130 cm^{-1} and 920, 1188 cm^{-1} were observed on 20% LaF_3/BaF_2 and 50% LaF_3/BaF_2, respectively, and only one relatively weak peak at 898 cm^{-1} was detected on 80% LaF_3/BaF_2. The peaks at 898 and 920 cm^{-1} may be tentatively assigned to O_2^{2-} species,[9,10,11,19,20] though the probable existence of surface-Madelung-potential-stabilized O_3^{2-} species can not be ruled out as to be elaborated elsewhere,[21] and the peaks at 1130 and 1188 cm^{-1} may be assigned to the O_2^- species[12,13] in different micro-environments. No peak was observed between 600 and 1500 cm^{-1} over the "blank" sample. Comparing the results of above Raman analysis and the results of catalytic performance evaluation (Table 3), It is interesting to see that, with the increase of LaF_3/BaF_2 ratio in the catalysts, both the relative content of O_2^- species (compared with the content of O_2^{2-} or O^- species) on the catalysts surface and the catalyst selectivity to C_2H_4 decrease. These phenomena suggest that the O_2^- may be a more favorable species than the O_2^{2-} or O^- to the selective oxidative-dehydrogenation of C_2H_6 to C_2H_4.

in situ FTIR Spectroscopy Characterization

Figure 1a shows the IR spectrum of 20%SrF_2/La_2O_3. After 20 min. of treatment with H_2 at 700°C. A broad band centered at c.a. 900 cm^{-1} was observed, which could be accounted for by overtone and combination of La-O fundamental modes.[22] When the above sample was exposed to O_2 at 650°C, a new band at 1113 cm^{-1} appeared. This band may be assigned to O-O stretching vibration of adsorbed superoxide species (O_2^-).[14,23] Injection of 20 ml of CO_2 with the flow of O_2 into the IR cell maintained at 650°C; the band intensity of O_2^- species decreased slightly, and three intense bands attributable to the CO_3^{2-} species (1451, 856 cm^{-1}) on the surface of La_2O_3[24] and the gas phase CO_2 (2354 cm^{-1})[25] were detected. Since no other oxygen adspecies were observed on the surface of 20%SrF_2/La_2O_3 at 650°C, O_2^- adspecies may be considered to be the active oxygen species for the oxidative coupling of methane. When the temperature of above O_2-adsorbed sample was gradually cooling down from 650°C to 25°C under O_2 atmosphere (Figure 1b~e), the bands at 1113 and 900 cm^{-1} shifted to higher wavenumbers (1123 cm^{-1} and 917 cm^{-1}), and two new bands at 1483, 1402 cm^{-1} and a shoulder peak at 863 cm^{-1} were detected, these bands were attributable to the surface CO_3^{2-} species[24] which might be generated by the gradual reaction of oxygen with the surface carbonaceous impurity reduced previously in the H_2 treatment procedure.

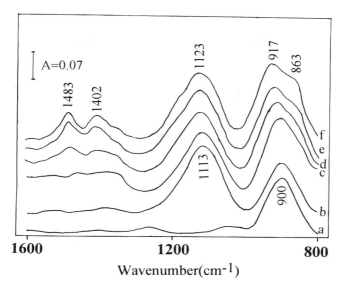

Figure 1. IR spectra of 20% SrF$_2$/La$_2$O$_3$ Catalyst, (a) After treatment with H$_2$ at 700°C. (b) After exposure to O$_2$ at 650°C, followed by cooling down successively under O$_2$ to (c) 500°C, (d) 300°C, (e) 100°C and (f) 25°C, respectively.

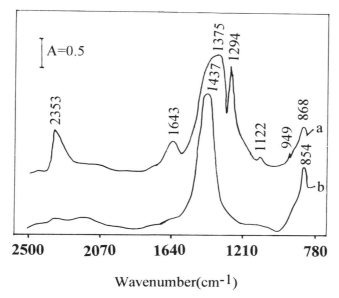

Figure 2. IR spectra of 20% SrF$_2$/La$_2$O$_3$ catalyst. (a) After treatment of O$_2$-pre-adsorbed 20% SrF$_2$/La$_2$O$_3$ with CH$_4$/O$_2$ (3/1) at 650°C, (b) followed by purging with He at 650°C.

After the treatment of O_2-pre-adsorbed 20%SrF_2/La_2O_3 sample with a stream of CH_4/O_2 (3:1) at 650°C for 45 min., the IR spectrum of the sample is shown in Figure 2a. The superoxide (O_2^-) band became weaker and shifted to 1122 cm^{-1}, at the same time, the absorption bands of gas phase CO_2 (2353 cm^{-1}),[25] adsorbed H_2O (1643 cm^{-1}), gaseous CH_4 (1294 cm^{-1})[26] and surface CO_3^{2-} (868 cm^{-1})[24] were observed, it was also interesting to note that a weak peak at 949 cm^{-1}, which was assignable to gaseous C_2H_4,[27] was also observed on the working 20%SrF_2/La_2O_3 catalyst. These observations indicated that, under the above spectra recording conditions, the mathane oxidative coupling reaction do happened on the 20%SrF_2/La_2O_3 catalyst in the IR cell, probably due to the lower yield of C_2H_6 (7.5%) than that of C_2H_4 (12.5%) under this reaction conditions, the bands of C_2H_6 were not detected. The broad band with maximum around 1375 cm^{-1} might have resulted from the overlapping peaks of CH_4 (1349 cm^{-1})[26] and CO_3^{2-} (1437 cm^{-1}),[24] and the band at 900 cm^{-1} might have been buried under the intense CO_3^{2-} band centered at 868 cm^{-1}. The peaks due to adsorbed CH_4 were not detected, suggesting that under the reaction conditions, the adsorption of CH_4 on the catalyst surface must be very weak. After purging the above sample with a flow of He at 650°C for 5 min., only the peaks of surface CO_3^{2-} species (854, 1437cm^{-1})[24] were shown in the IR spectrum (Figure 2b).

ACKNOWLEDGMENT

This work has been supported by the National Science Foundation of China.

CONCLUSIONS

Based on above results, we may conclude that, with the addition of metal fluoride to metal oxide and treatment in air at high temperature, anionic and/or cationic exchange between oxides and fluorides lattices took place to some extent, leading to the formation of new phases, anion vacancies, O^- ions, quasi-free electrons as well as expansion/contraction of lattices. These factors should be responsible to the significant improvement of the catalytic performance. In consideration of stronger electronegativity of F than O, the catalyst containing F^- might be more conducive to the formation of some oxygen species with fewer negative charges that would be favorable to the selective conversion of alkanes to alkenes. On the other hand, the dispersion of "inert" fluorides on the catalyst surface will be also beneficial to the isolation of the surface active center and decrease of CO_2-inhibition as shown by the smaller carbonate peaks versus F^--free composite metal-oxide catalyst system, and will therefore be favorable to the lowering of the "light-off" temperature and the improvement of selectivity.

REFERENCES

1. a. R. Burch, G.D. Squire, and S.C. Tsang, Comparative study of catalysts for the oxidative coupling of methane, *Appl. Catal* 43:105 (1988).
    b. S.J. Conway, D.J. Wang, and J.H. Lunsford, Selective oxidation of methane and ethane over Li+-MgO-Cl- catalysts promoted with metal oxides, *Appl. Catal. A : General* 79:L1 (1991).
    c. V.D. Sokolovskii, and E.A. Mamedov, Oxidative dehydrodimerization of methane, *Catal. Today* 14 (3/4):415 (1992).

2. T.R. Baldwin, R. Burch, E.M. Carbb, G.D. Squire, and S.C. Tsang, Oxidative coupling of methane over chloride catalyst, *Appl. Catal.* 56:219 (1989).

3. Fan-Cheng Wang, Z.L. Zhang, C.T. Au, and K.R. Tsai, Methane oxidative coupling to C_2 hydrocarbons over MnO-NaCl/TiO$_2$ catalyst, *Preprints of 3B Symp. on Methane Activation, Conversion and Utilization* 1989 International Chemical Congress of Pacific Basin Societies, Honolulu, Hawaii, 2 (1989).

4. a. Xiao-Ping Zhou, Wei-De Zhang, Hui-Lin Wan, and Khi-Rui Tsai, Methane oxidative coupling over metal oxyfluoride catalysts, *Chinese Chem. Lett.* 4:603 (1993).

 b. Xiao-Ping Zhou, Shui-Qin Zhou, Shui-Ju Wang, Wei-Zheng Weng, Hui-Lin Wan, and Khi-Rui Tsai, Methane oxidative coupling over fluoride-promoted cerium oxide catalysts, *Chem. Res. Chinese Univ.* 9:264 (1993).

 c. Xiao-Ping Zhou, Shui-Qin Zhou, Fu-Chun Xu, Shui-Ju Wang, Wei-Zheng Weng, Hui-Lin Wan, and Khi-Rui Tsai, Investigation on high efficient catalyst for the oxidative dehydrogenation of ethane, *Chem. Res. Chinese Univ.* 9:269 (1993).

 d. Xiao-Ping Zhou, Shui-Ju Wang, Wei-Zheng Weng, Hui-Lin Wan, and Khi-Rui Tsai, Activation of O_2 and catalytic properties of CeO_2/CaF_2 catalysts for methane oxidative coupling, *J. Nat. Gas Chem.* (China) 2:280 (1993).

 e. Xiao-Ping Zhou, Wei-De Zhang, Shui-Ju Wang, Hui-Lin Wan, and Khi-Rui Tsai, Methane oxidative coupling over alkaline-earth fluoro-oxide catalysts, *J .Nat .Gas Chem.* (China) 2:344 (1993).

 f. Shuiqin Zhou, Xiaoping Zhou, Huilin Wan, and K. R. Tsai, Oxidative coupling of methane over BaF_2-TiO_2 catalysts, *Catal. Lett.* 20:179 (1993).

5. A. Sher, R. Solomon, K. Lee, and M.W. Muller, Transport properties of LaF_3, *Phys. Rev.* 144:593 (1966).

6. L. Eyring, Chapter 27. The binary rare earth oxides, *in* "Handbook on the Physics and Chemistry of Rare Earths Vol. 3", K.A. Gschneidner, Jr and L.R. Eyring, ed., North-Holland Physics Publishing, (1979).

7. M. Che, and A.J. Tench, Characterization and reactivity of mononuclear oxygen species on oxide surfaces, *Adv. Catal.* 31:77 (1982).

8. Y. Inoue, and I. Yasumori, Catalysis by alkaline earth metal oxides. III. X-ray photoelectron spectroscopic study of catalytically active MgO, CaO and BaO surfaces, *Bull. Chem. Soc. Jpn.* 54:1505 (1981).

9. A. Metcalfe, and S. Ude Shankar, Interaction of oxygen with evaporated silicon films, *J. Chem. Soc. Faraday Trans.* I 76:630 (1980).

10. J.L. Gland, B.A. Sexton, and G.B. Fisher, Oxygeninteractions with the Pt(111) surface, *Surf. Sci.*, 95:587 (1980).

11. R.D. Jones, D.A. Summerville, and F. Basolo, Synthetic oxygen carriers related to biological system, *Chem. Rev.* 79:139 (1979).

12. D. McIntosh, and G.A. Ozin, Metal atom chemistry and surface chemistry. 1. Dioxygen-silver, Ag^+,O_2^-, and tetraoxygensilver, Ag^+,O_4^-, reactive intermediates in the silver atom-dioxygen system relevance to surface chemistry, *Inorg. Chem.* 16:59 (1977).

13. a. L. Andrews, Matrix reactions of K and Rb atoms with oxygen molecules, *J. Chem. Phys.* 54:4935 (1971).

 b. L. Andrews, J. T. Hwang, and C. Trindle, Matrix reactions of cesium atoms with oxygen molecules. Infrared spectrum and vibrational analysis of $Cs^+O_2^-$. Infrared observation of $Cs^+O_2^{2-}Cs^+$ and $Cs^+O_4^-$. Theoretical structure elucidation of $M^+O_4^-$, *J. Phys. Chem.* 77:1065 (1973).

14. L. Vaska, Dioxygen-metal complexes:toward a unified view, *Acc. Chem. Res.* 9:175 (1976).

15. C. Li, K. Domen, K. Maruya, and T. Onishi, Dioxygen adsorption on well-outgassed and partially reduced cerium oxide studied by FT-IR, *J. Am. Chem. Soc.* 111:7683 (1989).

16. M. Che, and A.J. Tench, Characterization and reactivity of molecular oxygen species on oxide surface, *Adv. Catal.* 32:82 (1983).

17. a. A.P. Baddorf, and B.S. Itchkawitz, Identification of oxygen species on single crystal K(110), *Surf. Sci.* 264:73 (1992).

b. J.P.S. Badyal, X. Zhang, and R.M. Lambert, A model oxide catalyst system for the activation of methane:Lithium-dopted NiO on Ni(111), *Surf. Sci.* 225:L15 (1990).

c. J.L. Dubois, M.Bisiaux, H. Mimoun, and C.J. Cameron, X-ray photoelectron spectroscopic studies of lanthanum oxide based oxidative coupling of methane catalysts, *Chem. Lett.* 967 (1990).

18. W.H. Zachariasen, Crystal Chemical studies of the 5f-series of elements. XIV. Oxyfluorides, XOF, *Acta Cryst.* 4:231 (1951).

19. J.S. Valentine, Dioxygen ligand in mononuclear group VIII transition metal complexes, *Chem. Rev.* 73:237 (1973).

20. F. Al-Mashta, N. Sheppard, V. Lorenzelli, and G. Busca, Infrared study of adsorption on oxygen-covered α-Fe_2O_3: Bands due to adsorbed oxygen and their modification by co-adsorbed hydrogen or water, *J. Chem. Soc. Faraday Trans.* I 78:979 (1982).

21. K.R. Tsai, H.L. Wan, H.B. Zhang, and G.D. Lin, Catalyst design in oxidative dehydrogenation of C_1-C_4 alkanes to alkenes over rare-earth-based catalysts *Proc. 34th IUPAC Cong.* Beijing, China, 671 (1993).

22. J.H. Denning, and S.D. Ross, The vibrational spectra and structures of rare earth oxides in the a modification, *J. Phys. C: Solid State Phys.* 5:1123 (1972).

23. A.B.P. Lever, G.A. Ozin, H.B. Gray, Electron transfer in metal-dioxygen adducts, *Inorg. Chem.* 19:1823 (1980).

24. T.L. Van, M. Che, J.M. Tatibouet, and M. Kermarec, Infrared study of formation and stability of $La_2O_2CO_3$ during the oxidative coupling of methane on La_2O_3, *J. Catal.* 142:18 (1993).

25. S.C. Bhumkar, and L.L Lobban, Diffuse reflectance infrared and transient studies of oxidative coupling of methane over Li/MgO catalyst, *Ind. Eng. Chem. Res.* 31:1856 (1992).

26. "Standard Infrared Grating Spectra, V43-44", published by Sadtler Research Laboratories, Inc, 42923P(1974).

27. C. Brecher, and R.S. Halford, Motions of molecules in condensed systems. XI. Infrared spectrum and structure of a single crystal of ethylene, *J. Chem. Phys.* 35:1109 (1961).

OXIDATIVE COUPLING OF METHANE OVER SULFATED SrO/La_2O_3 CATALYSTS

János Sárkány, Qun Sun, Juana Isabel Di Cosimo, Richard G. Herman, and Kamil Klier

Zettlemoyer Center for Surface Studies and Department of Chemistry
Sinclair Laboratory, 7 Asa Drive
Lehigh University
Bethlehem, PA 18015

INTRODUCTION

Saturated linear hydrocarbons, particularly methane, are major components of natural gas and of the gas produced by certain coal gasifiers. While methane makes an excellent gaseous fuel, it is desirable to convert it *via* 1-step processes to higher molecular weight products for transportation, storage, and utilization as chemical feedstocks. The desired primary reactions are shown in Equations 1-4, while Equation 5 is also a reaction of interest:

$$CH_4 + 0.5\ O_2 \rightarrow CH_3OH \tag{1}$$

$$CH_4 + O_2 \rightarrow CH_2O + H_2O \tag{2}$$

$$2CH_4 + O_2 \rightarrow C_2H_4 + 2H_2O \tag{3}$$

$$2CH_4 + 0.5\ O_2 \rightarrow C_2H_6 + H_2O \tag{4}$$

$$2CH_4 + 2CO_2 \rightarrow C_2H_4 + 2CO + 2H_2O. \tag{5}$$

After Keller and Bhasin published their research results for methane coupling over a decade ago,[1] many laboratories have been striving to develop efficient methane conversion catalysts and technologies. Various aspects of the state of the art of methane oxidation, including early developments, have been reviewed by Foster in 1985,[2] Gesser et al. in 1985,[3] Pitchai and Klier in 1986,[4] Scurrell in 1987,[5] Lee and Oyama in 1988,[6] Hutchings et al. in 1989,[7] Amenomiya et al. in 1990,[8] Lunsford in 1990 and 1991,[9,10] Mackie in 1991,[11] Hamid and Moyes in 1991,[12] Krylov in 1993,[13] and Zhang et al. in 1994.[14] Therefore, the literature will not be extensively reviewed here.

While the catalytic oxidative coupling paths represented by Equations 3 and 4 show considerable promise, it is evident from patent examples that the process conditions are still quite severe, i.e. the usual reaction temperature of 650-800°C is still too high. Reactions leading to oxygenates (Equations 1 and 2) are more difficult to conduct selectively, but they have been identified as being very desirable, particularly the oxidation to methanol.[15] At the same time, the standard free energy of all the oxidations, Equations 1-4, is negative over a wide range of temperatures, establishing a thermodynamic driving force for these reactions even at room temperature should an effective catalyst be found. More practical considerations led us to seek a desirable temperature range of 350-650°C. The lower limit of 350°C is based on experience with the dehydration of most oxide catalysts, which lose water at temperatures $\leq$350°C. The upper limit of 650°C is in the range of temperatures at which uncontrolled free radical reactions will occur and often will lower the selectivity by driving the oxidation process to CO and CO_2.

Oxide catalysts were chosen here, in particular the very basic SrO/La_2O_3 catalyst that is active in the formation of methyl radicals, and therefore of C_2 hydrocarbons.[16,17] However, the methyl radicals formed over SrO/La_2O_3 can be trapped by a subsequent catalyst bed, e.g. MoO_3/SiO_2[18] but not SiO_2,[19] and at least partially converted to CH_2O. Since the formation of surface carbonates might be at least partially responsible for deactivation of basic catalysts under some methane conversion reaction conditions, the catalytic behavior and/or stability of the very basic SrO/La_2O_3 might be enhanced by acidic doping. With analogy to earlier findings that sulfate ion strongly enhanced the acidic properties of iron oxide,[20] as well as alumina and titania,[21] SO_4^{2-} was used as an acidic surface dopant to improve the catalytic performance of the 1 wt% SrO/La_2O_3 catalyst.

EXPERIMENTAL

The 1 wt% SrO/La_2O_3 catalyst used in this study was prepared by AMOCO Oil Co. using the procedure described by DeBoy and Hicks.[16,17] The sulfated SrO/La_2O_3 catalysts were produced by the incipient wetness impregnation technique. The appropriate amount of $(NH_4)_2SO_4$ was dissolved in deionized water, the measured quantity of SrO/La_2O_3 was added, and the slurry was continuously stirred with a magnetic stirrer until dryness was achieved. This was followed by drying the solid overnight at 120°C and calcination in air at 600°C for 6 hr. Prior to catalytic testing, the samples were activated *in situ* under air (or O_2) flow at 500°C for 1 hr unless stated otherwise. The gases used in this study were zero grade purity and were used without further purification.

Catalytic testing was carried out in a fixed-bed continuous-downflow 9 mm OD (7 mm ID) quartz reactor that narrowed just below the catalyst bed to 4 mm OD (2 mm ID). In nearly all tests, 0.1000 g of catalyst was used. The catalytic tests of the catalysts included the standard conditions of employing a CH_4/air = 1/1 reactant mixture at 1 atm (0.1 MPa) and with gas hourly space velocity (GHSV) = 70,000 ℓ/kg catal/hr over the range of temperatures of 500-700°C. Usually each test was carried out at steady state methane conversion at each designated set of reaction conditions for 1-3 hr. The principal products analyzed by on-line automated sampling of the exit gas using gas chromatography (Hewlett-Packard 5890 PC-controlled GC) were CO_2, C_2 hydrocarbons (C_2H_6 + C_2H_4), CO, and H_2O. Usually not included were C_3 compounds that were present in only small traces. Formaldehyde was collected in two water-filled scrubbers in series, the first was kept at room temperature and the second at 0°C, and then quantitatively determined by the modified Romijn's iodometric titration method.[22] In the present research, the carbon mass balance was always better than 90% and usually better than 95%.

RESULTS AND DISCUSSION

Activity and Selectivity of the SrO/La₂O₃ Catalyst

To determine the degree of reproducibility of catalyst testing of the AMOCO Oil Co. SrO/La₂O₃ catalyst, three different researchers tested four different portions of the catalyst. It was found that this catalyst was active at 500°C, and it is shown in Figure 1 that catalytic behavior of the SrO/La₂O₃ catalyst was reproducible and that no deactivation was observed. In this figure, the open data points were obtained while increasing the temperature, while the solid data points were obtained while the temperature was decreased in a stepwise manner. The conversion at 500°C is very sensitive to the reaction temperature since the methane oxidation reaction, once initiated, is highly exothermic. Thus, the data points indicated for 500°C have appreciable scatter, and the data points indicating ≈10-13% conversion could actually be for a slightly higher temperature than 500°C.

It is notable that methane was converted to products over the SrO/La₂O₃ catalyst at temperatures as low as 500°C. This contrasts with the typical steam reforming reaction of methane over commercial Ni-based catalysts that is carried out at ≈850°C. At temperatures higher than 550°C under the reaction conditions employed here, over 70% of the oxygen was consumed. Thus, at the higher temperatures studied, the conversion level of CH_4 was limited by the availability of the O_2 reactant. The %yields of total C_2 products as a function of temperature under these reaction conditions are shown in Figure 1. The %yield is defined as the product of the total C_2 selectivity (C-mol% ethane + ethene) and the total conversion of methane (mol%). Once again, the data indicated that the SrO/La₂O₃ catalyst was quite reproducible and stable.

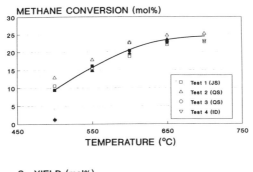

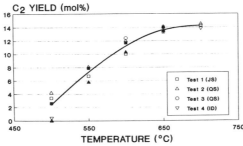

Figure 1. Methane conversion and C_2 hydrocarbon yield (mol%) over the 1 wt% SrO/La₂O₃ catalyst with CH_4/air = 1/1 at 1 atm and GHSV = 70,000 ℓ/kg catal/hr. The open symbols represent sequentially increasing reaction temperatures, while the filled symbols represent decreasing reaction temperatures.

Effect of Sulfate Concentration

The effect of sulfate content on the catalytic behavior of the 1 wt% SrO/La_2O_3 catalyst was examined at 500°C under the standard testing conditions. The content of sulfate added to the 1 wt% SrO/La_2O_3 varied as 0, 0.5, 1.0, 2.0, and 4.0 wt% SO_4^{2-} of the total weight of catalyst. The 1 wt% SO_4^{2-}/1 wt% SrO/La_2O_3 showed the largest promotional effects on the methane conversion (Figure 2) and C_2^+ selectivity (Figure 3). In each case, as the activity increased, the C_2 selectivity also increased.

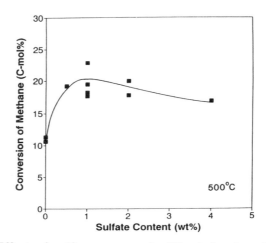

Figure 2. Effect of sulfate content (wt%) of the 1 wt% SrO/La_2O_3 catalyst on methane conversion (carbon mol%) to all products from a CH_4/air = 1/1 reactant mixture at 500°C, 1 atm, and GHSV = 70,000 ℓ/kg catal/hr.

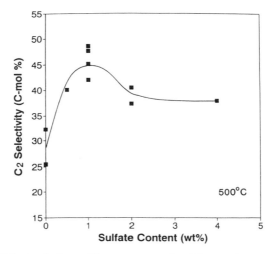

Figure 3. Effect of the sulfate content (wt%) on the C_2 hydrocarbon selectivity observed at 500°C over the 1 wt% SrO/La_2O_3 catalyst. The experimental parameters are given in Figure 2.

From these figures, it can be seen that adding an acidic sulfate promoter to the catalysts to a 1 wt% level increased the overall conversion of methane and selectivity to C_2 hydrocarbons by factors of 1.5-2.0. The sulfate doping of the catalyst also increased the %yields of the C_2 hydrocarbon products and the total conversion of methane [mol fraction]. Thus, adding 1 wt% SO_4^{2-} to the catalyst increased the %yield at 500°C from 3% to 7.5-11%, e.g. the catalyst showing 45% C_2 selectivity in Figure 3 exhibited a %yield for the C_2 hydrocarbons of 9%. At this reaction temperature, all sulfated catalysts exhibited higher methane conversions and C_2 selectivities than the non-sulfated SrO/La_2O_3 catalyst.

Effect of Reaction Temperature

The effect of temperature on the catalytic activity of the 1 wt% SO_4^{2-}/1 wt% SrO/La_2O_3 catalyst was determined and compared with the nonsulfated catalyst. The sulfate-promoted catalyst, compared to the nonsulfated one, again showed a large increase for methane conversion at 500°C, as indicated in Figure 4. Upon increasing the temperature stepwise to 700°C, the conversion levels increased. However, the extent of the promotional effect decreased with increasing temperature, and practically no effect at reaction temperatures >650°C was observed. At the high CH_4 conversion levels, the reactant oxygen was nearly or totally depleted (>85-90%), which limited the oxidative reactions with methane. This, as well as other factors, contributed to the converging catalytic behavior.

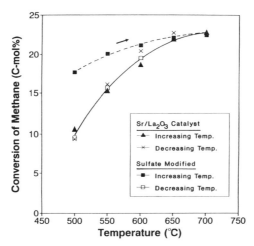

Figure 4. Effect of temperature on methane conversion in the range of 500-700°C over 1 wt% SrO/La_2O_3 and 1 wt% SO_4^{2-}/1 wt% SrO/La_2O_3 under standard testing conditions.

Upon sequentially decreasing the reaction temperature after the 700°C experiment, the former promoting effect was not observable even at 500°C. Plots of the C_2 selectivities and C_2 %yields exhibited the same behavior as the conversion curves in Figure 4, i.e. the C_2 selectivity increased as the CH_4 conversion increased. In those plots, not shown here, the presence of the sulfate anion caused a large promoting effect at 500°C, where the C_2 selectivity and C_2 yield increased from 32.2 to 45.1% and from 3.4 to 8.0%, respectively. However, the promotional behavior disappeared after completing the temperature cycle. In both cases, the %yield of C_2 hydrocarbons was $\approx 2.4\%$ for the final 500°C tests. A new pretreatment of the tested catalyst in air at 500°C for 1 hr did not restore the former enhanced activity obtained at this temperature. With regard to the C_2 hydrocarbon products, the ethene/ethane ratio increased with methane conversion and with total C_2 selectivity, reaching approximately 1/1 at the higher reaction temperatures.

Possible explanations of the instability of the promotional effect observed over the sulfate doped catalyst include the high O_2 conversion (>85-90%) at the higher reaction temperatures, resulting in a reducing environment that could transform the sulfate to a volatile component. This would result in the loss of the acidic surface dopant. In addition, the thermal treatments utilized during the catalytic testing could have altered the surface areas of the catalysts. Both of these possibilities were examined as causes of the loss of the enhanced catalytic properties of the sulfated catalyst that were observed for the moderation temperature reaction conditions. The pretreated sulfated catalyst had an initial BET surface area of 16.2 m²/g and a sulfate content of 0.93%. After the catalytic testing sequence that reached 700°C, the catalyst had a lower surface area of 8.2 m²/g and a sulfate content of 0.36% (Table 1). The corresponding surface areas of the pretreated and tested SrO/La_2O_3 catalyst were 10.4 and 7.0 m²/g, respectively. Thus, there was a loss of both sulfate and surface area induced by the high temperature testing. The final surface areas of the catalysts after high temperature testing were similar, as shown in Table 1. Subsequent studies showed that the loss of initial surface area and stabilization at 7-9 m²/g occurred for both catalysts during the first hour of testing under methane conversion conditions, even at 500°C, e.g. to 7.5 and 9.2 m²/g for the SrO/La_2O_3 and $SO_4^{2-}/SrO/La_2O_3$ catalysts, respectively. Further chemical analyses are underway, and analyses by XPS and laser Raman spectroscopy, including *in situ* experiments, are being carried out, especially with respect to the presence and role of carbonate on the catalysts.

Table 1. Chemical[1] and Surface Analysis[2] of SrO/La_2O_3-Based Catalysts

Sample	Sulfate (wt%)	Carbonate (wt%)	BET Area (m²g⁻¹)
(A) Before Reaction			
1 wt% SrO/La_2O_3	---	3.30	10.4
1 wt% $SO_4^{2-}/SrO/La_2O_3$	0.93	4.35	16.2
(B) Pretreatment only at 800°C (for 1 hr in Air or He)			
1 wt% $SO_4^{2-}/SrO/La_2O_3$ (Air)	0.96	---	15.2
1 wt% $SO_4^{2-}/SrO/La_2O_3$ (He)	0.93	---	13.2
(C) After Reaction (500 → 700 → 500°C; GHSV = 70,000 ℓ/kg/hr; CH_4/Air = 1/1)			
1 wt% SrO/La_2O_3	---	7.50	7.0
1 wt% $SO_4^{2-}/SrO/La_2O_3$	0.36	5.35	8.2
(D) Reaction at 550°C with Low GHSV (70,000 → 5,400 ℓ/kg/hr; CH_4/Air = 1/1)			
1 wt% $SO_4^{2-}/SrO/La_2O_3$	0.87	15.35	

[1]The chemical analyses were carried out by Galbraith Laboratories, Inc.
[2]The BET areas were measured with a Micrometrics Gemini-2360 instrument.

Stability at 550°C

It was found that decreasing the GHSV of the reactant from ≈70,000 ℓ/kg catal/hr to 5,400-17,000 ℓ/kg catal/hr resulted in the lost of activity (to ≈8% CH_4 conversion) and selectivity (to 1-5 C-mol% C_2). Upon returning to the original reaction conditions, only ≈50% of the activity and selectivity was recovered. As indicated in Table 1, a dominant factor appeared to be build-up of carbonate on the catalyst during this test. Indeed, increasing the reaction temperature in small increments to 600°C at GHSV = 70,175 ℓ/kg catal/hr regenerated the original activity and selectivity of this catalyst. A sudden jump in both parameters was noted as the temperature passed through 585°C. Upon testing a fresh catalyst under the standard reaction conditions at 550°C for 25 hr, no deactivation was observed, and the CH_4 and O_2 conversions were maintained at ≈20 and ≈85 mol%, respectively, while the selectivities of C_2 hydrocarbons and of CO_x were both ≈50 C-mol%. These experimental results correspond to a space time yield of the C_2 hydrocarbon products of 2 kg/kg catal/hr at 550°C.

These changes in activity and selectivity appear to be interrelated to the partial pressure of CO_2 over the catalyst. It is now being established by a number of research groups (see other papers in this volume) that basic catalysts are susceptible to carbonate formation on the catalysts and that this carbonate layer can directly affect the resultant catalytic properties. This appears to be surface poisoning by blockage of the active sites, which can be reversed under some reaction conditions by decomposition of the carbonate. This is being investigated further.

CONCLUSIONS

Acidic doping of the strongly basic 1 wt% SrO/La_2O_3 catalyst with sulfate promoted both the conversion and selectivity in the oxidative coupling of CH_4 to C_2 hydrocarbons. At temperatures that are moderate for oxidative coupling, e.g. 550°C with a CH_4/air = 1/1 reactant gas mixture and with GHSV = 70,000 ℓ/kg catal/hr, the sulfate doping showed a maximum effect at 1 wt% SO_4^{2-} concentration. For this catalyst, the space time yield at 550°C was 72 mol/kg catal/hr of C_2 hydrocarbons with 20% CH_4 conversion and 50% C_2 selectivity. The sulfated catalysts revealed high stability in the catalytic performance at 550°C with CH_4/air = 1/1 at GHSV = 70,000 ℓ/kg catal/hr; no deactivation was observed during a 25 hr test. The C_2 hydrocarbon selectivity correlated in a general fashion with CH_4 conversion for catalysts of all compositions, both sulfated-doped and undoped catalysts, used under both increasing and decreasing temperature testing sequences between 500 and 700°C, supporting the same principal mechanism for all samples. At lower GHSV values (down to 5,400 ℓ/kg catal/hr), it appeared that carbonate build-up partially poisoned the 1 wt% SO_4^{2-}/SrO/La_2O_3 catalyst at 550°C, but the deactivation was reversed by increasing the reaction temperature above 580°C at GHSV = 70,000 ℓ/kg catal/hr.

ACKNOWLEDGEMENTS

This research was partially supported by the U.S. Department of Energy-Morgantown Energy Technology Center and the AMOCO Corp. under the AMOCO University Natural Gas Conversion Program.

REFERENCES

1. Keller, G. E. and Bhasin, M. M., J. Catal., 73, 9 (1982).
2. Foster, N. R., Appl. Catal., 19, 1 (1985).
3. Gesser, H. D., Hunter, N. R., and Prakash, C. B., Chem. Rev., 85(4), 235 (1985).
4. Pitchai, R. and Klier, K., Catal. Rev.-Sci. Eng., 28, 13 (1986).
5. Scurrell, M. S., Appl. Catal., 32, 1 (1987).
6. Lee, J. S. and Oyama, S. T., Catal. Rev.-Sci. Eng., 30, 249 (1987).
7. Hutchings, G. J., Scurrell, M. S., and Woodhouse, J. R., Chem. Soc. Rev., 18, 251 (1989).
8. Amenomiya, Y., Birss, V. I., Goledzinowski, M., Galuska, J., and Sanger, A. R., Catal Rev.-Sci. Eng., 32, 163 (1990).
9. Lunsford, J. H., Catal. Today, 6, 235 (1990).
10. Lunsford, J. H., in "*Natural Gas Conversion*" (Studies in Surface Sci. and Catal., Vol. 61), ed. by A. Holmen, K.-J. Jens and S. Kolboe, Elsevier, New York, 3 (1991).
11. Mackie, J. C., Catal. Rev.-Sci. Eng., 33, 169 (1991).
12. Hamid, H. B. A. and Moyes, R. B., Catal. Today, 10, 267 (1991).
13. Krylov, O. V., Catal. Today, 18, 209 (1993).
14. Zhang, Z., Verykios, X. E., and Baerns, M., Catal. Rev.-Sci. Eng., 36, 507 (1994).
15. Dautzenberg, F. M., in "*Preprints, Symp. on Methane Activation, Conversion, and Utilization, PACIFICHEM '89*," Intern. Chem. Congr. of Pacific Basin Societies, Honolulu, HI, Paper No. 170, pp 163-165 (1989).
16. DeBoy, J. M. and Hicks, R. F., Ind. Eng. Chem. Res., 27, 1577 (1988).
17. DeBoy, J. M. and Hicks, R. F., J. Catal., 113, 517 (1988).
18. Sun, Q., Di Cosimo, J. I., Herman, R. G., Klier, K., and Bhasin, M. M., Catal. Lett., 15, 371 (1992).
19. Sun, Q., Herman, R. G., and Klier, K., Catal. Lett., 16, 251 (1992).
20. Kayo, A., Yamaguchi, T., and Tanabe, K., J. Catal., 83, 99 (1983).
21. Saur, O., Bensitel, M., Mohammed Saad, A. B., Lavalley, J. C., Tripp, C. P., and Morrow, B. A., J. Catal., 99 104 (1986).
22. Walker, J. F., "Formaldehyde," 3rd Ed., ACS Monograph 159, Reinhold Publ. Co., New York, 489 (1964).

THE OXIDATIVE COUPLING OF METHANE OVER ZrO$_2$ DOPED Li/MgO CATALYSTS

G.C. Hoogendam[1], K. Seshan[1], J.G. van Ommen[1], J.R.H. Ross[2]

1) University of Twente, Faculty of Chemical Technology, P.O. Box 217. 7500AE Enschede, The Netherlands (phone +31-53-892860; fax +31-53-339546)
2) University of Limerick, Plassey Technology Park, Limerick, Ireland (phone +353-61-333644)

INTRODUCTION

Since the early reports of Keller and Bhasin (1) the Oxidative Coupling of Methane (OCM) has been extensively studied (2,3). A lot of catalyst systems have been proposed with widely ranging activities and selectivities. Most studies are aimed at finding the highest possible yields for ethane and ethylene. However, in several economic evaluations it has been shown that a high selectivity is more important than a high activity(3). Since its discovery by Lunsford and co-workers (4), Li/MgO has been one the most studied catalysts (5,6,7).

The first model for active site formation in Li/MgO catalysts was made by Lunsford et al. (4). ESR results had shown the presence of [Li$^+$O$^-$] centers. The following reactions were proposed for their formation:

$$2Li^+O^{2-} \;+\; \square \;+\; 1/2\,O_2 \;\Leftrightarrow\; 2\,Li^+O^- + O^{2-}$$

with $\square$ denoting an oxygen vacancy. The number of active sites is mainly determined by the oxygen partial pressure and the temperature. It was also observed that the rate of methyl radical formation behaved in a parallel manner to the Li loading. thus, it was concluded that that CH$_3$· was produced according to the following reaction:

$$Li^+O^- \;+\; CH_4 \;\rightarrow\; Li^+OH^- \;+\; CH_3$$

The active site was regenerated with molecular oxygen in a two-step mechanism.

The disadvantage of Li/MgO lies in the poor stability. However, it is still one of the most selective catalysts. Previous studies have shown that adding a third metal oxide to the Li/MgO system can significantly improve the stability and activity while maintaining a high selectivity (8). Of the tested oxides, SnO_2 was found to be the most promising (9). A kinetic study on a Li/Sn/MgO catalyst showed that, in comparison with Li/MgO, CH_4 and O_2 are more strongly adsorbed. This observation can explain the increased activity at relatively low temperatures. However, it was not understood why the Li/Sn/MgO was more stable than undoped Li/MgO. A later study showed that the Li/Sn/MgO catalysts contained a mixed phase, $Li_2Mg_3SnO_6$. Like MgO, this phase has a cubic structure. Due to the small difference in lattice parameters, XRD reflections of $Li_2Mg_3SnO_6$ always appear immediately next to or under the MgO reflections, making detection of the mixed phase more difficult. The discovery of $Li_2Mg_3SnO_6$ has led to a more detailed study on mixed oxide formation in doped Li/MgO catalysts (10, 11). In this paper, the effect of ZrO_2 as dopant for Li/MgO is studied. ZrO_2 doped Li/MgO was previously identified as a active and stable catalyst for OCM (10).

EXPERIMENTAL

Details of catalyst preparation are given elsewhere (8). Two series of ZrO_2 doped Li/MgO catalysts were prepared. In the first series the lithium concentration varied between 0.8 and 9 wt%, while the amount of ZrO_2 was constant (ca. 2wt%). The second series consisted of catalysts with increasing ZrO_2 concentration (0.3 to 21wt%) while the lithium concentration was constant. All ZrO_2 doped Li/MgO catalysts contained three phases: Li_2CO_3, Li_2MgZrO_4 and MgO. In the first series, MgO was the main phase in all catalysts, while catalysts with less than 4 wt% Li showed the Li_2MgZrO_4 as the most significant phase. Catalysts that contained a higher wt% lithium contained Li_2CO_3 as the second most predominant phase and Li_2MgZrO_4 as the least significant phase. Details on the second series are given in (9). In short, an increasing amount of ZrO_2 resulted in an increased amount of Li_2MgZrO_4, and a decreasing amount of Li_2ZrO_2. Catalyst testing was done in a quartz reactor (5 mm id). 375 mg of catalyst was placed in the reactor in between two quartz wool plugs. The temperature inside the reactor was measured at the end of the catalyst bed. Typically, the feed consisted of CH_4:O_2:He = 67:7:26, with a total flow rate of 100 ml/min. The products were analysed by a Varian 3400 gas chromatograph equipped with a molsieve column for separation of H_2, O_2, N_2, CO and CH_4 and a haysep N column for separation of all other products, including H_2O. Measurements were performed at temperatures in the range 600-825 ° C.

RESULTS AND DISCUSSION

A series of ZrO_2 doped Li/MgO catalysts with a varying lithium concentration but a fixed amount of ZrO_2 was used to study the effect of lithium loading. Figure 1 shows the methane conversion and C_2 selectivity as a function of lithium concentration at 700°C. The methane conversion decreases from 13% at a lithium loading of 0.8 wt% Li to less than 1% at a lithium loading of 9 %. The selectivity is approximately constant at 75%. It is clear that an increase in the lithium concentration does not automatically result in an increase in activity. In fact, the activity decreases if there is more than 4 wt% Li.

The effect of Zr loading was studied by using a series of catalysts containing similar amounts of lithium but with increasing ZrO_2 concentration. The methane conversion and C_2 selectivity as a function of zirconium concentration at 700°C are

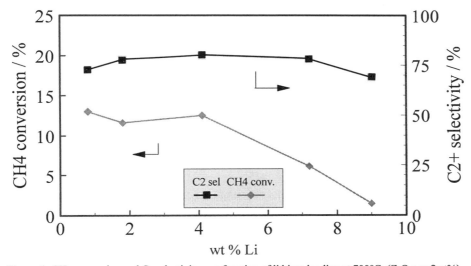

Figure 1. CH_4 conversion and C_2 selectivity as a function of lithium loading at 700°C, (ZrO_2: ca 2wt%).

shown in Figure 2. The methane conversion increases from 4% for undoped Li/MgO to 8.8% at a zirconium loading of 11.9 %. Again, the selectivity is approximately constant at 75%.

The activation energy for methane conversion was calculated for the catalysts with increasing zirconia loading. The results are presented in Table 1, together with some data on the composition of the catalysts.

Table 1. $E_{act,app}$ and catalysts composition for a series of Li/Zr/MgO catalysts.

Catalyst	wt% Zr	wt% Li	$E_{act,app}$ (kJ/mol)
Zr-0.3	0.3	4.5	191
Zr-2	2.9	4.3	189
Zr-3	6.1	4.4	180
Zr-6	11.9	4.3	175
Zr-21	20.8	3.9	165

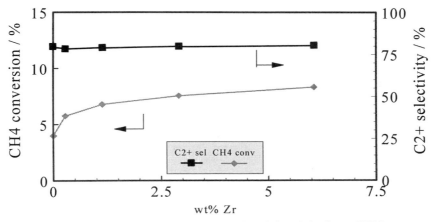

Figure 2. CH$_4$ conversion and C$_2$ selectivity as a function of zirconia loading at 700°C

Apparently, an increase in ZrO$_2$ loading results in a decrease in the apparent activation energy. However, it is important to realise that the apparent activation energy is the sum of the true activation energy plus the enthalpy of adsorption of CO$_2$. The fact that CO$_2$ plays an important role can be clearly demonstrated by the model for active surface species developed by Korf et al. (11).

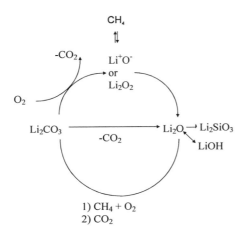

Figure 3. Schematic model of surface species as reported in reference 11.

Based on the observation that Li/MgO catalysts prepared in the presence of CO$_2$ (carbonate rich Li/MgO catalysts) were much more active than catalysts prepared in the absence of CO$_2$, Korf et al. suggested a mechanism in which the active site is formed via the decomposition of Li$_2$CO$_3$. This schematic model for the formation of surface species is shown in Fig.3. The decomposition of Li$_2$CO$_3$ in the presence of O$_2$ results in the formation of an active site: Li$^+$O$^-$ or Li$_2$O$_2$. When methane reacts with the active site, Li$_2$O is formed Li$_2$O can react with water to LiOH or with qaurtz to Li$_2$SiO$_3$, or, it can react with with CO$_2$ to regenerate Li$_2$CO$_3$. This model immediately shows that the partial pressure of CO$_2$ plays an important role in the activity of the catalyst, and that CO$_2$ is a poison for the OCM reaction. Since then, it has been shown many times for different, mostly alkali containing catalysts, that CO$_2$ acts as a poison (12,13,14). It was also recognised that the effect of CO_2 poisoning should be taken into account in kinetic equations (15,16). A simple rate equation should contain the adsorption constant of CO_2 in the denominator, if the adsorption enthalpy of CO_2 is large it affects the apparent energy of activation:

$$E_{act,app} = E_{true} + \Delta H_{ads(CO2)}$$

Based on the results in Table 1, in combination with the observation that the selectivity is the same for all catalysts, it is possible that the effect ZrO_2 is to facilitate the decomposition of lithium carbonate. On the other hand it is clear that the presence of ZrO_2 results in the formation of a mixed phase: Li_2MgZrO_4. Apparently the lithium in the catalyst rather takes part in the formation of this phase than in the formation of Li_2CO_3. This indicates that lithium carbonate is less stable. Therefore, increasing the Zr concentration results in less Li_2CO_3 in the catalyst. The model of Korf shows that an increase in the decomposition of lithium carbonate results in an increase in activity. However, the model also suggests that an increase in activity results in an increase in deactivation through the formation of Li_2SiO_3 or volatile LiOH. This effect has indeed been observed.

CONCLUSIONS

Varying concentrations of lithium and zirconium in Li/MgO catalysts doped with ZrO_2 result in varying methane conversions at 700 °C but in a constant selectivity. The effect of ZrO_2 can be explained with the model of Korf et al. (11), assuming that ZrO_2 facilitates the decomposition of Li_2CO_3, thus creating more active sites and resulting in more active catalysts.

Acknowledgement

This research was funded in part by the European Union Joule I programme.

LITERATURE CITED

1. G.E. Keller, M.M. Bhasin, J. Catal. 73 (1982) 9
2. A.M. Maitra, Appl. Catal. A, 104 (1993) 11
3. J.H. Lunsford, Stud. Surf. Sci.Catal., 75 (New Frontiers in Catalysis), 1993, 103
4. T. Ito, J. Wang, C.H. Lin, J.H. Lunsford, J. Am. Chem. Soc., 107 (1985) 5062
5. X.D. Peng, D.A. Richards, P.C. Stair, J. Catal. 121 (1990) 99
6. J.H. Edwards, R.J. Tyler, S.D. White, Energy Fuels, 4 (1990) 85
7. J.S. Hargreaves, G.J. Hutchings, R.W. Joyner, C.J. Kiely, Proc. 2nd Anlo-
    Dutch Conf. on Catalysis, Hull, (1991)
8. S.J. Korf, J.A. Roos, L.J. Veltman, J.G. van Ommen, J.R.H. Ross, Appl. Catal. 56
    (1989) 119

9. S.J. Korf, J.A. Roos, J.A. Vreeman, J.W.H.C. Derksen, J.G van Ommen, J.R.H. Ross, Catal. Today 6 (1990) 417

10. G.C. Hoogendam, A.N.J. van Keulen, K.Seshan, J.G. van Ommen, J.R.H. Ross, Stud.Surf.Sci.Catal. 81 Nat. Gas Conv. II) ,1994, 187

11. S.J. Korf, J.A. Roos, N.A. de Bruijn, J.G. van Ommen, J.R.H. Ross, Catal. Today 2 (1988) 535

12. S.J. Korf, J.A. Roos, N.A. de Bruijn, J.G. van Ommen, J.R.H. Ross, J.Chem.Soc., Chem.Comm. 19 (1987) 1433

13. K.P. Peil, J.G. Goodwin, G. Marcelin, Stud. Surf. Sci.Catal. 61 (Nat. Gas Conv.) 1991, 73

14. J.L. Dubois, C.J. Cameron, Prepr. ACS, Div. Pet. Chem. 37 (1992) 85

15. J.A. Roos, S.J. Korf, R.H.J. Veehof, J.G. van Ommen, J.R.H. Ross, Appl. Catal. 52 (1989) 131

16. M. Xu, C.Li. X. Yang, M.P Rosynek, J.H. Lunsford, J. Phys. Chem. 96 (1992) 6395

EVIDENCE FOR THE PRODUCTION OF METHYL RADICALS ON THE Na2WO4/SiO2 CATALYST UPON INTERACTION WITH METHANE

Zhi-Cheng Jiang* , Liang-Be Feng, Hua Gong and Hong-Li Wang**

Lanzhou Institute of Chemical Physics
Chinese Academy of Sciences
Lanzhou, 730000, China

INTRODUCTION

Methane activation on the Na_2WO_4/SiO_2 catalyst has been studied by high-temperature quenching ESR. A paramagnetic resonance signal with g = 2.002 for methyl radicals and one with g = 1.910 for W^{5+} were simultaneously appeared in the ESR spectrum collected at 77K after the reaction of methane with the catalyst at 750 ^{0}C in the absence of gas-phase oxygen. In conjunction with previously published results of the surface characterizations of the catalyst system, a surface reconstruction-associated redox mechanism for selective oxidation of methane to methyl is proposed, which involves heterolytic dissociation of methane on the catalyst surface and it is very likely that, the unsaturated lattice oxygen, adjacent to the Na^+ vacancy created at the high-temperature of OCM reaction is the active oxygen species responsible for methane activation.

It has been taken for granted that the reaction of oxidative coupling of methane (OCM) to ethylene or other C_2 products is a hetero-homogeneous catalysis process and the production of methyl radicals at the catalyst surface is the initial step for the reaction, which takes place at a high temperature, usually 700 ^{0}C and higher. However, the surface chemistry for the OCM catalysts has not yet been sufficiently explored and a general consensus on methane activation has not yet been reached so far. The well-known pioneering work by Lunsford and his associates demonstrated the presence of methyl radicals in the gas stream flowing over Li^+/MgO catalyst by Matrix Isolation Electron Spin Resonance (MIESR), and they proposed surface adsorbed O^- ions as the active oxygen and [Li^+O^-] as the adtive sites for methane activation.[1] Since then, there has been considerable discussions and varied opinions on the active species of oxygen on the catalyst surface which abstracts a hydrogen atom from methane in OCM reaction. The oxygen species proposed to be active for OCM include O^-,[1,2] O_2^-,[3,4] and O_2^{2-} [5,6]. It appears that the active species of oxygen may differ for different systems and different species of oxygen may even be interconvertible. Lattice oxygen has also been proposed as the active oxygen for some systems. [7,8] In particular, the unsatuarated surface lattice oxygen or F centers, rather than that of [Li^+O^-] centers, were recently proposed to be responsible for methane activation. [9,10]

---

*   To whom correspondence should be addressed.

**   Associated with Dalian Institute of Chemical Physics, Chinese Academy of Sciences, Dalian, 116023, China.

Among the many catalyst systems published so far for OCM reaction, a novel type of catalyst comprised of Na_2WO_4 supported on silica with the promotion of Mn oxide has shown the significantly better catalytic performances (methane conversion 30%, C_2 selectivity 81%) than that of the previously reported systems[11]. We have reported the results of a surface characterization study on the catalyst in a previously published article.[12] A strong interaction between Na_2WO_4 and silica support was found, which brought about a reconstructive phase-transformation of the support, silica gel, to cristobalite by annealing at 850 °C. A surface restructuring, associated by the depletion of surface lattice oxygen, were also found to happen at the temperature region of 700-800 °C. In this communication we report for the first time the direct experimental evidence from Electron Spin Resonance (ESR) measurements for the methyl radical formation on the catalyst, and, in conjunction of the previous experimental findings of the surface restructuring property, a possible surface reconstruction associated redox mechanism for methane activation to methyl is discussed briefly.

The high-temperature quenching ESR measurements were made on a Varion E-115 spectrometer operated at X-band (9.5 G Hz) with 100 kHz modulation. ESR parameters were calibrated with DPPH (g=2.0036). The catalyst sample tube was connected to a vacuum system for pretreatment including evacuation and adsorption/ reaction at elevated temperature up to 800 °C. It was then sealed and quenched in liquid nitrogen for ESR measurements.

The Na_2WO_4/SiO_2 catalyst sample was first heated from room temperature to 700 °C in the pretreatment /reaction vacuum system (2.4 x 10^{-3} Torr) to remove gas phase and any adsorbed surface oxygen, then exposed to pure methane (99.99%) and further

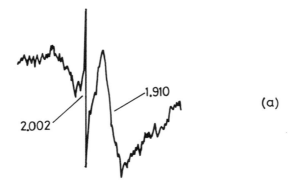

2.002   1.910   (a)

a. Quenching with the presence of methane in the sample tube.

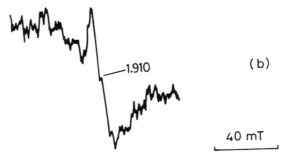

1.910   (b)

40 mT

b. Quenching with prior evacuation of methane from the sample tube.

Figure 1.   ESR spectra for Na_2WO_4/SiO_2 after reaction with pure methane at 750°C. Data collected at 77K.

heated to 750^0C for 20 min. After this reaction, the sample tube was immediately quenched in liquid nitrogen, when the reactant methane was still in the reaction tube, for ESR measurement. The collected ESR spectrum is shown in Fig.1a. In this spectrum, a ESR signal with g = 2.002 and another with g = 1.910 were simultaneously observed. The former signal was disappeared when the gas-phase reactant in the sample tube was first evacuated from the sample tube before quenching, in this case only the signal with g = 1.910 was obtained (Fig. 1b).

Furthermore, the former signal is in excellent agreement with that reported for the methyl radicals in the gas phase (g=2.003)[1]. This signal is therefore attributed to the methyl radical produced at 750^0C on the catalyst. The absence of hyperfine structure of the signal in our sample is due to that, the gas-phase reactant methane in the reaction tube was frozen as a condesed layer on the catalyst surface at 77K and the produced methyl radicals were trapped in the condesed methane. However, these methyl radicals were not stable on the catalyst surface at the temperature of 750 ^{0}C and escaped into the gas-phase, so that they could not be observed after evacuation. These methyl radicals were produced as the consequence of reaction of methane with a pretreated sample of Na_2WO_4/SiO_2 catalyst at 750 ^{0}C, and there were not any gas-phase or adsorbed oxygen which were involved in the reaction, except the surface lattice oxygen of the catalyst. Thus, it is apparent that the active oxygen species for selective oxidation of methane to methyl must be the lattice oxygen in the system. It is also clear that, upon an H atom is abstracted from methane by the surface lattice oxygen to form methyl radical and OH^-, the catalyst should be reduced. The ESR signal with g = 1.910 was thus attributed to W^{5+}, the reduction product from W^{6+} in Na_2WO_4 due to the depletion of lattice oxygen. This assignment is in agreement with several published ESR data for W^{5+} formation.[13,14] Furthermore, it is consistent with the previously reported results of surface characterizations of the catalyst.[12] We have proposed that a surface cluster compound, $WSi_3O_{8.5}$, was formed on restructured Na_2WO_4/SiO_2 catalyst upon heating at 700-850^0C. It was produced by the WO_4 tetrahytra in Na_2WO_4 attaching onto the SiO_2 via depletion of the lattice oxygen at elevated temperature:

$$700\text{-}850\ ^0C$$
$$Na_2WO_4 + 3SiO_2 (WSi_3O_{8.5} + 2Na^+ + 3/4\ O_2 \qquad (1)$$

In this process , W^{6+} in Na_2WO_4 is reduced to W^{5+}, while Na^+ ions migrate from the lattice of Na_2WO_4 into the support to take part in the reconsturctive phase-transformation of silica gel to cristobalite.[15] It is interesting to note that the temperature for this surface reconstruction and lattice oxygen depletion is just in the same range with that of OCM reaction, so that it is reasonable to assume that this surface reconstruction process might be incorporated in the methane activation. A surface restructuring assisted redox mechanism for selective oxidative of methane to methyl radicals on the Na_2WO_4/SiO_2 catalyst may be schematically expressed as follows:

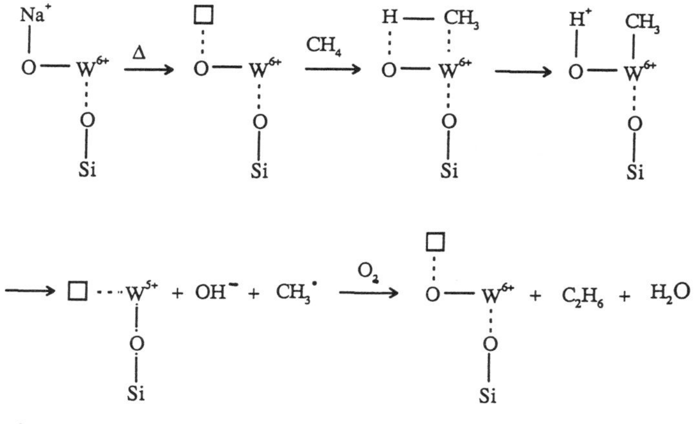

Scheme 1. Mechanism for methane activation over Na_2WO_4/SiO_2 catalyst.

Methane activation may start from adsorption on the surface of Na_2WO_4/SiO_2 catalyst, with H^+ at the Na^+ vacancy, created by the migration of mobile Na^+ ion at the elevated temperature, and CH_3^- at the W^{6+} ion site. Upon the attack by the surface lattice oxygen, a C-H bond in methane is heterolytically dissociated between the two sites. H^+ is abstracted by the neighboring lattice oxygen of Na^+ vacancy to form OH^- and a methyl radical is produced from an adsorbed CH_3^- by transfer an electron to W^{6+} ion. The methyl radicals escape into the gas-phase for further coupling and W^{6+} reduce to W^{5+}, as we have shown the direct ESR evidence for the simultaneous formation of methyl radicals and W^{5+} in the reaction of methane over Na_2WO_4/SiO_2 catalyst in the absence of gas-phase or adsorbed surface oxygen. This intermediate reduction state of W ion is not stable at the oxidative environment and easy to be reoxidized, as it has been proved by the disappearance of the ESR signal with g =1.910 when exposing the reduced catalyst to gas-phase oxygen. Thus, an effective redox cycle might be established for the catalyst in the reaction of oxidative coupling of methane. As to the role of Mn oxide in the reaction, evidences have shown that it enhances the rapid exchange between gas-phase and lattice oxygen and will be discussed later in a subsequent article.

In summary, by high temperature quenching ESR measurements we have demonstrated the simultaneous production of methyl radicals and reduction of W^{6+} to W^{5+} in the reaction of methane over Na_2WO_4/SiO_2 catalyst, in the absence of gas phase oxygen or any adsorbed species of surface oxygen. This points to the participation of surface lattice oxygen of the catalyst in the abstraction of hydrogen from methane. It is very likely that the coordinately unsaturated lattice oxygen, adjacent to the Na^+ vacancy, which was creacted at the elevated temperature of OCM reaction, should be responsible for methane activation. The catalyst operates in a redox mechanism and it is associated by a surface reconstruction process.

ACKNOWLEDGMENT

This work was supported by the National Science Foundation of China (NSFC) and the State Key Laboratory for Oxo Synthesis and Selective Oxidation (OSSO). The authors thank Prof. Shu-Ben Li and his graduate student Xue-Ping Fang for kindly providing the catalyst samples used in this work.

REFERENCES

1. Driscoll, D.J.; Martir, W.; Wang, J.-X.; Lunsford, J.H. *J. Am. Chem. Soc.* 1985, 102, 58
2. Wang, J.-X.; Lunsford, J.H. *J. Phys. Chem.* 1986, 90, 5883
3. Lin, C.-H.; Campbell, K. D.; Wang, J.-X; Lunsford, J.H. *J. Phys. Chem.* 1986, 90, 534
4. Jiang, Z.-C.; YU, Z.-Q.; Zhang B.; Shen, S.-K.; Li, S.B.; Wang H.-L., in *Novel Production Methods for Ethylene, Light Hydrocarbons and Aromatics* (Albright, L.F., Crynes, B.L. and Nowak, S. eds.), Marcel Dekker, New York, 1992, pp. 75-83
5. Kharas, K.C.C.; Lunsford, J.H. *J. Am. Chem. Soc.* 1989, 111, 2336
6. Yamashita, H.; Machida, Y.; Tomita, A. *Appl. Catal.* 1991, 79, 203
7. Hatano, H.; Otsuka, K. *J. Chem. Soc., Faraday Trans.* 1988, 85, 189
8. Choudhary,V.R.; Rane, V.H. *J. Catal.* 1991, 130, 411
9. Matsuura, I.; Utsumi, Y.; Doi, T.; Yoshida, Y. *Appl. Catal.* 1989, 47, 299
10. Wu, M.-C.; Truong, C.M.; Coulter, K.; Goodman, D.W. *J. Catal.* 1993, 140, 344
11. Fang, X.-P.; Li, S.-B.; Lin, J.-Z.; Gu, J.-F.; Yang D.-X. *J. Mol. Catal.* (China), 1992, 6, 427
12. Jiang, Z.-C.; Yu, C.-J.; Fang, X.-P.; Li, S.-B.; Wang, H.-L. *J. Phys. Chem.*, 1993, 97, 12870
13. Konings, A.J.A.; van Dooren, A.M.; Koninsberger, D.C.; de Beer, V.H.J.; Farragher, A.L; Schuit, G.C.A. *J. Catal.* 1978, 54, 1
14. Dimonie, M.; Coca, S,; Dragutan, V. *J. Mol. Catal.* 1992, 76, 79
15. Wells, A. F. *Structural Inorganic Chemistry*, 5th ed.; Oxford University Press: New York, 1984

TWO-STEP, OXYGEN–FREE ROUTE TO HIGHER HYDROCARBONS FROM METHANE OVER RUTHENIUM CATALYSTS

Manoj M. Koranne and D. Wayne Goodman

Department of Chemistry
Texas A & M University
College Station, TX 77843-3255

ABSTRACT

The direct conversion of methane to higher hydrocarbons over a Ru/SiO_2 catalyst as well as $Cu-Ru/SiO_2$ catalysts has been investigated via a two-step, oxygen-free route. The reaction consists of decomposition of methane over supported Ru catalysts at temperatures (T_{CH4}) between 400 to 800K to produce surface carbonaceous species followed by rehydrogenation of these species to higher hydrocarbons at T_{H2} of 368K. A 3% Ru/SiO_2 catalyst exhibited high initial ethane yields, but accumulation of "inactive" carbon led to a complete deactivation of this catalyst in few reaction cycles. For a $Cu-Ru/SiO_2$ catalyst (Cu/Ru=0.1), about 95% carbon deposited could be hydrogenated at 368K thus improving the net ethane yield. However, when tested for multiple reaction cycles, the $Cu-Ru/SiO_2$ catalyst (Cu/Ru=0.1) exhibited a rather slow deactivation, which was mainly attributed to gradual accumulation of inactive carbon on the surface of the catalyst. These results are consistent with the known tendency for Cu to deposit on low coordination Ru sites and with previous suggestions that carbon is predominantly formed on such sites.

INTRODUCTION

Conversion of natural gas (mainly methane) to easily transportable liquid hydrocarbons/oxygenates has acquired immense importance in recent years due to a world-wide surplus of natural gas. During the past decade, emphasis has shifted to direct routes for conversion of methane into higher hydrocarbons or oxygenates. The two main catalytic routes which have been proposed for direct conversion of methane, oxidative coupling of methane[1-3] and direct partial oxidation of methane to methanol or formaldehyde[4-6], have not yet proved to be economically attractive[7]. It is therefore necessary to explore alternate routes for direct methane conversion.

The direct conversion of methane to ethane is not thermodynamically allowed. In order to circumvent this thermodynamic limitation, van Santen and coworkers[8-9] have proposed a two-step, low temperature, oxygen free route for methane conversion. In the first step, the decomposition of methane is carried out at high temperatures (400-800K) where the enthalpy change is positive but free energy change is negative. The surface carbonaceous species formed in the first step are subsequently rehydrogenated to higher hydrocarbons at lower temperatures (350-450K).

A similar two-step route has been proposed by Amariglio and coworkers[10-11]. They propose to perform both the steps at a unique low temperature (350-450K) wherein formation of higher hydrocarbons is thermodynamically allowed[11].

Recent spectroscopic studies have demonstrated that methylidyne, vinylidyne, and graphitic carbonaceous species exist on single-crystal Ru catalysts following methane decomposition[12]. In related studies performed at elevated pressures, the ethane/propane yields from methane conversion via a two-step process over single-crystal Ru catalysts, correlated well with the hydrocarbon intermediates identified using spectroscopic techniques[12-14].

The activity/selectivity behavior for the two-step methane conversion process has also been examined on supported Ru catalysts[15]. The results of the kinetics measurements carried out at industrially relevant pressures (1-10 atm) show that while the supported Ru catalysts offer great promise, the formation of inactive carbon on these catalysts cannot be avoided. This both lowers the ethane yields over supported Ru catalysts and limits the number of reaction cycles that can be performed without intermediate catalyst regeneration step[15]. It was suggested that low coordination Ru sites (which are highly active) are responsible for the formation of inactive carbon[15]. Recent reports on bimetallic Cu-Ru system indicate that it is possible to preferentially deposit Cu onto low coordination Ru sites[16-18]. Here, we report results for activity/selectivity behavior over Cu-doped Ru catalysts. We aim to selectively poison highly active Ru sites with Cu, in an effort to improve ethane yield and lifetime of supported Ru catalysts for the two-step methane conversion process.

EXPERIMENTAL

Catalysts

A 3 wt% Ru supported on SiO_2 (Cab-O-Sil, HS5, Cabot Corp., Surface Area = 300 m^2/g) base catalyst was prepared using standard incipient wetness impregnation technique starting from ruthenium nitrosyl nitrate (99.999% Pure, Alfa Chemicals). The catalyst was dried overnight at 80°C, ground to a powder and reduced at 700K for 12 hours in pure hydrogen. Portions of the reduced catalysts were impregnated with required levels of copper (Cu/Ru atomic ratio = 0.05, 0.1, and 0.5) using an aqueous solution of copper nitrate (Alfa Chemicals). The catalysts were then re-reduced in pure H_2 at 700K for 4 hours.

Catalyst Characterization

The Ru surface area was measured using standard hydrogen chemisorption technique[19]. The catalyst was reduced in situ for 14 hours at 700 K prior to adsorption measurements. The system was evacuated at 700K for 4 hours, and then cooled to room temperature. In order to account for the reversible H_2 chemisorption, the system was evacuated at room temperature for 2 minutes and adsorption isotherms were obtained at room temperature in the pressure range of 60-400 torr[19].

TEM micrographs of the silica-supported Ru base catalyst were obtained using a high resolution Joel 2010 instrument. The catalyst powder was mounted on a 200 mesh Cu grid by simple dusting. Several bright field TEM micrographs of different portions of the sample were obtained at magnifications up to 1,000,000. An average particle size was determined from these micrographs.

XRD pattern of the catalyst was obtained with a Scintag diffractometer using CuKα radiation. No peaks corresponding to Ru could be detected indicating that either the particle size is less than 40 Å (detection limit of the instrument) or the loading is too low for detection.

Reaction System

A typical microreactor flow system was used to study the activity/selectivity of the supported catalysts. The reaction system consisted of a stainless steel reactor (1/4" in diameter) in a tube furnace. The temperature in the reactor was controlled ($\pm$ 1K, computer controlled, home built) using a thermocouple located inside the reactor. A complete description of the reaction system is given elsewhere[15].

A premixed, diluted mixture of methane in helium was allowed to flow through the reactor by switching a three-way valve located just above the reactor. The amount of methane added to the system could be controlled by varying either the concentration of methane or the time for which methane was admitted in the system.

Reaction Procedure and Product Analysis

Due to the inherent unsteady nature of the reaction (two steps; viz. methane decomposition at 500-800 K, followed by hydrogenation at 360-380 K), it is important to account for all the carbon fed to the system. The procedure used to determine both total amount of carbon retained on the catalyst during decomposition step and the amounts of hydrocarbons formed during the hydrogenation step was as follows. A dilute mixture of methane in He (0.3-5 %) was allowed to flow over a catalyst bed (0.3 g in most cases) held at temperatures ranging from 500-800 K for a specific time interval (0.5-5 min) and at a GHSV between 6000-12000 h^{-1}. Methane conversion, defined as percentage of methane retained on the catalyst was determined by monitoring the effluent concentration of methane using a flame ionization detector. The reactor was then rapidly cooled below 150°C to avoid "aging" of the carbonaceous species formed during methane decomposition. The surface carbonaceous species were hydrogenated at temperatures ranging from 350-380 K using pure hydrogen (15-20 cc/min). Since the concentration of the products change with time, small portions of the effluent stream were trapped (using a multi-position Valco sampling valve) at various time intervals for further analysis. The concentration of methane and ethane/propane were then determined by injecting these aliquots in a GC (Carle; Series 400) equipped with a Porapak Q column and a flame ionization detector. The total amounts of methane and ethane/propane formed during hydrogenation were determined from the concentration-time curves and hydrogen flow rate. Thus, ethane selectivity, defined as 2 x moles of ethane/total moles of product (i.e., on a C-atom basis) could be determined. The amounts of propane formed were, at best, an order of magnitude lower than ethane and hence are not reported here. A simple mass balance yielded the amount of "inactive carbon", i.e., carbon that could not be hydrogenated at temperatures less than 380K.

RESULTS AND DISCUSSION

Catalyst Characterization

The Ru dispersion determined from the results of irreversible hydrogen chemisorption in a manner similar to that suggested by Yang and Goodwin[19] was 23%. The average particle size determined from the results of irreversible hydrogen chemisorption was 40Å using an average area of 0.0817 nm^2 per Ru atom and assuming the particle to be cubic with 5 sides exposed.

TEM micrographs showed near spherical particles uniformly distributed on the support. The average particle size was estimated from TEM was 30Å, in agreement with the particle size estimated by hydrogen chemisorption and consistent with the XRD analysis.

Sinfelt[20] used hydrogen chemisorption at room temperature for determination of the number of Ru atoms exposed in a bimetallic Cu-Ru/SiO$_2$ system. Subsequently, it was shown that hydrogen spillover from Ru to Cu may lead to an overestimation of the total Ru atoms exposed[21-23]. King and coworkers[16-18], using H-NMR to delineate the hydrogen atoms associated with Ru, and also those associated with spillover from Ru to Cu, were able estimate the total Ru surface area in Cu-Ru/SiO$_2$ catalysts. More recently, Chen and Goodwin[23], taking advantage of the fact that hydrogen spillover from Ru to Cu is suppressed at low temperatures[21], performed hydrogen chemisorption at 77K to estimate the Ru surface areas in a Cu-Ru/SiO$_2$ bimetallic system. Chen and Goodwin[24] concluded that copper blocks Ru on a one-to-one basis so long as only two dimensional Cu islands are formed and further that three dimensional islands are formed long before a monolayer coverage of Cu by Ru is complete. Assuming that copper blocks ruthenium on one-to-one basis, the estimated Ru dispersions for Cu-Ru/SiO$_2$ catalysts with Cu/Ru ratios of 0.05 and 0.1 are 18% and 13% respectively.

Reaction Studies

3% Ru/SiO$_2$ Catalyst. The activity/selectivity behavior of a 3% Ru/SiO$_2$ catalyst (~16 % dispersion; 50Å particle size) for the conversion of methane to higher hydrocarbons via the two-step process has been reported in detail elsewhere[15]. Here, the activity/selectivity behavior of a similar 3% Ru/SiO$_2$ catalyst, but with a dispersion of 23%, was reexamined to allow easy comparison with Cu-doped Ru/SiO$_2$ catalysts using this batch of base catalyst. In general, the results obtained for the 3% Ru/SiO$_2$ catalyst (~23 % dispersion), are consistent with those reported earlier. For example, the results indicate that the ethane yield (defined as moles of ethane/site) increased initially with increase in temperature from 550K to 725K (see Figure 1). A further increase in temperature, however, led to a decrease in the ethane yield. This trend is consistent with that observed over single-crystal Ru catalysts at elevated pressures[13] which exhibited maximum at around 500-600K (see Figure 1). The temperature at which the maxima occurs for supported Ru, however, appears to have shifted to a higher temperature. In our previous publication[15], this behavior was explained in detail and is summarized below for clarity.

Methane conversion, ethane selectivity, and carbon hydrogenated at 368K as a function of methane decomposition temperature are displayed in Figure 2. For the unpromoted Ru/SiO$_2$ base catalyst, the methane conversion (defined as percent methane decomposed) increased monotonically with increase in temperature. This is consistent with the trend observed by van Santen and coworkers[9] over Ru/SiO$_2$ and by Wu et al.[13] over single-crystal Ru catalyst. However, at temperatures between 400-600K, the supported catalysts did not show significant activity for ethane formation, in spite of the fact that some methane was retained on the catalyst (not shown in Figure 1). Unlike single-crystal Ru catalysts, the supported catalysts consist of different types of sites (low and high coordination) which likely exhibit different activities to methane decomposition. It is likely that highly active sites preferentially chemisorb methane, leading to intermediates similar to those formed over single-crystal Ru catalysts. These rapidly transform to "inactive" carbon, which cannot be hydrogenated at lower temperatures (<400K)[15]. This is also supported by the fact that only 30-50% of the carbon decomposed was hydrogenated at 368K regardless of decomposition temperature (see Figure 2c). Once these highly active sites are covered (and poisoned) additional surface carbon species lead to the formation of ethane, resulting in higher ethane yields at higher temperatures. At temperatures greater than 725K, however, the inactive carbon is probably formed even on high coordination Ru sites resulting in lower ethane yields[15]. If indeed the highly active sites are responsible for the formation of inactive carbon, selective poisoning of these sites by an inactive element (like copper) may lead to improved performance. The activity/selectivity behavior for the Cu-Ru/SiO$_2$ catalysts is given below.

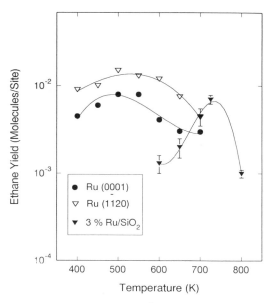

Figure 1. Comparison of ethane yield for single crystal[5] and supported Ru catalyst (conditions used were: $GHSV_{CH4}$ = 6000h[-1], total methane input = 4.8μmol, T_{H2} = 368K and $GHSV_{H2}$ = 2000h[-1]).

Cu-Ru/SiO₂ Catalysts. For a Cu-Ru/SiO₂ catalyst with Cu/Ru=0.1, the methane conversion, ethane selectivity, and carbon hydrogenated at 368K as a function of methane decomposition temperature are displayed in Figure 2. For the base Ru/SiO₂ catalyst, methane conversion increases with increase in methane decomposition temperature. However, the methane conversion is significantly lower for Cu-Ru/SiO₂ (Cu/Ru=0.1) catalyst than that for the Ru/SiO₂ catalyst. It is well known that addition of small amounts of copper to Ru/SiO₂ catalysts leads to blockage of some active Ru sites[17,20,24]. Since copper is inactive for methane decomposition in the temperature range studied, such a blockage of Ru sites leads to lower methane conversions. The ethane selectivity for the Cu-Ru/SiO₂ catalyst (Cu/Ru=0.1), on the other hand, is comparable to that for the Ru/SiO₂ catalysts, suggesting that copper does not significantly affect the nature of surface carbon intermediates formed on methane decomposition.

A comparison of the percentage of methane hydrogenated at 368K for the Cu-Ru/SiO₂ and the Ru/SiO₂ catalysts is illustrated in Figure 2c. For lower methane decomposition temperatures, the percentage of methane hydrogenated at 368K is significantly higher for the Cu-Ru/SiO₂ catalyst (Cu/Ru=0.1) than that for the Ru/SiO₂ catalyst. This means that lower amount of "inactive" carbon is formed on the Cu-Ru/SiO₂ catalyst than on the base Ru/SiO₂ catalyst. These observations are consistent with the previously cited evidence that addition of small amounts of copper to Ru crystallites will preferentially block low coordination sites[16,17], which in turn are suggested to be the sites responsible for the formation of "inactive" carbon[15].

Methane conversion, ethane selectivity, and carbon hydrogenated at 368K as a function of copper content are illustrated in Figure 3. In general, the methane conversion decreases with increase in copper content. The percentage of carbon hydrogenated at 368K increases with increase in copper content, up to about 10% Cu. These observations are again consistent with the proposition that, for catalysts with low copper contents, copper preferentially blocks the low coordination sites, which are suggested to be responsible for the formation of inactive carbon. It should be noted that these differences are more

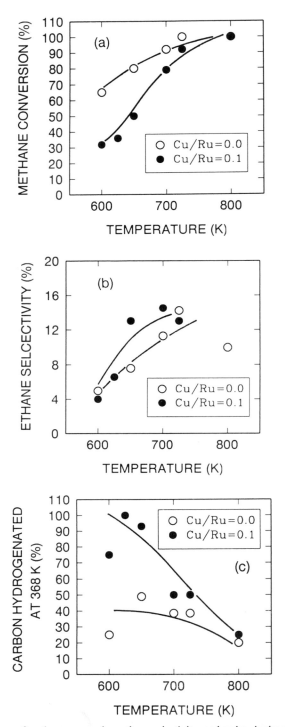

Figure 2. Comparison of methane conversion, ethane selectivity, and carbon hydrogenated at 368K as a function of methane decomposition temperature for Ru/SiO$_2$ and Cu-Ru/SiO$_2$ (Cu/Ru=0.1, see Figure 1 for conditions used).

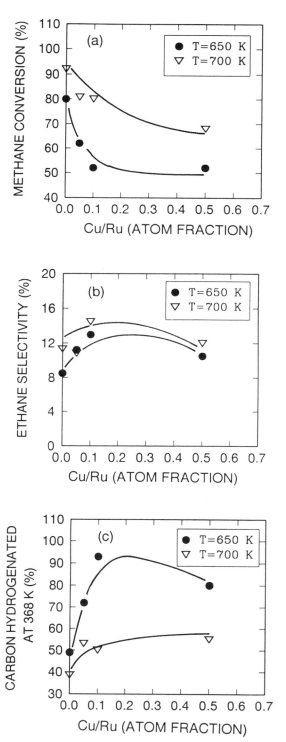

Figure 3. Methane conversion, ethane selectivity, and carbon hydrogenated at 368K as a function of copper content for a series of Cu-Ru/SiO$_2$ catalysts (see Figure 1 for conditions used).

pronounced at lower methane decomposition temperatures. At higher methane decomposition temperatures, the inactive carbon may be formed even on high coordination Ru sites as is evidenced by STM studies of single-crystal Ru catalysts[25].

For a Cu/Ru ratio of 0.5 all the Ru atoms should be completely covered by copper if copper follows only a two dimensional growth pattern. However, the Cu-Ru/SiO$_2$ catalyst with Cu/Ru=0.5, exhibited significant activity for methane decomposition (see Figure 3) indicating that copper does not cover all surface Ru atoms, even at this high Cu/Ru ratio. This can be attributed to either the formation of three dimensional copper islands at high copper contents[18,21], or to the formation of isolated copper crystallites[24].

For the Cu-Ru/SiO$_2$ catalyst with Cu/Ru ratio 0.1 (where presumably only the edge and corner atoms are covered by copper) the ethane yield (based on estimated surface Ru surface areas) at 650K is about 0.012. Considering the uncertainty involved in the calculation of total Ru sites for bimetallic supported catalysts, this value is in agreement with that for the single-crystal Ru catalysts.

Multiple Reaction Cycles. Although the Cu-Ru/SiO$_2$ catalysts show desirable ethane yields for one reaction cycle, from an industrial standpoint, it is important to determine the lifetime and regeneration capabilities of these catalysts. In our previous study, we reported on the effect of multiple reaction cycles on the ethane yields for Ru/SiO$_2$ catalysts[15]. It was found that the catalyst could be operated only for a few reaction cycles (about 3 to 4), after which the catalyst exhibited negligible activity. This necessitates an intermediate high temperature hydrogenation step for regeneration. The rapid decrease in the activity was attributed to accumulation of "inactive" carbon on the catalyst.

The effect of multiple reaction cycles on the ethane yield, methane conversion, and ethane selectivity for Cu-Ru/SiO$_2$ catalyst (Cu/Ru=0.1) is illustrated in Figure 4. In contrast to the behavior for the Ru/SiO$_2$ catalyst, the methane conversion did not fall to zero even after 5 reaction cycles. Since the ethane selectivity decreased rapidly with number of reaction cycles (Figure 4c) the ethane yield decreased to about a fifth of the initial yield after about 5 reaction cycles. At the same time, the percentage of carbon hydrogenated at 368K decreased from 93% to about 75%, indicating that some inactive carbon accumulates on the catalyst. Although the amount of inactive carbon on the Cu-Ru/SiO$_2$ catalyst is low (as compared to base Ru/SiO$_2$ catalyst), as is also evident from a only a slight decrease in methane conversion, it is likely that the nature of intermediates may be affected by the presence of inactive carbon, leading to decrease in ethane selectivity. The catalyst activity could be completely recovered after a high temperature rereduction (700K for 4 hours) in pure H$_2$. Further studies on the nature of surface intermediates on Ru single-crystal catalysts in the presence of copper are in progress.

While significant improvements in the catalyst performance have been realized, further improvements in the catalyst lifetime are necessary in order for this process to be commercially attractive. Studies using other dopants, such as sulfur and palladium, are in currently in progress.

CONCLUSIONS

Methane conversion to higher hydrocarbons via a two-step, oxygen free route was investigated for Ru/SiO$_2$ and Cu-Ru/SiO$_2$ catalysts. The Ru/SiO$_2$ catalyst exhibited high initial yields but accumulation of "inactive" carbon led to a complete deactivation of this catalyst in few reaction cycles. For the Cu-Ru/SiO$_2$ catalysts although the methane conversion decreased with increase in copper content, the percentage of carbon hydrogenated at 368K increased with increase in copper content (upto Cu/Ru = 0.1), thus improving the net ethane yield. The Cu-Ru/SiO$_2$ catalyst with Cu/Ru = 0.1 exhibited a slow

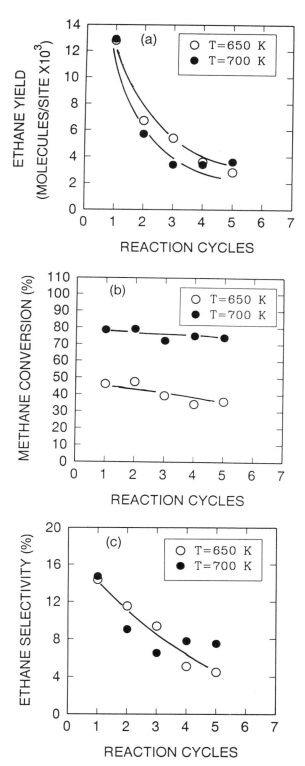

Figure 4. Ethane yield, methane conversion, and ethane selectivity as a function of reaction cycles for the Cu-Ru/SiO$_2$ catalyst with Cu/Ru = 0.1 (see Figure 1 for conditions used).

deactivation behavior due to gradual accumulation of inactive carbon on the catalyst. These results are consistent with the known tendency of Cu to deposit preferentially on low coordination Ru sites, and with previous suggestions that inactive carbon is predominantly formed on such sites.

ACKNOWLEDGEMENTS

Financial support from AMOCO Corporation is gratefully acknowledged.

REFERENCES

1. G.E. Keller and M.M. Bhasin, *J. Catal.*, 73:9 (1982).
2. T. Ito and J.H. Lunsford, *Nature*, 314:721 (1985).
3. Y. Amenomiya, V.I. Birss, M. Goledzinowski, J. Galuszka, and A.R. Sanger, *Catal. Rev. Sci. Eng.*, 32:163 (1990).
4. R. Pitchai and K. Klier, *Catal. Rev. Sci. Eng.*, 28:13 (1986).
5. N.D. Spencer and C.J. Pereira, *J. Catal.*, 116:399 (1989).
6. M.M. Koranne, J.G. Goodwin, Jr., and G. Marcelin, *J. Phys. Chem.*, 97:673 (1993).
7. J.M. Fox, *Catal. Rev. Sci. Eng.*, 35:169 (1993).
8. T. Koerts and R.A. van Santen, *J. Chem. Soc. Chem. Commun.*, 1281 (1991).
9. T. Koerts, M.J. Deelen, and R.A. van Santen, *J. Catal.*, 138:101 (1992).
10. M. Belgued, P. Pareja, A. Amariglio, and H. Amariglio, *Nature*, 352:789 (1991).
11. M. Belgued, H. Amariglio, P. Pareja, A. Amariglio, and J. Saint-Just, *Catal. Today*, 13:437 (1992).
12. M.-C. Wu. and D.W. Goodman, *J. Am. Chem. Soc.*, 116:1364 (1994).
13. P. Lenz-Solomun, M.-C. Wu., and D.W. Goodman, *Catal. Lett.*, 25:75 (1994).
14. M.-C. Wu, P. Lenz-Solomun, and D.W. Goodman, *J. Vac. Sci. Tech.*, in press (1994).
15. M.M. Koranne and D.W. Goodman, *Catal. Lett.*, in press (1994).
16. M.W. Smale and T.S. King, *J. Catal.*, 119:441 (1989).
17. T.S. King, X. Wu, and B.C. Gerstein, *J. Am. Chem. Soc.*, 108:6056 (1986).
28. X. Wu, B.C. Gerstein, and T.S. King, *J. Catal.*, 121:271 (1990).
19. C.H. Yang and J.G. Goodwin, Jr., *J. Catal.*, 78:182 (1982).
20. J.H. Sinfelt, *J. Catal.*, 29:308 (1973).
21. J.T. Yates, Jr., C.H.F. Peden, and D.W. Goodman, *J. Catal.*, 94:576 (1985).
22. D.W. Goodman and C.H.F. Peden, *J. Catal.*, 95:321 (1985).
23. J.E. Houston, C.H.F. Peden, D.S. Blair, and D.W. Goodman, *Surf. Sci.*, 167:427 (1986).
24. G.L. Haller, D.E. Reasasco, and J. Wang, *J. Catal.*, 84:477 (1983).
25. M.-C. Wu, Q. Xu, and D.W. Goodman, *J. Phys. Chem.*, in press (1994).

HOMOLOGATION REACTIONS OF ALKANES ON TRANSITION-METAL SURFACES

R.H. Cunningham, R.A. van Santen, A.V.G. Mangnus and J.H.B.J. Hoebink

Schuit Institute of Catalysis
Eindhoven University of Technology
P.O. Box 513
5600 MB Eindhoven
The Netherlands

Abstract

A combination of experiments using TAP, and isotopic labelling of hydrocarbons using ^{11}C over metal catalysts has provided useful information about the formation of carbon surface species and their activity. The formation of non-reactive surface species following high temperature CO or CH_4 adsorption was shown to be more likely at high temperatures, and that this was a fairly rapid process. The activation energy for C-C coupling reactions was found to be higher following promotion of a Ru catalyst with V. Differences in n-/iso-ratio between labelled and non-labelled products indicated that different reaction pathways led to the formation of n-hexane and iso-hexane.

INTRODUCTION

Work by Koerts et al.[1] showed that carbon atoms from [13]CH_4 could be inserted into ethylene or propylene in a surface reaction carried out on supported Co and Ru catalysts. Due to the thermodynamics, this reaction cannot be carried out in one step, and so a procedure was developed where the CH_4 activation step was separated from the associative carbon insertion step[1]. Figure 1 illustrates the devised reaction scheme. A CH_4/He stream was passed over the Ru catalyst at 450 °C, resulting in the formation of dehydrogenated surface CH_x species. The catalyst was then cooled rapidly in He to 50 °C at which point pulses of ethylene were passed over the metal surface. This resulted in the formation of several types of dehydrogenated surface species, including C_3H_y. These species could be hydrogenated at 50 °C resulting in the formation of propane. The use of ^{13}C in these experiments provided important mechanistic information about the formation of longer alkanes via C-C bond formation.

TPR and IR-spectroscopy studies have proven the existence of different types of

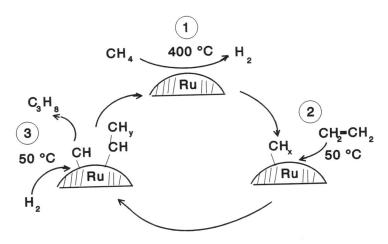

Figure 1. Reaction cycle for the stepwise conversion of methane and an olefin to a higher homologue[1].

adsorbed carbon species following the dissociation of CH_4 or CO over metals such as Ru or Co, and the species formed from CO have been shown to have a higher methanation activity than the non-dissociated molecule. The activity of species formed from CO dissociation has been found to be similar to that from methane decomposition on transition metals between 350 °C and 550 °C[2,3]. It is generally recognised that there are three types of surface carbon species formed, and these are usually referred to as C_α, C_β and C_γ. C_α is usually thought to be carbidic in nature and formed at low temperatures[4-6], and can be hydrogenated to form methane at temperatures ≤ 50 °C. Calculations have shown this carbon to be adsorbed in hollow sites bounded to three or more surface metal atoms[7]. A less reactive surface carbon type, usually referred to as C_β, can be hydrogenated at temperatures between 100 °C and 300 °C, and has been described as being amorphous in nature[8,9]. Carbon deposited at higher temperatures via methane or CO dissociation is characterized as being graphitic in nature, C_γ. Once deposited it is believed that the carbidic carbon can be converted to the graphitic form by heating, however this transformation is generally regarded as being irreversible. It has been shown[3] that only the C_α species have reasonable selectivities for the production of higher hydrocarbons. Recently we have carried out experiments using the TAP[10] apparatus here in Eindhoven to study in more detail the nature of the carbon species formed on transition metals as a result of high temperature CO and methane treatment. Further evidence has been obtained for the formation of non-reactive surface species at high temperatures, and we have observed that this process can be fairly rapid.

We have recently used a procedure similar to the two-step reaction described above for the production of ^{11}C-labelled n-hexane from [11]CO and 1-pentene over supported Ru catalysts[11]. The production of labelled n-hexane is part of a larger project where n-hexane isomerization over zeolite catalysts will be studied using Positron Emission Profiling (PEP); a similar technique has been used previously for the study of CO oxidation over automotive exhaust catalysts[12,13]. Over Ru, a C_n olefin can undergo cleavage, mainly at the terminal bond position. It has been shown that the resulting C_1 surface species can recombine with the alkene to produce C_{n+1} alkanes[14]. The incorporation of ^{11}C into the reaction products therefore allows one to distinguish this self-homologation process from that involving the reaction of a surface C_5 species with a C_1 species produced from high-temperature CO adsorption. Differences in the n-/iso- product ratios between labelled and non-labelled products also gave information about the pathway for n- and iso-product formation.

TAP EXPERIMENTS

Experimental

Three catalysts: 5%Ru/SiO$_2$, 5%Pt/SiO$_2$ and 4.5%Rh/SiO$_2$ were prepared in this laboratory using incipient wetness impregnation of Grace silica 332 with aqueous solutions of the appropriate metal salts. A fourth catalyst, 2.3%Pt-2.7%Rh/SiO$_2$, was supplied for us by a group in Germany[15].

For each experiment approximately 50 mg of catalyst was placed in the centre section of the TAP micro-reactor. On either side of the catalyst bed the reactor was filled with inert Grace silica. Two types of experiment were carried out using the TAP apparatus: multi-pulse, and pump-probe. In the case of multi-pulse experiments, many pulses of CO (typically about 1 x 10^{-7} mol per pulse) were passed over a pre-reduced catalyst. Product analysis by a mass spectrometer in the analysis chamber of the TAP equipment provided information on the CO adsorption process, and surface reactions to produce CO$_2$. Usually multi-pulse experiments were used to deposit a large quantity of reactive species on the catalyst surface. Pump-probe experiments involved a short single pulse of CO followed 3 seconds later by a single pulse of H$_2$ of similar size. From this it was possible to obtain information about product formation and the production of non-reactive surface species over relatively short time intervals. During each experiment this pump-probe cycle was repeated several times for better accuracy. Using these experiments the quantity of non-reactive carbon with respect to the CO adsorption temperature was measured over Ru, and a comparison of the reactivity of methane-produced surface C$_1$ towards ethane production was made between Ru, Pt and Pt-Rh. Finally, a comparison was made between the reactivity of Ru and Rh for methane formation from CO adsorption.

Results and discussion

The reaction of CO and H$_2$ over Ru resulted in the formation of CO$_2$ and methane. At lower temperatures CO adsorption was indicated by a low CO signal. At these temperatures it was observed that the production of CO$_2$ and CH$_4$ was not great. It can be assumed therefore that at temperatures of $\leq$250 °C CO is adsorbed on the Ru catalyst without dissociation. At higher reaction temperatures an increased production of both CO$_2$ and CH$_4$

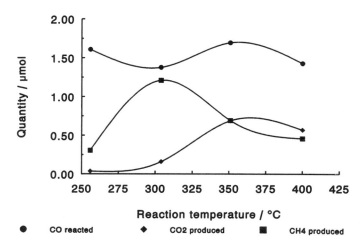

Figure 2. Carbon species produced at various reaction temperatures over 5%Ru/SiO$_2$

was observed. The production of CO_2 is a result of the Boudouard reaction whereby surface C_1 species are also produced[16,17]. The formation of CH_4 at these higher temperatures therefore indicates that C_1 surface species must be reactive intermediates in this reaction rather than non-dissociated adsorbed CO. At reaction temperatures above about 310 °C a sharp reduction was observed in the CH_4 production although the quantity of CO adsorbed was found to be almost independent of temperature. This indicated that at these temperatures a large quantity of non-reactive surface carbon species was formed following CO adsorption. The quantity of CO reacted, and, CO_2 and CH_4 produced over Ru at four different reaction temperatures is illustrated in figure 2. The quantity of non-reactive carbon (C_γ) remaining on the surface can be calculated by carrying out a carbon mass-balance (1) and (2):

$$n_{C,\ in} = n_{CO,\ in} \tag{1}$$

$$n_{C,\ out} = n_{CO,\ out} + n_{CH_4,\ out} + n_{CO_2,\ out} \tag{2}$$

where: $n_{i,in}$ = the number of moles of component i at the reactor inlet
          $n_{i,out}$ = the number of moles of component i at the reactor outlet

The percentage of carbon which remains on the catalyst as a non-reactive species (C_γ) is then given by (3):

$$\%C_\gamma = \frac{n_{C,\ out} - n_{C,\ in}}{n_{C,\ in}} \times 100\% \tag{3}$$

At low temperatures it is assumed that no C_γ is formed as there is no evidence for CO dissociation below 250 °C. It is assumed that the deficit in the carbon mass-balance at low temperatures comes as a result of adsorbed molecular CO. The percentage carbon remaining as non-reactive species at various temperatures is illustrated in figure 3.

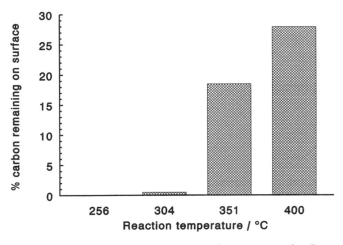

Figure 3. % carbon remaining on Ru surface as non-reactive C_γ

When the time between the CO pulse and the subsequent H_2 probe-pulse was short (3-5 seconds) the production of non-reactive surface carbon was still observed, indicating that its formation is an extremely rapid process. Evidence for the formation of these surface species has also been obtained for Rh. Over both Ru and Rh it is found that there is a decrease in the conversion of CO to methane with increasing temperature, this is illustrated in figure 4. At all temperatures studied however, the conversion of CO to methane is higher over the Ru catalyst than over the Rh catalyst. This seems logical as Ru is the better of the two metals for CO dissociation.

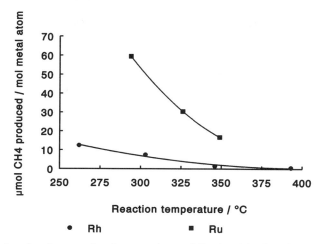

Figure 4. Quantity of methane produced per metal atom following CO adsorption over Ru and Rh

The formation of higher hydrocarbons (namely ethane), from C_1 produced from methane, was also studied. Results of pump-probe experiments carried out at 250 °C are given in table 1, where the reactant was methane, and the probe-pulse gas was hydrogen. All the results were normalised with respect to the number of metal atoms present in each catalyst sample, as the three catalysts had different metal loadings. Of the three catalysts studied (Ru, Pt-Rh and Pt), Pt was found to be the best metal for C-C coupling, Pt-Rh was found to have only about 50% of the activity of the pure Pt catalyst, and no C-C coupling reaction was observed over Ru. In the case of Ru this does not mean that C-C coupling does not take place, but just that with the quantity of reactant used, the product yield is below the detectable threshold. The effluent gas from the reactor was also checked for higher hydrocarbons, e.g. propane or butane, however with such small quantities of reactants used there was no evidence for the formation of these products.

Table 1. Ethane production following a single pulse of CH_4 (5.05×10^{-8} mol) and a subsequent single H_2 pulse (4.0×10^{-8} mol) over Ru, Pt-Rh and Pt catalysts.

Catalyst	μmol C_2H_6 produced / mol metal atom
5%Ru/SiO$_2$	-
2.3%Pt-2.7%Rh/SiO$_2$	1.03
5%Pt/SiO$_2$	2.13

The results of the TAP experiments show clearly that the formation of non-reactive carbon species does occur to a greater extent at elevated temperatures. Based upon a significant yield of CO_2, and little non-reactive surface-carbon formation, it is found that the optimum temperature for CO adsorption over Ru for reactive surface-carbon production is 300 °C. The TAP experiments also demonstrate that of the three metals Pt, Rh and Ru, Pt is the most active for C-C coupling when C_1 species have been formed via CH_4 adsorption. However as Pt is inactive for CO dissociation, this is a poor choice of metal for the formation of labelled *n*-hexane via ^{11}CO and 1-pentane. On the bases of the TAP experiments alone, the Ru and Ru-Rh are more likely to be of use for the labelling experiments as both can dissociate CO and form hydrocarbons (methane or ethane) from surface carbon species.

LABELLED *n*-HEXANE PRODUCTION

Experimental

For the production of ^{11}C-labelled *n*-hexane two sorts of catalyst have been used: a pure Ru/SiO$_2$ catalyst and Ru catalysts promoted with V. In all, three catalysts have been used for this work, all of which were produced in this laboratory via incipient wetness impregnation of Grace silica with aqueous solutions of the metal salts. The catalysts studied were 5.0%Ru/SiO$_2$, 5.3%Ru-O.7%V/SiO$_2$ and 5.3%Ru-1.4%v/SiO$_2$. The experimental set-up consisted of a quartz micro reactor, approximately 6 ml in volume, and containing approximately 400 mg of catalyst. During experiments He was continuously passed over the catalyst sample. The introduction of He and H_2 to the system was controlled by 24-volt solenoid valves, and the gas flow rates were controlled by Brooks 5850TR thermal mass-flow controllers. Non-labelled carbon monoxide could be introduced to the system via an electrically operated six-way valve (Valco Instrument Company), fitted with a 1 μmol injection loop. 1-pentene was introduced in a similar manner via a saturator operated at 21 °C. ^{11}C-labelled carbon monoxide could be introduced to the catalyst via a second electrically operated six-way valve. Products of the hydrocarbon reaction were separated by an on-line gas chromatograph (Packard model 427) fitted with a TCD and a 2 m squalane/alumina column operated at 70 °C. Labelled compounds could also be detected by the use of NaI scintillation crystals connected to photo-multiplier tubes. By positioning the TCD of the gas chromatograph and the NaI detectors in series it was possible to determine the quantity of both non-labelled and labelled products obtained. The set-up is illustrated in Figure 5.

^{11}C is a positron emitter with a half life of 20.39 minutes. Positrons only travel a short distance in solid material, and after being slowed down, the positron annihilates with an electron (usually in a neighbouring atom) producing two γ-photons, each of 0.511 MeV, travelling in opposite directions. ^{11}C was produced by bombarding high purity nitrogen with 12 MeV protons accelerated in the Eindhoven University cyclotron. A 25 minute irradiation period of 3.5 l of target gas resulted in the production of 300 MBq of activity; this is equivalent to slightly less than 1 pmol of ^{11}C. Trace impurities of oxygen present in the target vessel were sufficient to convert all ^{11}C produced to $^{11}CO_2$. The $^{11}CO_2$ was converted with 100% selectivity to ^{11}CO over metallic zinc granules at 390 °C. This was confirmed by $^{11}CO/^{11}CO_2$ separation using a Chromosorb 102 column.

The reaction procedure for labelled *n*-hexane production was similar to that described in the introduction for the production of propane from methane and ethylene. ^{11}CO was dissociated at high temperature (300 °C - 400 °C), the catalyst was then cooled rapidly to a much lower temperature (100 °C - 150 °C) where 1-pentene was adsorbed. The resulting products were then removed from the catalyst surface by a hydrogen pulse at the same temperature as the 1-pentene adsorption.

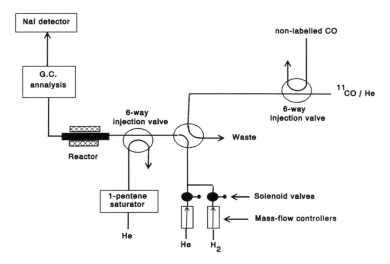

Figure 5. Simplified diagram of the set-up used for ^{11}C-labelled *n*-hexane production.

Results and discussion

On both the Ru catalyst and the V-promoted catalysts a range of labelled alkanes from C_1 to C_6 (including isomers) was produced, rather than labelled *n*-hexane alone. This came as a result of the high activity of Ru as a hydrogenolysis catalyst at the same temperatures used for product hydrogenation, i.e., ^{11}C was reacting with shorter fragments of 1-pentene. The selectivity for *n*-hexane could be defined in two ways. Firstly, the selectivity for *n*-hexane production in the product fraction isolated during the hydrogenation step **(4)**. Secondly, this selectivity could be multiplied by the percentage of α-carbon which was formed on the catalyst, to give an overall product selectivity based on the total quantity of adsorbed ^{11}C **(5)**. Where:

$S_{i,H}$ = selectivity for product i (i carbon atoms) during hydrogenation step
$S_{i,T}$ = total selectivity for product i
P_i = product i peak area
$X_{C\alpha}$ = fraction of labelled carbon adsorbed as the α form

Product selectivities for the hydrogenation step are given in table 2 for the three catalysts. In each case the values quoted are results where the optimum conditions were used to maximise *n*-hexane selectivity (this is discussed in greater detail below). Table 3 presents the percentage α-carbon formed on each catalyst following high temperature CO

$$S_{i,H} = \frac{P_i}{\sum\limits_{i=1}^{6} P_i} \times 100\% \qquad (4)$$

$$S_{i,T} = S_{i,H} \times X_{C_\alpha} \qquad (5)$$

dissociation, and total *n*-hexane selectivities based on the total quantity of adsorbed ^{11}C. In no experiment was there evidence for the formation of γ-carbon. It was also found that the β-carbon which was formed on the catalyst yielded only methane upon hydrogenation.

Table 2. Product selectivities hydrogenation at 110 °C

Catalyst	Product selectivity / %					
	methane–propane	butane	iso-pentane	n-pentane	iso-hexane	n-hexane
5%Ru/SiO$_2$	69.0	3.2	0.8	7.3	3.1	16.6
5.3%Ru–0.7%V/SiO$_2$	76.1	3.6	0.3	4.7	2.1	13.2
5.3%Ru–1.4%V/SiO$_2$	82.6	4.2	0.7	3.5	2.7	6.3

Table 3. Fraction of adsorbed carbon in the form of C$_\alpha$ and total n-hexane selectivities

Catalyst	Fraction C$_\alpha$	Total n-hexane % selectivity
5%Ru/SiO$_2$	0.091	1.51
5.3%Ru-0.7%V/SiO$_2$	0.161	2.13
5.3%Ru-1.4%V/SiO$_2$	0.234	1.47

An optimum temperature for CO adsorption was observed at 325 °C where a maximum selectivity for n-hexane production occurred. A change in the adsorption temperature of ± 50 °C led to a reduction in the n-hexane selectivity by as much as 50%. During the high temperature adsorption period the effluent gas from the reactor was analyzed and found to be mainly CO_2; at the same time a sharp increase in radioactivity was observed in the vicinity of the reactor. Both of these observations suggest that the formation of active carbon takes place via the Boudouard reaction. The influence of the CO adsorption temperature on the labelled n-hexane selectivity is shown in figure 6 for 5%Ru/SiO$_2$. A selectivity maximum was also observed in the case of the other two catalysts at about the same temperature. The selectivity maximum can be explained by thinking about the CO adsorption and dissociation process. As the adsorption temperature increases, the rate of the Boudouard reaction also increases, however the chance of CO adsorption will decrease with increasing temperature. The chance of forming non-reactive carbon also increases with temperature, and it is known that these non-reactive species do not lead to hydrocarbon production. There will therefore be a temperature where these three parameters are at an optimum to produce a maximum in n-hexane selectivity, and this is indeed what was found. In the case of the pure Ru catalyst it was found that it was essential to co-adsorb non-labelled CO with the labelled CO in order to adsorb a sufficient quantity of ^{11}C on the catalyst for labelled n-hexane production. As with the variation in CO adsorption temperature, the quantity of non-labelled CO which was co-adsorbed was also found to have a significant effect on the labelled n-hexane selectivity. This is illustrated in figure 7.

The non-labelled CO plays an important role in the reaction process. At low concentrations it blocks some of the Ru sites, thus reducing the hydrogenolysis activity: the chance for C-C coupling reactions therefore increases. At higher concentrations too many of the Ru sites become blocked, reducing not only the hydrogenolysis activity, but also the activity for C-C bond formation.

It was possible to calculate the activation energy for the C-C coupling reaction as the quantity of labelled n-hexane produced in each experiment was directly proportional to a

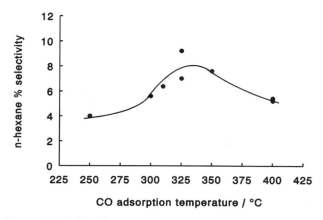

Figure 6. Influence of CO adsorption temperature on labelled *n*-hexane selectivity (5%Ru/SiO₂)

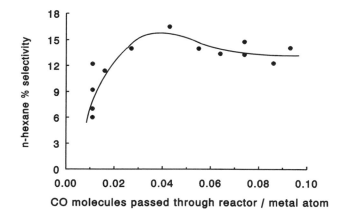

Figure 7. Influence of non-labelled CO on labelled *n*-hexane selectivity (5%Ru/SiO₂)

reaction rate. This was done for 5%Ru/SiO$_2$ and 5.3%Ru-0.7%V/SiO$_2$. Due to the small quantities of product obtained it was not possible to measure such an activation energy for *iso*-hexane formation directly from an Arrhenius-style plot, however it was noted that the *n*-/*iso*- product ratio was a useful parameter as this was also temperature dependent. From this the activation energy difference between *n*-hexane and *iso*-hexane formation could be calculated. Table 4 presents activation energies for *n*- and *iso*-hexane production over the pure Ru and the V-promoted catalyst.

It can be seen that promotion of the Ru catalyst with V raised the activation energy for hexane formation. It was observed that it was possible to use a higher 1-pentene adsorption / hydrogenation temperature in the case of the V-promoted catalyst without seriously affecting the *n*-hexane selectivity, i.e. the activation energy for the hydrogenation of all surface species from C$_1$ to C$_6$ must be increased by V promotion. The same effect was observed in earlier work where the effect of V promotion on Rh catalysts was studied[18]. In that study it was observed that although the activity of the surface species towards H$_2$ was far lower in the case of the promoted catalyst, the probability of chain growth was higher.

The conclusion was reached that the carbon species were more strongly bonded to the catalyst surface, and therefore more likely to take place in C-C coupling reactions. The results suggest that the same is true in the case of the V-promoted Ru catalysts.

Table 4. Activation energies for *n*- and *iso*-hexane formation over 5%Ru/SiO$_2$ and 5.3%Ru-0.7%V/SiO$_2$

Catalyst	Product	E_a / kJ mol^{-1}
5%Ru/SiO$_2$	*n*-hexane	40.12 ± 1.24
"	*iso*-hexane	81.02 ± 1.31
5.3%Ru-0.7%V/SiO$_2$	*n*-hexane	69.63 ± 2.93
"	*iso*-hexane	120.07 ± 34.24[*]

At all reaction temperatures studied there was a difference observed in the *n-/iso-* product ratio between the labelled and non-labelled product fraction. This is best illustrated in figure 8 where it can be seen that labelled *n-/iso-* hexane ratio is approximately twice that

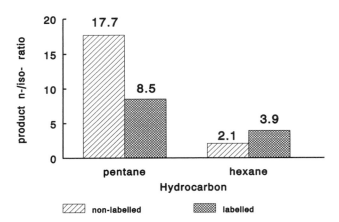

Figure 8. *n-/iso-* ratio for labelled and non-labelled hydrocarbons hydrogenated at 110 °C. 5%Ru/SiO$_2$

of non-labelled hexane. This strongly suggests that C$_1$ addition takes place preferentially at the terminal position of the growing hydrocarbon chain, and that *iso*-alkanes are mainly formed from a reaction pathway other than C$_1$ addition, i.e. via the recombination of cracked pentene fragments (C$_2$, C$_3$ and C$_4$ species). If this was not the case one would expect the *n-/iso-* product ratio to be the same for both labelled and non-labelled products. In figure 8 the very high non-labelled *n-/iso-* ratio for pentane can be explained simply from hydrogenation of 1-pentene. The greater likelihood of carbon addition occurring at a terminal position has also been concluded in earlier studies[19-21].

---

[*] Error in activation energy high due to the small number of data points.

CONCLUSIONS

From TAP it is clear that over metals such as Ru and Rh, the rate of formation of active surface C_1 species increases with temperature, however so does the chance of non-reactive surface carbon formation. Of the three metals Pt, Ru and Rh, TAP shows that Pt is by far the best metal for C-C coupling, however its inability to dissociate CO means that it can only be used for the production of higher hydrocarbons if CH_4 is used as the source of C_1.

From the difference in the *n-/iso-* ratio for labelled and non-labelled products it is clear that there are different reaction pathways leading to *n-* or *iso-* hydrocarbons. Promotion of a ruthenium catalyst with V raises the overall selectivity for *n*-hexane production. This comes about partly through an increase in the metal-carbon bond strength in the case of the promoted catalyst.

REFERENCES

1. T. Koerts, P.A. Leclercq, and R.A. van Santen, *J. Am. Chem. Soc.*, 114 (1992) 7272
2. F. Solymosi and A. Erdöhelyi, *Surf. Sci.*, 110 (1983) L630
3. T. Koerts and R.A. van Santen, *J. Chem. Soc., Chem. Commun.*, (1991) 1281
4. J.G. McCarty and H. Wise, *J. Catal.*, 57 (1979) 406
5. J.W. Niemantsverdriet and A.D. van Langeveld, *Catalysis*, (1987) 769
6. A.G. Sault and D.W. Goodman, *Advances in Chemical Physics*, 76 (1989) 153
7. A. de Koster and R.A. van Santen, *J. Catal.*, 127 (1991) 141
8. T.M. Duncan, P. Winslow, and A.T. Bell, *J. Catal.*, 93 (1985) 1
9. T.M. Duncan, J.A. Reiner, P. Winslow, and A.T. Bell, *J. Catal.*, 95 (1985) 305
10. J.T. Gleaves, J.R. Ebner, and T.C. Keuchler, *Catal. Rev. Sci. Eng.*, 30 (1988) 49
11. R.H. Cunningham, R.A. van Santen, J. van Grondelle, A.V.G. Mangnus, and L.J. van IJzendoorn, *J. Chem. Soc., Chem. Commun.*, (1994) 1231
12. K.A. Vonkeman, G. Jonkers, and R.A. van Santen, *Stud. Surf. Sci. Catal.*, 71 (1991) 239
13. G. Jonkers, K.A. Vonkeman, S.W.A. van der Wal, and R.A. van Santen, *Nature*, 355 (1992) 63
14. E. Rodriguez, M. Leconte, and J.M. Basset, *J. Catal.*, 132 (1991) 472
15. H. Bönnemann, W. Brijoux, R. Brinkmann, E. Dinjus, T. Joussen, and B. Korall, *Angew. Chem.*, 103 (1991) 1344
16. A.T. Bell, *Catal. Rev. Sci. Eng.*, 23 (1981) 203
17. P. Biloen and W.M.H. Sachtler, *Adv. Catal.*, 30 (1981) 165
18. T. Koerts and R.A. van Santen, *Catal. Lett.*, 6 (1990) 49
19. C. O'Donohoe, J.K.A. Clarke, and J.J. Rooney, *J. Chem. Soc., Faraday Trans. I.*, 76 (1980) 345
20. Z. Paál, M. Dobrovolszky, and P Tétényi, *J. Chem. Soc., Faraday Trans. I*, 80 (1984) 3037
21. M.A. Dobrovolszky, Z. Paál, and P. Tétényi, *Acta Chim. Hung.*, 119 (1985) 95

A COMPARATIVE STUDY OF CATALYTIC BEHAVIOURS OF Sr-Ti,Sr-Zr,Sr-Sn PEROVSKITES AND CORRESPONDING LAYERED PEROVSKITES FOR THE OXIDATIVE COUPLING OF METHANE

W. M. Yang, Q. J. Yan and X. C. Fu

Department of Chemistry, Nanjing University,
Nanjing 210008, P. R. China

The catalytic performance of the perovskites $SrTiO_3$, $SrSnO_3$, and $SrZrO_3$, and corresponding layered perovskites Sr_2TiO_4, Sr_2SnO_4, and Sr_2ZrO_4 was studied for the oxidative coupling of methane. Much higher C_2 selectivity and yield (35-41% and 11-15%, respectively) were obtained over layered perovskites under the reaction condition of 1073K, $CH_4/O_2/N_2$=2:1:4 and F/W=60000 ml/min. Postreaction characterization by XRD and XPS reveals that the layered perovskites undergo significant bulk transformation during the reaction, which leads to the formation of $SrCO_3$ and related perovskites, and the enrichment of Sr in the surface as well. No bulk and surface transformation is observed for the perovskites during the reaction. The O 1s and C 1s spectra show that only two kinds of oxygen species, namely lattice oxygen and surface CO_3^{2-} exit on the surface of the perovskites. In contrast, on the layered perovskite surface, significant amount of O_2^{2-} species is obtained. At 1073K, under reaction atmosphere, $SrCO_3$ react with O_2 to produce surface SrO_2, which plays an important role for the oxidative coupling of methane.

INTRODUCTION

Oxidative coupling of methane (OCM) to give C_2 hydrocarbons has attracted much attention in the last decade because of the increasing interest of the direct utilization of natural gas. A large number of study show that complex oxide with single phase crystal structure is a kind of promising catalyst for OCM reaction. Kaddouri[1] compared the catalytic property of complex compound $LnLiO_2$ possessing rock salt structure with alkali earth oxide Ln_2O_3 (Ln=Sm,Nd,La) and found that $LnLiO_2$ has higher C_2 selectivity and yield than relative oxides and the rock salt structure is very stable in the reaction process. Nagamato[2] studied a series of perovskite type oxides and found that they are active and stable for OCM. Machida and Enyo[3] reported that a 31.6% C_2 yield was obtained over $SrCe_{0.9}Yb_{0.1}O_{2.95}$ under certain reaction conditions. Q.J. Yan[4] and Campbell[5] reported that some layered perovskites are also active for OCM. Although many work reported the catalytic property of complex oxide but the detail study of oxygen species on the surface of this kind of catalyst is not much. Ito and Lunsford[6] suggested that [Li$^+$-O$^-$] center is responsible for methane activation for Li/MgO (as Li doped in rock salt structure MgO forming solid solution). Lambert et al.[7] studied the surface oxygen species of rock salt structure LiNiO$_2$ and noted that lattice oxygen O^{2-} is responsible for methane oxidative coupling. Kharas and Lunsford[8] suggested that a non-classical peroxide species in BaPbO$_3$ perovskite oxide is the active oxygen species. W.P. Ding et al.[9] found that over ATiO$_3$ (A=Ca,Sr,Ba) perovskites type oxides, O$_2^-$ and O$_2^{2-}$ is the selective oxidation surface oxygen species for OCM.

XPS has been a particularly effective technique for characterizing various surface oxygen species on the catalyst surface and determine the active species responsible for OCM[10-14]. The oxygen species present on Li/MgO catalyst surface have been studied by Peng and Stair[10] using XPS. They have delineated three kinds of oxygen species related to MgO, Li$^+$O$^-$ and Li$_2$CO$_3$, respectively and found that [Li$^+$O$^-$] is the active center responsible for OCM. Yamashita et al.[11] investigated a series of Ba-La-O mixed oxides and suggested that the precursor of active oxygen species responsible for methane activation is O$_2^{2-}$. They further pointed out that under reaction conditions, these peroxide species may decompose to O$_2^-$ or O$^-$ which directly react with methane. Using XPS, Ganguly[12] studied the oxygen species in perovskite oxides and confirmed that two kinds of oxygen species e.g. O^{2-} and O$_2^{2-}$ existed in BaPbO$_3$, BaBiO$_3$ and SrPbO$_3$. Recently, Lunsford and co-workers[13] investigated the catalytic property of BaPbO$_3$ and BaBiO$_3$, and found that the BaO$_2$ formed in the reaction process played an important role for methane conversion to form C$_2$ hydrocarbons.

In the current investigation, the catalytic behaviors of SrBO$_3$ and related layered perovskite Sr$_2$BO$_4$ (B=Ti,Sn,Zr) for OCM were examined, the surface oxygen species was studied by using XPS technique and the reaction mechanism was discussed.

EXPERIMENTAL

A.R. grade SrCO$_3$, TiO$_2$, SnO$_2$ and ZrO$_2$ were used as reagents. The perovskites SrBO$_3$ (B=Ti, Sn, Zr) and the related layered perovskites Sr$_2$BO$_4$ were prepared by solid state reaction at 1223K-1623K for 36-48 hr. Catalytic performance was carried out in a quartz fix-bed microreator with 4 mm i.d. at 1 atm. CH$_4$, O$_2$, and N$_2$ were purified to remove CO$_2$ and water prior to use. 0.1 g of 20-40 mesh catalyst particles were loaded into the reactor and the reaction was carried out under following conditions: 1003K or 1073K, CH$_4$/O$_2$/N$_2$=2:1:4, total flow rate=100ml/min. The blank experiments showed that at 1073K, the methane conversion can be neglected. The reactor effluent was first passed through MgClO$_4$ to remove water and then analyzed by G.C. using TCD detector. CO$_2$, C$_2$H$_4$, C$_2$H$_6$ were separated from O$_2$, N$_2$, CH$_4$, CO on a Porapark Q column. The latter four components were separately analyzed using a molecular sieve 13X column.

XRD analyses were performed on a Shimadzu XD-3A diffractometer before and after exposing to the reactant gas , with Cu Kα source. XPS spectra were acquired on VG ESCALAB MKII spectrometer with Mg Kα X-ray source. All measured binding energies were adjusted with respect to the C 1s peak at 284.6 eV due to adventitious carbon. The samples were treated prior to XPS analysis in flowing O$_2$ or in the same flow of CH$_4$/O$_2$/N$_2$ reaction mixture as those employed for the reaction studies and then swept out the adsorbed water and/or surface hydroxyls with pure N$_2$. The samples were cooled to room temperature and transported under N_2 to the XPS introduction chamber, without exposure to the atmosphere.

RESULTS AND DISCUSSION

Catalytic Behaviors

XRD results showed that the desired perovskites or layered perovskites crystalline structure were formed for all the samples (except Sr_2ZrO_4). Significant amount of $SrZrO_3$ and SrO crystalline phase were found in Sr_2ZrO_4 layered perovskite samples. This may be due to the lower temperature used for preparation compared to the literature[14].

Fig. 1-2 present the catalytic behaviors of all the samples run at 1073K within 14 hr. It can be seen that the three perovskite type samples maintained constant activity and selectivity over the time period investigated. Methane conversion over $SrTiO_3$, $SrSnO_3$ and $SrZrO_3$ reached 30%, 32% and 27%, respectively, with C_2 selectivities around 20-23%, while, over layered perovskite Sr_2TiO_4 and Sr_2SnO_4, methane conversion and C_2 selectivity increased markedly within the first 4-5 hr. Methane conversion and C_2 selectivity over Sr_2TiO_4 increased from 34% to 39% and 23% to 41%, respectively. Similarly, over the Sr_2SnO_4, methane conversion and C_2 selectivity increased from 35% to 38% and 22% to 39%, respectively in the first 4-5 hr. By contrast, over the layered perovskite Sr_2ZrO_4 containing SrO and $SrCO_3$, methane conversion and C_2 selectivity decreased in the first 3 hr and then maintained at constant. Apparently, the layered perovskites were more active and selective than the corresponding perovskite for OCM at 1073K. It is interesting that after on stream at 1073K for 14 hr and then cooled to 1003K, the catalytic activities and selectivities of all the samples were very low and the difference between perovskites and corresponding layered perovskites became very small.(Table 1)

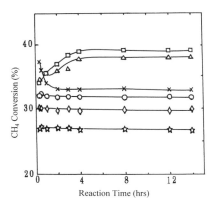

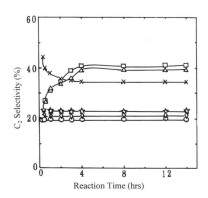

Fig. 1 Variation of CH_4 conversion with time.
◊ $SrTiO_3$; □ Sr_2TiO_4; ⊙ $SrSnO_3$; _ Sr_2SnO_4; ★ $SrZrO_3$; × Sr_2ZrO_4. (T=1073K;$CH_4/O_2/N_2$=2:1:4)

Fig. 2 Variation of C_2 selectivity with time.
◊ $SrTiO_3$; □ Sr_2TiO_4; ⊙ $SrSnO_3$; _ Sr_2SnO_4; ★ $SrZrO_3$; × Sr_2ZrO_4. (T=1073K;$CH_4/O_2/N_2$=2:1:4)

Table 2 shows the XRD results of used catalysts. It is shown that the perovskites structure did not change during the reaction at 1073K but the layered perovskite crystalline phase underwent extensive transformation. Part or all of the crystalline structure transformed into $SrCO_3$ and related perovskites. Since the C_2 selectivity of $SrTiO_3$ ($SrSnO_3$) were small and only 6% methane conversion was obtained over pure $SrCO_3$, the relation between phase transformation and the increase of the catalytic activity and selectivity is still unknown. We assume that some kinds of surface active phase may be generated during bulk phase transformation.

Table 1 Catalytic properties of the perovskites and the layered perovskites (after 14 hr on stream at 1073K and then cooled to 1003K for catalytic test)

Catalyst	CH_4 Conv.(%)	C_2 Sel.(%)	C_2-yield(%)
$SrTiO_3$	10.1	29.0	2.9
Sr_2TiO_4	12.6	33.3	4.1
$SrSnO_3$	6.2	24.2	1.5
Sr_2SnO_4	8.0	27.4	2.2
$SrZrO_3$	12.3	26.1	3.2
Sr_2ZrO_4	10.4	30.7	3.2

Table 2 Bulk phases of the catalysts identified by XRD

Fresh	Used[a]	Used[b]
Sr_2TiO_4	Sr_2TiO_4(trace)	Sr_2TiO_4 (70%)
	$SrTiO_3$	$SrTiO_3$
	$SrCO_3$	$SrCO_3$
Sr_2SnO_4	Sr_2SnO_4(trace)	Sr_2SnO_4 (78%)
	$SrSnO_3$	$SrSnO_3$
	$SrCO_3$	$SrCO_3$
Sr_2ZrO_4[c]	$SrZrO_3$	$SrZrO_3$
	$SrCO_3$	$SrCO_3$
$SrTiO_3$	$SrTiO_3$	$SrTiO_3$
$SrSnO_3$	$SrSnO_3$	$SrSnO_3$
$SrZrO_3$	$SrZrO_3$	$SrZrO_3$

[a] 14 h at 1003K; CH_4:O_2:N_2=2:1:4
[b] 14 hr at 1073K; CH_4:O_2:N_2=2:1:4
[c] Beside Sr_2ZrO_4, significant amount of $SrZrO_3$ and SrO also exit.

XPS Characterization

In order to investigate the surface active sites for the catalysts, the surface composition and oxygen species were examined using XPS.

The surface composition of the fresh and used catalyst is shown in Table 3. It can be seen that before and after exposing the sample to the reactant gas mixture, Sr/Ti or Sr/Sn ratio of perovskites is unchanged while that of layered perovskites increased markedly. This fact indicated the enrichment of $SrCO_3$ on the surface of layered perovskites and is consistent with the XRD data.

The O 1s and C 1s XPS spectra of fresh and used $SrTiO_3$ and Sr_2TiO_4 are shown in Fig. 3-4. Fig. 3(a) shows after treated with oxygen at 1073K and cooled to room temperature of the samples only one O 1s peak at 529.2 eV assigned to O^{2-} lattice oxygen of $SrTiO_3$ appears, while after exposing in CH_4/O_2/N_2=2:1:4 flow at 1073K for 10 hr and swept with pure N_2, a single O 1s peak with a shoulder at higher B.E. side appears (Fig.3b).

It can be deconvoluted into two peaks at 529.3 eV and 531.1 eV. There is also only one O 1s peak at 529.0 eV appears for the fresh catalyst Sr_2TiO_4. After reaction, a strong shoulder at higher B.E. side appears and the O 1s peak can be fitted by two peaks at 529.3 eV and 531.1 eV. The peak at 529.3 eV can be attributed to lattice oxygen of $SrTiO_3$ and/ or Sr_2TiO_4 while the assignment of the peak at 531.1 eV is still uncertain because the O 1s peak of both CO_3^{2-} and O_2^{2-} are located in this region[10,13,15-17]

Table 3 Surface composition of the catalysts determined by XPS

Catalyst	Surface Composition(atomic%)						
	C*	O	Sr	Ti	Sn	Sr/Ti	Sr/Sn
$SrTiO_3$ (fresh)	0	57	21	22	--	1.0	--
$SrTiO_3$ (used)	6	53	21	20	--	1.1	--
Sr_2TiO_4 (fresh)	0	59	29	14	--	2.1	--
Sr_2TiO_4 (used)	10	68	17	5	--	3.4	--
$SrSnO_3$ (fresh)	0	61	19	--	20	--	0.9
$SrSnO_3$ (used)	8	55	19	--	18	--	1.1
Sr_2SnO_4 (fresh)	0	58	29	--	13	--	2.2
Sr_2SnO_4 (used)	12	68	15	--	5	--	3.0

* Carbon in CO_3^{2-} species only; not include adventitious carbon;
"Fresh": Treated with O_2 at 1073K for 5 hr;
"Used": Reacted with $CH_4/O_2/N_2$ at 1073k for 10 hr.

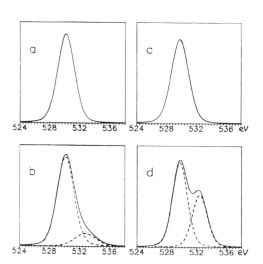

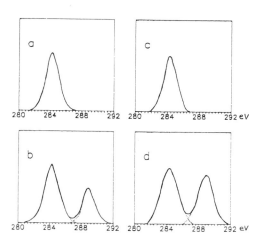

Fig. 3 XPS(O 1s) of $SrTiO_3$ and Sr_2TiO_4. (a,c) treated with O_2 at 1073K; (b,d) treated with O_2 at 1073K, then reacted with $CH_4/O_2/N_2$ mixture at 1073K for 10 hr and finally sweep out adsorbed water and/or surface hydroxyls with N_2 for 1 hr.

Fig. 4 XPS(C 1s) of $SrTiO_3$ and Sr_2TiO_4. (a,c) treated with O_2 at 1073K; (b,d) treated with O_2 at 1073K, then reacted with $CH_4/O_2/N_2$ mixture at 1073K for 10 hr and finally sweep out adsorbed water and/or surface hydroxyls with N_2 for 1 hr.

Fig. 4(a-d) display the C 1s spectra of $SrTiO_3$ and Sr_2TiO_4 both before and after the reaction. The fresh sample contains only adventitious carbon represented by the peak at 284.6 eV and is free of surface carbonate species which should appear at 289.0 eV. This is consistent with the conclusion deduced by the O 1s spectrum. However, the C 1s spectra of the used samples show two distinct features having B.E. of 284.6 and 289.0 eV represent the formation carbonate of the surface. The areas of the O 1s peak at 531.1 eV and the C 1s peak at 289.0 eV for the used Sr_2TiO_4 correspond to an O:C surface composition ratio of 3.7, indicating that 81% of the peak at 531.1 eV is due to CO_3^{2-} of $SrTiO_3$ and/or Sr_2TiO_4, with the remainder due to O_2^{2-}. By contrast, for the used $SrTiO_3$ the areas of the O 1s peak at 531.1 eV and the C 1s peak at 289.0 eV correspond to an O:C surface composition ratio of 3.1, which is close to the value of 3.0 for CO_3^{2-}, indicating that it is free of O_2^{2-} on the surface of used $SrTiO_3$. Based on XRD and XPS results, we suggest that O_2^{2-} are formed by following steps during the reaction at 1073K:

$$Sr_2TiO_4+CO_2=SrTiO_3+SrCO_3 \quad 1)$$
$$SrCO_3+O_2=SrO_2+CO_2 \quad 2)$$

The SrO_2 phase was not shown up in the XRD patterns, which suggested that its formation may be limited only on the surface of the layered perovskites, hence, there is no bulk SrO_2 phase detected by XRD (see Table 2).

All the results described above indicated that the bulk phase of the perovskites is stable and the surface composition is almost unchanged (except trace amount of CO_3^{2-}) during the OCM process, and the catalytic activity and selectivity remained constant in the time period investigated. By contrast, the initial catalytic activity and selectivity of Sr_2TiO_4 and Sr_2SnO_4 are relatively low. The phase transformation during the first 4-5 hr on stream leads to the generation of surface SrO_2 and increasing of the catalytic activity and selectivity. XPS results suggested that only O^{2-} species existed on the fresh layered perovskites (Sr_2TiO_4 and Sr_2SnO_4) surface. The decomposition of the layered perovskites phases were accompanied by the generation of surface O_2^{2-} and CO_3^{2-} species, and by parallel increases in the catalytic activity and selectivity for OCM during the first 4-5 hr of reaction. Apparently, the O_2^{2-} species are more active than O^{2-} for OCM. Compare to the perovskite catalysts, the C_2 selectivity and yield on the related layered perovskites are much higher, this may be attributed to the presence of O_2^{2-} oxygen species on layered perovskite surface. Layered perovskite Sr_2ZrO_4 contains considerable amount of SrO, hence its catalytic activity varies in a different tendency during the first 4-5 hr of reaction. The initial activity and selectivity are the highest and declined with time to reach a steady state. Since this catalyst contains considerable SrO which can react with O_2 to form active surface SrO_2 during the pretreatment prior to exposure in the reaction mixture, the active SrO_2 content is the highest at the beginning of the reaction and undergo some decrease of activity due to CO_2 poisoning. When the reaction temperature decreased from 1073K to 1003K, the discrepancy of the catalytic activity between perovskites and related layered perovskites became rather small. This may be due to the decrease of O_2^{2-} species on the surface deduced by the back shift of reaction (2) at lower temperature.

The foregoing discussion leads to the conclusion that both O^{2-} and O_2^{2-} species are active for methane activation to produce C_2 hydrocarbons, but O_2^{2-} present higher C_2 selectivity and yield. According to this assumption, we agree with Stair et al.[18] that two kinds of C-H activation mechanism must be existed e.g. insertion and abstraction mechanism. Methane reacts with O_2^{2-} through abstraction mechanism produces gas phase methyl radicals while it reacts with basic (O^{2-}) and acidic pairs (metal ions) through insertion mechanism produces surface-bonded methyl. Compare to the gas phase methyl radicals, the oxidation probability of surface-bonded methyl by gas phase oxygen is larger, hence O_2^{2-} present higher C_2 selectivity than O^{2-}.

ACKNOWLEDGEMENT

The financial support of the NSFJ and NSFC is gratefully acknowledged.

REFERENCES

1. A. Kaddouri, R. Kieffer, A. Kienneman, P. Poix and J.L. Rehspringer, Appl. Catal., 51, L1(1989)
2. H. Nagamoto, K. Amanuma, H. Nobutomo and H. Inoue, Chem. Lett., 237 (1988)
3. K. Machida and M. Enyo, J. Chem. Soc. Chem. Commun., 1639 (1987)
4. Yan, Q., Jin, Y., Wang, Y., Chen, Y. and Fu, X., Proceed. of the 10th Inter. Congr. on Catal., Part C, P2292 (1992)
5. Campbell, K. D., Catal. Today, 13, 245 (1992)
6. T. Ito and J.H. Lunsford, Nature, 314, 721 (1985)
7. G.D. Moggridge, J.P.S. Badyal and R.M. Lambert, J. Catal., 132, 92 (1991)
8. K.C.C. Kharas and J.H. Lunsford, J. Am. Chem. Soc., 111, 2336(1990)
9. W.P. Ding, Y. Chen and X.C. Fu, Appl. Catal., 104, 61(1993)
10. X.D. Peng, D.A. Richards and P.C. Stair, J. Catal., 121, 99(1991)
11. H. Yamashita, Y. Machida and A. Tomita, Appl. Catal., 79, 203 (1991)
12. P. Ganguly and M.S. Hegde, Phys. Rev. B, 37, 5107 (1988)
13. D. Dissanayake, K.C.C. Kharas, J.H. Lunsford and M.P. Rosynek, J. Catal., 139, 652 (1993)
14. R. Scholder, D. Rade and H. Schwarz, Z. Anorg. Allgen. Chem., 362, 149 (1968)
15. D.D. Sarma, K. Sreedhar, P. Ganguly and C.N.R. Rao, Phys. Rev. B, 36, 2371 (1987)
16. D.D. Sarma, K. Sreedhar, P. Ganguly, and C.N.R. Rao, Phys. Rev. B, 39, 6194 (1989)
17. J.P.C. Baryal, X. Zhang and R.M. Lambert, Surf. Sci. Lett., 225, L15 (1990)
18. X.D. Peng and P.C. Stair, J. Catal., 128, 264 (1991)

PROMOTION BY DICHLOROMETHANE OF THE
OXIDATIVE COUPLING OF METHANE ON
LiMn/MgO CATALYSTS

R. Mariscal, M.A. Peña, and J.L.G. Fierro

Instituto de Catálisis y Petroleoquímica, CSIC
Campus UAM, Cantoblanco, 28049 Madrid, Spain

INTRODUCTION

It has been observed that by introduction of halomethane into the gas stream during the testing of catalysts for the methane coupling reaction, the selectivity for the formation of C_2 products is greatly enhanced[1-4]. The presence of gaseous HCl in the feed has also been reported to promote the OCM reaction[5]. However, it is interesting to note that the highest C_2 selectivity was achieved when the corresponding chloride salts were used to introduce alkali metals in the catalyst[1,3,7-9]. Burch et al.[1,2] have shown that addition of alkali halides to manganese dioxide can convert its activity from total oxidation to coupling, and that exposure to chlorine can also modify the behavior of manganese catalysts. Despite the studies related to catalytic systems containing chloride ions, important factors such as the nature of surface phase, its stability and the enhancement of C_2 yields still remain unclear.

In this paper, magnesia-supported Mn oxide catalysts promoted by Li have been investigated in the methane coupling reaction at 1023 K. The changes in both surface and bulk properties as a function of pretreatments have been revealed by photoelectron spectroscopy and X-ray diffraction. Moreover, the influence of injecting a single pulse of dichloromethane on the catalytic and structural properties was carefully studied.

EXPERIMENTAL

Magnesium oxide (UCB, 99.6 wt%) powder, calcined at 1073 K, was used as carrier. A 0.3 wt% Mn/MgO catalyst was prepared by adding MgO to an aqueous solution of manganese nitrate ($Mn(NO_3)_2.4H_2O$, Riedel-de Haën AG) in the appropriate concentration and evaporating to dryness. The precursor was kept at ambient temperature for 16 h, then dried at 403 K for 10 h and subsequently calcined in air at 1073 K for 4 h. An aliquot of the Mn/MgO catalyst was then impregnated with an aqueous solution of Li_2CO_3 (Merck). In order to increase the solubility of Li_2CO_3, CO_2 was bubbled into the solution and the flask maintained under ice. Once solutions become transparent, the required amount of 0.3 wt% Mn/MgO was added to the flask solution and then the excess water removed. Drying and calcination steps were the same as given above. A 2.5 wt% Li/MgO catalyst was prepared according to the same procedure of the ternary (Li/Mn/MgO) system.

X-ray diffraction patterns were recorded using a Philips PW 1010 vertical diffractometer using nickel-filtered CuKα radiation, under constant instrumental parameters. For each sample, Bragg angles between 5 and 70° were scanned at a rate of 2°/min.

Photoelectron spectra were recorded from quenched catalysts using a Fisons ESCALAB MkII 200R spectrometer equipped with a MgKα X-ray excitation source (1253.6 eV) and a hemispherical electron analyzer. C1s, O1s, Mn2p, Mg2p and Cl2p peaks were accumulated in order to improve the signal-to-noise ratio. The intensities were estimated by calculating the integral of each peak after subtraction of the "S-shaped" background and corrected using the cross sections quoted in literature. Accurate binding energies ($\pm$ 0.1 eV) (BE) were obtained by charge referencing to the adventitious C1s peak at 284.9 eV.

The activity of the catalysts was determined using a 5 mm i.d. quartz fixed-bed flow reactor at atmospheric pressure and 1023 K. The 0.1 g of catalyst used in each case was mounted into the reactor between quartz wool plugs. The flows were controlled by Brooks mass flow controllers to give a final mixture $CH_4:O_2 = 20:1$. Dichloromethane (10 μL) was injected through a port in the O_2 supply line. At fixed intervals the effluents were sampled using a gas sampling valve and on-line analyzed by GC equipped with TCD.

Table 1. Catalytic behavior of magnesia-supported catalysts after adding a CH_2Cl_2 pulse into the feed stream.

Catalyst	t(min)	X_{CH4}(%)	S_{CO}(%)	S_{CO2}(%)	S_{C2H4}(%)	S_{C2H6}(%)
Li/MgO	0	1.69	46.0	27.5	2.1	24.4
	2	2.09	47.6	30.5	2.1	19.8
	20	1.76	45.6	41.4	-	13.0
Mn/MgO	0	2.51	22.4	72.8	-	4.9
	2	2.80	35.8	54.4	1.3	8.5
	20	2.43	30.4	64.4	-	5.2
Li/Mn/MgO	0	0.20	31.4	29.1	-	39.6
	2	3.66	6.0	2.0	29.1	62.9
	12	0.37	22.3	28.4	-	49.3
	20	0.27	26.0	28.7	-	45.3

Reaction conditions were: temperature = 1023 K; $CH_4:O_2$ = 20 (molar ratio); catalyst weight = 0.1 g.

RESULTS AND DISCUSSION

Activity Changes after Dosing Dichloromethane

To investigate the relative selectivities of the Mn/MgO and Li/Mn/MgO catalysts, they were heated in the CH_4/O_2 mixture at 1023 K and the products analysed until steady state conditions were reached. Then a pulse of 10 μL of CH_2Cl_2 was injected in the feed stream and the product distribution analyzed as function of the time on stream. The activity data and product distributions under both stationary state conditions and after pulsing CH_2Cl_2 are compiled in Table 1 for the Li/MgO, Mn/MgO and Li/Mn/MgO catalysts using fixed experimental conditions. For the sake of comparison, C_2 selectivities are plotted in Figure 1 as a function of the time on stream. The points at zero time represent the steady state C_2 selectivity for the formation of various products before injection. The Mn/MgO catalyst generated no ethylene but had low selectivity to ethane (4.9%). Conversely, the parent Li/MgO counterpart yielded both ethylene and ethane, although the latter was dominant, being ca. five times greater than for the Mn/MgO catalyst. The behavior of the ternary Li/Mn/MgO system was between that of the binary catalysts; only ethane was observed with a maximum selectivity of 39.6%, although the methane conversion was much lower. When CH_2Cl_2 was added to the feed stream no significant changes in product distributions were found. For the Mn/MgO catalyst, only small improvement in C_2 selectivity was observed, selectivity increased over a period of two minutes to ca. 10%, but decreased quickly. The promotion effect of CH_2Cl_2 was much more marked for the ternary Li/Mn/MgO system in both activity and C_2 selectivity: the methane conversion was ca. 18 times higher than in the absence of halocompound and the maximum C_2 selectivity was 92% (ca. one third was ethylene), with the concomitant inhibition of CO_x formation. Although this effect is transitory, some enhancement of ethane selectivity is still observable after 20 minutes on stream.

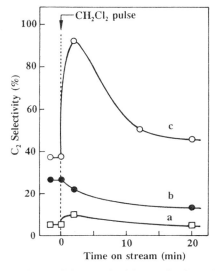

Figure 1. Dependence of the C_2 selectivity on the time of stream after dosing a small CH_2Cl_2 pulse into the feed stream. ($\square$), Mn/MgO; ($\bullet$), Li/MgO; ($\bigcirc$), Li/Mn/MgO. Stationary state data are also included for comparison.

XPS of Quenched Catalysts after Dosing CH_2Cl_2

Cl2p and $Mn2p_{3/2}$ core level spectra of the quenched Mn/MgO and Li/Mn/MgO catalysts and the reference Li/MgO homologue after injection of a single pulse of CH_2Cl_2 over the catalyst bed are display in Figure 2. Judging from these spectra, no significant changes in the BEs values of these peaks were observed. What is very important is the observation of the same BE at 198.0 eV of the $Cl2p_{3/2}$ peak for all catalysts. This BE value is typical of Cl^- ions, but not of chlorine in covalent C-Cl bonds; this peak usually appears slightly above 199.0 eV. The $Mn2p_{3/2}$ peaks for both binary Mn/MgO and ternary Li/Mn/MgO catalysts were fitted to three components: the lower BE peak is due to Mn^{3+} species, the peak at intermediate BE would be ascribed to Mn^{4+} ions, and the third one placed at the highest BE was introduced to account for final states and energy relaxation phenomena involved in the photoelectron process. By comparing the intensity of the former two peaks, it is clear that more reduced Mn is present in the Li-free catalyst.

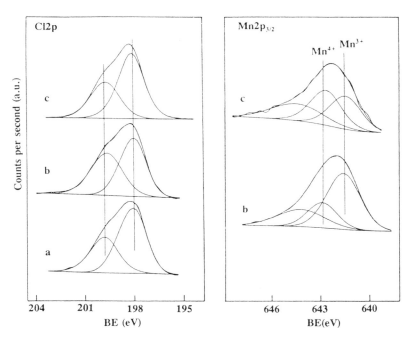

Figure 2. Cl2p and $Mn2p_{3/2}$ core level spectra of quenched catalysts after injection of a CH_2Cl_2 pulse: (a), Li/MgO; (b), Mn/MgO; (c), Li/Mn/MgO.

Quantitative analytical information obtained from the XPS measurements on the quenched samples is shown in Table 2. Injection of a pulse of CH_2Cl_2 resulted in the formation of Cl^- ions whose concentration changed from catalyst to catalyst. Thus, while the Cl/Mg ratio was rather low for the binary Mn/MgO and Li/MgO catalysts, it increased by almost an order of magnitude for the ternary Li/Mn/MgO catalyst. Substancial differences in the surface Mn/Mg ratios were also found when comparing promoted and unpromoted Mn/MgO catalysts. The Mn/Mg ratio was more than three times higher for the Mn/MgO than for its Li-doped homologue, suggesting formation of a highly dispersed Mn phase. Structural analysis by XRD revealed the presence of a Mg_6MnO_8 phase in calcined catalysts[10]. This phase is derived from that of MgO by replacing Mg^{2+} in every cation layer by alternately Mn^{4+} and a cation vacancy, in which Mn^{4+} is stabilized against reduction. The structure of the catalyst seems to be more complicated for the Li/Mn/MgO catalyst. The decrease of the XPS Mn/Mg ratio in this promoted catalyst compared to the unpromoted homologue suggests the formation of a certain proportion of Mn-Li solid solution, which may be confined to several subsurface regions, in complete agreement with other XRD and ESR data[10]. The strong enhancement of both activity and C_2 selectivity in the ternary Li/Mn/MgO catalysts led to the conclusion that the structure of the alkali-promoted Mn oxide catalysts plays a crucial role in determining the activity and selectivity in the OCM reaction. As pointed out by Burch et al.[1] and in agreement with our data[10], the non-selective MnO phase is formed under typical reaction conditions, i.e., 1023 K and CH_4/O_2 above 4. However, when some chlorine is introduced into the feed stream in the form of CH_2Cl_2 a redox mechanism occurs at catalyst surface:

$$CH_2Cl_2 + 2Mn^{2+} \longrightarrow 2Cl^- + 2Mn^{3+} + CH_2\cdot\cdot$$

According to this equation, formation of Mn^{3+} implies necessary the presence of chlorine. It can also be argued that the presence of Li in a Mn-Li solid solution stabilizes the intervalent Mn^{3+} species which acts selectively in the OCM reaction. As there is an excess of Li (Li/Mn = 66.3), the surface Li may stabilize the chloride against hydrolysis, although both activity and selectivity to C_2 hydrocarbons recover almost to stationary state levels 60 minutes after dosing the halocompound into the feed stream, indicating that such stabilization is transitory.

Table 2. Quantitative XPS data on Li/Mn/MgO catalysts after exposure to a pulse of CH_2Cl_2

Catalyst	Cl/Mg	Mn/Mg	Cl/Mn
Li/Mn/MgO	0.285	0.015	19.00
Mn/MgO	0.022	0.052	0.42
Li/MgO	0.038	-	-

CONCLUSIONS

The results presented in this work show that under typical condictions for the OCM reaction injecting a single pulse of dichloromethane into the CH_4/O_2 feed stream affects strongly both methane conversion and C_2 product distribution on the ternary Li/Mn/MgO catalyst system but only slightly in the binary Li/MgO and Mn/MgO counterparts. Since, in the absence of catalyst, no significant changes are observed in either catalytic activity and selectivity to C_2 hydrocarbons after dosing a CH_2Cl_2 pulse, it is inferred that the role of the halocompound is to alter the catalyst structure. Photoelectron spectroscopy on the quenched catalysts has shown that the enhanced selectivity is associated with a reduced Mn oxide. The CH_2Cl_2 is decomposed on catalysts surface through a redox process yielding chloride ions which are stabilized by the presence of lithium.

ACKNOWLEDGMENT

The authors wish to acknowledge Comisión Interministerial de Ciencia y Tecnología, Spain, for financial support of this work (Grant MAT91-0494).

REFERENCES

1. R. Burch, S. Chalker, G.D. Squire, and S.C. Tsang, Oxidative coupling of methane over manganese oxide catalysts, *J. Chem. Soc. Faraday Trans.* 86:1607 (1990).
2. R. Burch, S. Chalker, P. Loader, R. Mariscal, D.A. Rice, and G. Webb, The mechanism of ethylene formation on chloride and oxychloride catalysts, *Catal. Today* 13:301 (1992).
3. S. Ahmed, and J.B. Moffat, Effect of carbon tetrachloride as a feedstream additive in the oxidative coupling of methane an alkali oxide/SiO_2 catalysts, *Appl. Catal.* 63:129 (1990).
4. S. Ahmed, and J.B. Moffat, Promotion by tetrachloromethane of the oxidative coupling of methane on silica-supported alkaline earth oxides, *J. Catal.* 121:408 (1990).
5. H.P. Withers, Jr., C.A. Jones, J.J. Leonard, and J.A. Sofranko, Methane Conversion, US Patent No. 4,634,800, Atlantic Richfield Company (1987)
6. K. Otsuka, M. Hatano, and T. Komatsu, Synthesis of C_2H_4 by partial oxidation of CH_4 over transition metal oxides with alkali-chlorides, *in:* "Methane Conversion", D.M. Bibby, C.D. Chang, R.F. Howe, and S. Yurchak, ed., Stud. Surf. Sci. Catal. vol. 36, p. 383, Elsevier, Amsterdam (1988).
7. S.J. Conway, D.J. Wang, and J.H. Lunsford, Selective oxidation of methane and ethane over Li^+-MgO-Cl⁻ catalysts promoted with metal oxides, *Appl. Catal. A:General* 79:L1 (1991).
8. D.I. Bradshaw, P.T. Coolen, R.W. Judd, and C. Komodromos, Partial oxidation of methane over hopcalite and over manganese dioxide promoted by chlorine and alkali, *Catal. Today* 6:427 (1990).
9. W. Ueda, T. Isozaki, Y. Moriyama, and J.M. Thomas, A comparison of catalytic performance of the lithium and sodium analogues of bismuth oxyhalides in the oxidative dimerization of methane, *Chem. Lett.* 2103 (1989).
10. R. Mariscal, J. Soria, M.A. Peña, and J.L.G. Fierro, Features of Li-Mn-MgO catalysts and their relevance in the oxidative coupling of methane, *J. Catal.* 147:535 (1994).

KINETICS OF REACTION OF OXYGEN WITH LITHIUM NICKEL OXIDE, AND THE ROLE OF SURFACE OXYGEN IN OXIDATIVE COUPLING OF METHANE

Y. -K. Sun,[†] J. T. Lewandowski, G. R. Myers, A. J. Jacobson,[††] and R. B. Hall[*]

Corporate Research Laboratories, Exxon Research and Engineering Company, Annandale, New Jersey 08801;
[†] Merck Manufacturing Division, Merck & Co., Annandale, New Jersey 08801
[††] Department of Chemistry, University of Houston, Houston, Texas 77204

[*]to whom inquiries should be addressed

INTRODUCTION:

Alkali-promoted metal oxides have been widely investigated as catalysts for the selective reaction of methane with oxygen to form ethane and ethylene via oxidative coupling. A wide variety of oxides have been studied, including: basic oxides such as MgO (1), rare-earth oxides such as Sm_2O_3 (2,3), and transition metal oxides such as Mn oxides (4-7) and NiO (8-11). With all of these materials, deep oxidation of methane to CO_2 and H_2O competes with oxidative coupling. It is estimated that for the process to be economic, C_2^+ yields (methane conversion times selectivity to C_2^+ hydrocarbons) in excess of 40% are necessary (12). However, despite considerable research activity, yields in excess of about 25% have not been achieved in a single pass. Several concepts for achieving higher yields by means of novel engineering approaches have been discussed (13, 14). These approaches generally involve some sort of product separation in order to prevent secondary oxidation of the desired products, and a staging or recycle of unreacted methane. A key factor to the success of such approaches is the availability of catalytic materials that have very high intrinsic selectivity, even though it may be at relatively low conversion.

Lithium nickel oxide is of interest in this regard because it is one of the few materials for which an intrinsic C_2^+ selectivity of 100% has been reported. Hatano and Otsuka reported (8d) that, for limited conversions, 100% selectivity can be achieved in the absence of gas-phase oxygen, where the catalyst converts methane utilizing oxygen contained in the solid. Reports of 100% C_2^+ selectivity are generally suspect. It is well known (9, 8b) that the appearance of product CO_2 in the gas phase can be reduced by the formation of solid alkali-carbonates. This results in an overestimate of the selectivity when based on detection of gas-phase products alone. Hatano and Otsuka, however, calculated selectivity based on the ratio of C_2^+ products to methane conversion, which should have circumvented complications due to carbonate formation.

Lithium nickel oxide is rapidly reduced by reaction with methane alone. As the catalyst becomes partially reduced, the selectivity rapidly declines. It can, however, be reoxidized with gas-phase oxygen in the absence of methane, and the apparent 100% selectivity recovered. The structures that are relatively selective can only be stabilized under oxidative coupling conditions if gas-phase oxygen is continuously fed to the reactor (9,10). The selectivity achieved under these conditions, however, is significantly lower, roughly 60% at a methane conversion of 10%. It has been postulated (8b) that there is an additional form of oxygen on the catalyst in the presence of gas-phase oxygen, and that this oxygen may participate in the formation of CO_2. Understanding the nature of the interaction of gas-phase oxygen with the catalyst, the nature of oxygen species formed, and how each interacts with methane is important in determining whether, and how 100% C_2^+ selectivity might be achieved at low conversion.

Here we report results on studies of the kinetics of reaction of gas-phase oxygen and of methane with a structurally-ordered catalyst that has high C_{2+} selectivity (15-17). We confirm that lattice oxygen can be highly selective for methane activation and production of C_{2+} hydrocarbons. However, a proper accounting of CO_2 that remains on the catalyst as a carbonate, reveals that the selectivity is only 75%, in contrast to the earlier study (8d). In addition, we identify a surface oxygen state in equilibrium with gas-phase oxygen that participates in the production of CO_2. We determine the reaction kinetics of the surface and lattice oxygen states using transient isotope-switching techniques and steady-state rate measurements.

EXPERIMENTAL

The transient isotope-switching apparatus has been described previously (18). The technique of transient isotope-switching can provide fundamental information about the kinetics of catalytic reactions *under operating conditions*. A steady-state reaction condition is first established in a plug-flow, catalytic reactor. At some time, t, the isotopic composition of one or more of the reactant gases is switched. Products evolving after time t with the original label are from reactions of intermediates on/in the catalyst prior to the isotopic-label switch. The partial pressures and flow rates of the reactants are not altered by this switch so that the steady-state concentrations of reactants and products in the gas phase, *and on the catalyst surface*, are not disturbed. The area under the transients provides a direct measure of the steady-state concentrations of intermediates on/in the catalyst at the time of the switch, and the temporal shape of the transients provides information on the reaction kinetics of the intermediates.

The reactor in the current experiments was a plug-flow, fixed-bed quartz microreactor, 8 mm in diameter, with a total volume of 0.8 cc. Approximately 0.4 g of catalyst was supported in the middle of the reactor on a fused quartz frit. Temperature in the catalyst bed was measured with a chromel-alumel thermocouple shielded in a quartz jacket. The reactant gas stream was controlled by a mass flow control switching system capable of switching one or more of the reactants to its isotopically labelled counterpart in approximately 1 s.

The lithium nickel oxide catalyst used in this study was synthesized by reacting lithium carbonate and nickel oxide in air at 1073 K. The catalyst was characterized by x-ray diffraction to determine the purity of phase, and by microanalysis to determine the elemental composition. The nominal composition of the catalytic material used here was $Li_{0.44}Ni_{0.56}O$. The short and long range order of the material was characterized by small angle neutron scattering, EXAFS, and x-ray scattering (15, 17). The catalyst was black in color and had a surface area of approximately 2 m^2 g^{-1}, as determined by gas adsorption. To remove surface carbon contamination and any carbonate phase remaining from the synthesis or formed during methane conversion, the catalyst was heated to 923 K in a flow stream of 20% oxygen in either argon or helium.

Rates of consumption of reactants and formation of products were monitored continuously with a quadrupole mass spectrometer (Extranuclear) through a capillary sampling tube located at the outlet of the reactor, and periodically by a fast gas chromatography system (MTI M200). This GC system was capable of determining the entire product distribution through C_4 hydrocarbons every 60 s. Oxygen $^{16}O_2$ (99.997%) was obtained from Airco, Ar (99.999%) from Airco, and 20% $^{18}O_2$ (97%) in Ar (99.998%) from ICON.

RESULTS AND DISCUSSION

Isotope Exchange with Surface-Bound Oxygen

In Fig. 1. we show the time dependence of the mass spectrometric signals for gas-phase $^{16}O_2$, $^{16}O^{18}O$, and $^{18}O_2$ following a switch from 20% $^{16}O_2$ in Ar to 20% $^{18}O_2$ in Ar at 678 K and at a total flow rate of 40 cc min^{-1}. The mass spectrometric intensities are converted to the partial pressures of each species in the reactor. The calibration against absolute number density in the reactor is readily accomplished because the feed gas provides an accurate, internal reference point. In the upper panel of Fig. 1, it is shown that upon switching, the partial pressure of $^{16}O_2$ decreases rapidly with a complementary increase in the partial pressure of $^{18}O_2$, the result being a constant oxygen pressure. In the lower panel, it is shown that $^{16}O^{18}O$ evolves from the reactor, with a relatively rapid rise in rate followed by a slower decay. To check whether gas-phase reactions or reactions at the reactor wall contribute to the isotope exchange, a blank experiment was conducted with the catalyst replaced by quartz chips. No $^{16}O^{18}O$ was observed at temperatures up to 1123 K.

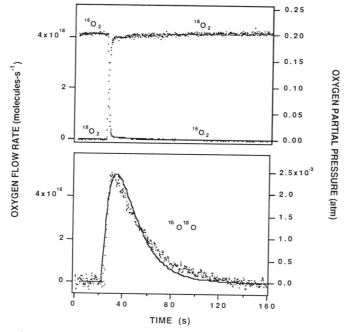

Fig. 1. Transient responses of mass spectrometric intensities of oxygen at 678 K upon switching from 20% $^{16}O_2$ in Ar to 20% $^{18}O_2$ in Ar. A background in the $^{16}O^{18}O$ signal due to $^{16}O^{18}O$ in the oxygen-18 feed has been appropriately subtracted. The flow rate was 50 cc-min^{-1}. The catalyst was $Li_{0.44}Ni_{0.56}O$ (weight = 0.36 g). The solid curve is a result from a model, see Sec. 3.1.2. for detail.

The full width at half maximum (FWHM) of the $^{16}O^{18}O$ evolution peak is ~ 29 s at 678 K, much longer than the ~ 1 s of the isotope switching transient. There is also a decay in the $^{16}O_2$ intensity, and a rise in the $^{18}O_2$ that occurs over a similar time. This is less obvious in Fig.1 because the amount of O_2 desorbing from the surface is a small fraction of the partial pressure of O_2 coming into the reactor. Subtracting the signals due to rapidly changing feed gas components following the switch, which decay/rise exponentially with a time constant of ~ 1 s, the area under the $^{16}O_2$ decay is found to be approximately equal to the area "missing" in the $^{18}O_2$ rise. Furthermore, the areas are roughly 1/2 that of the area under the $^{16}O^{18}O$ transient. This is what is expected for an exchange process involving dissociative adsorption of gas-phase O_2, and statistical, recombinatory desorption of an atomic state of oxygen in or on the catalyst (11, 19, 20).

An important characteristic of the exchange process at lower temperatures (below about 850 K) is that if the gas-phase label is switched from $^{16}O_2$ to $^{18}O_2$ and then back again to $^{16}O_2$, the shapes and amplitudes of the transient signals do not change, irrespective of the time the catalyst is exposed to $^{18}O_2$. This shows that the reservoir of exchangeable oxygen is fixed. Lengthy exposures to $^{18}O_2$, up to 400 s, do not lead to an increase in the amount of oxygen exchanged. This argues against the existence of a diffusion-limited exchange with lattice oxygen atoms near the surface, and suggests that the exchangeable oxygens are confined to the surface. The number of ^{16}O atoms contained in the $^{16}O^{18}O$ transient, obtained from the integral of the $^{16}O^{18}O$ signal over time, is 1.8×10^{18}. Hence the total number of exchangeable oxygen atoms is 3.6×10^{18}. This corresponds to a surface coverage of ~ 5×10^{14} cm^{-2}, or ~ 0.3 monolayers of ^{16}O. This number is estimated for a catalyst weight of 0.36 g with the assumptions that the surface area is ~ 2 m^2 g^{-1}, and the surface oxygen atom site density is 1.6×10^{15} cm^{-2}, i.e. the density of the (100) surface. Thus, the surface state is not a minority, defect site, as has been proposed (19). The density of populated surface sites under reaction conditions can be quite high in the presence of gas-phase O_2, and this chemisorbed oxygen can contribute significantly to the catalytic reactivity of these materials.

Another important characteristic in this temperature range is that the catalyst is saturated with oxygen for oxygen partial pressures ranging from 0.2 to less than 0.002 atmospheres. Consequently, the kinetics of $^{16}O^{18}O$ production in Fig. 1 is determined by the desorption kinetics of the surface oxygen, rather than by the the adsorption kinetics of the gas-phase oxygen. This is consistent with the observation by Lambert and coworkers (19) that oxygen chemisorbs on activated lithium nickel oxide even at room temperature.

With this, a simple kinetic model of the exchange process has been developed (11). This model has been used to obtain the kinetic parameters for desorption of atomic oxygen. Consistent results are obtained from an analysis of transients (as in figure 1) at several temperatures, and from the temperature dependence of exchange under steady-state conditions with a feed gas consisting of an equi-molar mixture of $^{16}O_2$ (10%) and $^{18}O_2$ (10%) in Ar (11). The activation energy obtained from these data for the recombinative desorption of the surface oxygen from $Li_{0.44}Ni_{0.56}O$ is 48 kcal mol^{-1}. The value for the pre-exponential factor, $k_d^{(0)}$, is 8×10^{-3} cm^2 s^{-1}. The latter is of the order expected for recombinative desorption involving species with moderate surface diffusivity.

An exchangeable-oxygen state is also observed with NiO. For NiO, however, higher temperatures are required to achieve the rate of exchange observed with the lithium substituted compound described above. NiO must be heated to 810 K in order to obtain Isotope-switching transients with half-widths similar to that shown in fig. 1. The rate constant, obtained from the temperature dependence of the $^{16}O^{18}O$ transients, for exchange of surface oxygen on NiO is $k_d = 5 \times 10^{-3} \exp(-56/kT)$ [cm^2 s^{-1}]. Thus the energy barrier to exchange of surface oxygen from NiO is significantly higher than it is for $Li_{0.44}Ni_{0.56}O$

Kinetics of Desorption of Surface and Lattice Oxygen

Under the conditions of the current experiments, the rate determining step in the isotope exchange is oxygen desorption. Thus, the kinetic parameters obtained for isotope exchange can be used to reproduce oxygen thermal desorption spectra measured in independent experiments. The thermal desorption spectrum of surface oxygen from lithium nickel oxide with a composition of approximately $Li_{0.5}Ni_{0.5}O$ has been reported previously by Lambert and coworkers (19). In these experiments, oxygen is dosed onto the catalyst powder at room temperature, and the powder heated linearly in time. Oxygen evolving from the powder into a vacuum chamber is monitored with a mass spectrometer. We have used a similar technique to measure TPD spectra at several different lithium loadings. The spectra obtained for a heating rate of 1 K/s, at several lithium loadings are shown in figure 2. The lower temperature features are due to desorption from the surface state, the higher temperature curves are due to evolution of lattice oxygen. The separation of the surface and lattice desorption curves in temperature is consistent with the observation that isotope exchange occurs exclusively with the surface oxygen at temperatures below 850 K. The peak temperatures for desorption of surface oxygen from $Li_{0.44}Ni_{0.56}O$ and NiO are 793 K, and 905 K, respectively. Using the rate coefficients k_d determined from exchange experiments above, we calculate that for a 1 K/s heating rate the maxima should occur at 785 K and 920 K.

It is clear from fig. 2 that lithium substitution into nickel oxide lowers the barrier for evolution of both surface and lattice oxygen. Similar results for lattice oxygen have been reported by Otsuka and coworkers (8d). Relatively, the effect on lattice oxygen is more pronounced than is the effect on surface oxygen. As will be shown below, reaction of methane with lattice oxygen has a high selectivity for production of ethane, while reaction with surface oxygen leads to CO_2. With this, the results in figure 2 suggest that lithium substitution increases selectivity by increasing the relative rate of reaction of lattice oxygen so that it is faster than reaction of surface oxygen at typical catalyst operating temperatures of around 1000 K.

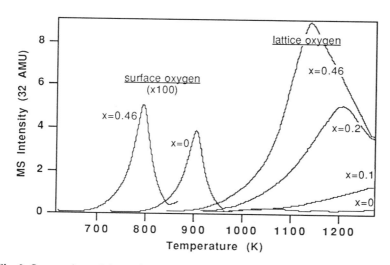

Fig. 2. Oxygen thermal desorption spectra for $Li_xNi_{1-x}O$ for several different values of lithium loading, x.

Isotope Exchange With Lattice Oxygen

Isotope exchange with $Li_{0.44}Ni_{0.56}O$ was measured at a number of temperatures above 850 K. The results are presented in Fig. 3. At 678 K, the $^{16}O^{18}O$ transient is substantially longer than the switching time (fig. 1). At 873 K, the O(s) desorption/exchange rate is significantly faster and the FWHM of the transient (~ 2 ms, calculated from k_d) becomes much shorter than the isotope switching time (~1 s). As a result, the $^{16}O^{18}O$ transient tracks the gas-phase switching rate, i.e. the $^{16}O^{18}O$ intensity is determined by exchange during the switching time when both $^{16}O_2$ and $^{18}O_2$ are still in the reactor. However, there is apparently still little exchange with lattice oxygen since the $^{16}O^{18}O$ signal following the switch decreases rapidly to zero. At 923 K, we start to see $^{16}O^{18}O$ at longer times. This signals the participation of lattice oxygen in the exchange. More and more ^{16}O from the catalyst participates in the exchange as temperature increases above 923 K.

At the higher temperatures, ^{18}O populates not only the surface state but also lattice sites, both at the surface and in the near surface region of the catalyst. The depth to which the catalyst becomes labeled depends on the details of the bulk diffusion kinetics, but in general, the longer the time to which the catalyst is exposed to $^{18}O_2$, the greater the depth of the labelling. Consequently, when the gas-phase oxygen is switched back to $^{16}O_2$, the $^{16}O^{18}O$ transient that marks the replacement ^{18}O lattice sites will depend on the previous exposure time. In the "reverse" switch, the $^{16}O^{18}O$ transient is observed to have a faster decay and a smaller time-integrated area compared to the initial switch from $^{16}O_2$ to $^{18}O_2$ (11). This is because only a limited depth has been labelled, and the fact that some of the near surface ^{18}O is exchanged, and some diffuses further into the bulk. Longer exposure times result in slower decay rates and more ^{18}O showing up the the second transient. A crude estimate of the effective energy barrier to isotope exchange with lattice oxygen <u>and</u> diffusion of lattice oxygen into the bulk, can be obtained from the extent (depth) of labelling as a function of exposure time. We obtain values of 58 kcal mol^{-1} for $Li_{0.44}Ni_{0.56}O$, and 62 kcal mol^{-1} for NiO (11, 20).

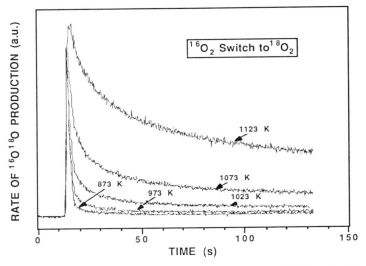

Fig. 3. Transient responses of mass spectrometric intensities of $^{16}O^{18}O$ from 873 to 1125 K upon switching from 20% $^{16}O_2$ in Ar to 20% $^{18}O_2$ in Ar. The flow rate was 40 cc min^{-1}. The catalyst was $Li_{0.44}Ni_{0.56}O$ (weight = 0.36 g).

Reactions of Surface and Lattice Oxygen with CH_4

Reaction of CH_4 with the lattice oxygen of lithium nickel oxide was studied first by reacting CH_4 with $Li_{0.44}Ni_{0.56}O$ in the absence of gas-phase oxygen. In order to determine the intrinsic reactivity and selectivity on a nearly stoichiometric material, the catalyst was exposed only to a small pulse of CH_4. The duration of the pulse was such that less than three layers of the lattice oxygen were reacted. A typical pulse was 15% CH_4 for 10 s. We find that the lattice oxygen has a high selectivity in converting CH_4 to C_2H_6. Between 973 and 1023 K, the selectivity to C_2H_6 was 75%. The remaining 25% was fully oxidized to

CO$_2$. However only 1% appeared in the gas phase, 24% remained on the catalyst as carbonate. The amount of carbonate left on the catalyst was determined by oxygen titration following each CH$_4$ pulse. We believe that the selectivities reported earlier (8d) are erroneously high, probably because of the difficulties of properly accounting for carbonate formation. The C$_2^+$ selectivity remains at 75% for methane exposures up to about 30 s. For exposures between 30 and about 1000 seconds, the selectivty remains relatively constant at about 60%. For these intermediate length exposures, the original selectivity can be recovered by reoxidizing the catalyst in oxygen. At exposure times in excess of 1000 seconds, the selectivity decreases precipitously, and it can no longer be recovered by a simple reoxidation of the catalyst (8, 9, 21).

In order to investigate the role played by the gas-phase and the surface oxygen, pulses of CH$_4$ and O$_2$ with fixed CH$_4$ (13.5%) but different O$_2$ (0 to 8.2%) partial pressures were reacted on Li$_{0.44}$Ni$_{0.56}$O at 1023 K, and conversion of methane and selectivity were determined (21). With increasing oxygen partial pressure, the production rate of C$_2$H$_6$ decreased, accompanied by an increase in the rate of CO$_2$. Importantly, the conversion of CH$_4$ did not increase significantly in the presence of gas-phase oxygen. These results suggest that the surface oxygen does not participate significantly in the activation of the C-H bond of methane.

To further study the reaction between methane and the surface oxygen, we used oxygen isotope labeling to preferentially populate the surface state with ^{18}O. Determining the isotope label of oxygen in carbon dioxide provides us with information on the relative rates of reaction of methane with lattice and surface oxygen. The results obtained at 973 K are shown in Fig. 4.

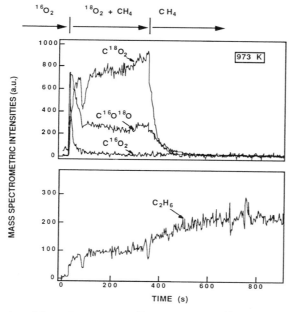

Fig. 4. Reaction of CH$_4$ with the surface ^{16}O and the lattice ^{18}O at 973 K. The reactant composition in the reactor is shown on the top of the figure: 20% ^{16}O$_2$ in Ar to 20% CH$_4$ + 10% ^{18}O$_2$ in Ar to 20% CH$_4$ in Ar. The low rate was 40 cc min^{-1}. The dips in the mass spectrometric signals at t ~ 80 s are caused by a transient decrease in pressure from the flow controller.

The reactant-gas compositions, and times at which they are switched, are indicated at the top of the figure. Mass spectrometric intensities of ethane, carbon dioxides, and water (not shown) were monitored continuously. At first, only ^{16}O$_2$ is fed to the reactor. Thus, both surface and lattice oxygen start with an ^{16}O label. When the feed is switched to $^{18}O_2$ and CH_4, the surface oxygen state is rapidly labelled with ^{18}O; labelling of lattice oxygen occurs on a much longer time scale. Fig. 4 shows that reaction of methane on this surface produces both ethane and carbon dioxide. The oxygen label of the CO_2, however, is mainly ^{18}O (87% ^{18}O and 13% ^{16}O). $C^{16}O_2$ is below the detection limit (<2%).

Several factors complicate the interpretation of the carbon dioxide labelling. First, CO_2 can exchange oxygen with the catalyst, proceeding through a carbonate intermediate involving either a surface ^{18}O or a lattice ^{16}O. Second, lattice oxygen may also be partially labelled by ^{18}O at 973 K. We attempted to assess the importance of these effects by measuring isotope scrambling of $C^{16}O_2$ in the presence of labelled oxygen For these experiments, the feed-gas composition was 5% $C^{16}O_2$, 20% $^{18}O_2$ and 75% Ar. Appreciable labelling of the CO_2 is observed. However, approximately 15 to 25% of the CO_2 retains the $C^{16}O_2$ labelling. In the experiments illustrated in fig. 4, $C^{16}O_2$ is below the detection limit (<2%). The combination of these results places an upper limit on the fraction of the CO_2 that can originate from reaction with lattice oxygen at 5-10%. At least 90-95% of the oxidation originates from reaction with surface and gas-phase oxygen. Detailed kinetic modelling of the gas-phase reaction network suggests that reaction with surface oxygen must account for roughly 1/2 of the CO_2 formation These results will be published elsewhere.

Another important observation in Fig. 4 is that when the gas-phase oxygen is eliminated, there is, in addition to the sharp decrease in the rate of CO_2 production, an increase in the rate of ethane production. This confirms that the selectivity is increasing since the rate of methane conversion does not change appreciably when the gas-phase oxygen is eliminated, and that the decrease in CO_2 is not solely due to formation of carbonates. This is consistent with the conclusion that lattice oxygen is responsible for production of ethane, that it is highly selective in doing so, and that CO_2 originates from reaction with surface and gas-phase oxygen. Of course, the current experiments do not identify what species are most rapidly oxidized by reaction with surface-oxygen, whether it is methane, methyl, ethyl or other products and intermediates involved in the coupling process.

SUMMARY

In summary, we conclude that lattice oxygen is highly selective for the oxidative coupling of methane to ethane, and that surface oxygen promotes full oxidation. Substitution of lithium into nickel oxides increasing C_2^+ selectivity by lowering the binding energy of the selective, lattice oxygen, allowing it to become more competitive with the surface oxygen. At high lithium loadings (x>0.25) these materials can achieve selectivies of up to 80%. However, this selectivity declines rapidly as lattice oxygen is removed, probably because of a phase segregation to Li_2O and NiO. Since the selectivity of NiO is essentially zero, and this phase segregation occurs so readily, it is unlikely that these materials will be useful catalysts.

LITERATURE CITED

1. (a) J.H. Lunsford, *Catal. Today* **6**, 235 (1990). (b) D.J. Driscoll, W. Martir, J.-X. Wang and J.H. Lunsford, *J. Am. Chem. Soc.* **107**, 58 (1985). (c) T. Ito and J.H. Lunsford, *Nature* **314**, 721 (1985). (d) C.-H. Lin, T. Ito, J.-X. Wang, and J.H. Lunsford, *J. Am. Chem. Soc.* **109**, 4808 (1987).
2. (a) K. Otsuka, K. Jinno and A. Morikawa, *J. Catal.* **100**, 353 (1986). (b) K. Otsuka and T. Nakajima, *J. Chem. Soc., Faraday Trans. 1* **83**, 1315 (1987).
3. (a) A. Ekstrom and A. Lapszewicz, *J. Am. Chem. Soc.* **110**, 5256 (1988). (b) A. Ekstrom and A. Lapszewicz, *J. Chem. Soc., Chem. Commun.* 797 (1988).
4. G.E. Keller and M.M. Bhasin, *J. Catal.* **73**, 9 (1982).
5. C.A. Jones, J.S. Leonard and J.A. Sofranko, *J. Energy and Fuels* **1**, 12 (1987).
6. (a) J.A. Labinger, S. Mehta, K.C. Otto, H.K. Rockstad and S. Zoumalan, in: *Catalysis*, Ed. J.W. Ward (Elsevier, Amsterdam, 1987) p. 513. (b) J.A. Labinger and K.C. Otto, *J. Phys. Chem.* **91**, 2682 (1987). (c) J.A. Labinger, K.C. Otto, S. Mehta, H.K. Rockstad and S. Zoumalan, *J. Chem. Soc., Chem. Commun.* 543 (1987).
7. Y. Amenomiya, V.I. Birss, M. Goledzinowski, J. Galuszka and A.R. Sanger, *Catal. Rev. - Sci. Eng.* **32**, 163 (1990), and references therein.
8. (a) K. Otsuka, Q. Lin and A. Morikawa, *Inorg. Chimica Acta* **118**, L23 (1986). (b) H. Hatano and K. Otsuka, *J. Chem. Soc. Faraday Trans. 1* **85**, 199 (1989). (c) K. Otsuka and T. Komatsu, *J. Chem. Soc., Chem. Commun.* 388 (1987). (d) M. Hatano and K. Otsuka, *Inorg. Chimica Acta* **146**, 243 (1988).
9. I.J. Pickering, P.J. Maddox and J.M. Thomas, *Angew. Chem. Int. Ed. Engl. Adv. Mater.* **28**, 808 (1989).
10. R.K. Ungar, X. Zhang and R.M. Lambert, *Appl. Catal.* **42**, L1 (1988).
11. Y.-K, Sun, J.T. Lewandowski, A.J. Jacobson, G.R. Myers and R.B. Hall, in *The Activation of Dioxygen and Homogeneous Catalytic Oxidation*, Eds.: H.R. Barton, A.E. Martell, D. Sawyer, Plenum Publishing, 1993
12. (a) S. Field, S.C. Nirula and J.G. McCarty, "An Assessment of the Catalytic Conversion of Natural Gas to Liquids", Report of SRI International Project No. 2352, (1987). (b) J.M. Fox, III, T.-P. Chen, and B.D. Degen, *Chem. Eng. Prog.* 42 (1990).
13. A.L. Tonkovich, R. Aris, and R.W. Carr, *Science*, **262**, 221 (1993)
14. R.B. Hall, G.R. Myers, *Proceedings, Division of Petroleum Chemistry: Symposium on Methane and Alkane Conversion Chemistry*, American Chemical Society Meeting, San Diego, CA, March 13-18, 1994
15. I.J. Pickering, J.T. Lewandowski, A.J. Jacobson and J.A. Goldston, *Solid State Ionics* **53-56**, 405 (1992).
16. W. Li, J.N. Reimers and J.R. Dahn, *Phys. Rev.* **B46**, 3236 (1992).
17. R.B. Hall, Y.-K. Sun, J.T. Lewandowski, G.R. Myers, I.J. Pickering, W.T.A. Harrison and A.J. Jacobson, (to be published).
18. C. A. Mims, R. B. Hall, A. J. Jacobson, J. T. Lewandowski, and G. Myers, in:*Surface Science of Catalysis - in situ Probes and Reaction Kinetics*, Eds.: D. J. Dwyer and F. M. Hoffman, ACS Symposium Series 482 (1991).
19. (a) G.D. Moggridge, J.P.S. Badyal and R.M. Lambert, *J. Catal.* **132**, 92 (1991). (b) X. Zhang, R.K. Ungar and R.M. Lambert, *J. Chem Soc., Chem. Commun.* 473 (1989).
20. P.J. Gellings and H.J.M. Bouwmeester, *Catal. Today* **12**, 1 (1992).
21. Y.-K, Sun, J.T. Lewandowski, G.R. Myers and R.B. Hall, (to be published).

CONVERSION OF METHANE TO HIGHER HYDROCARBONS OVER SUPPORTED TRANSITION METAL OXIDE CATALYSTS

Liwu Lin[*], Laiyuan Chen, Zhusheng Xu and Tao Zhang

Dalian Institute of Chemical Physics, Chinese Academy of Sciences, P.O.Box 110, Dalian 116023, P.R.China. Tel. (+86-411) 4671991, fax.(+86-411)4691570.

INTRODUCTION

Studies on the utilization of methane, the main component of natural gas, are of significant importance both in industry and catalytic science. Considerable interest has been concentrated on the production of higher hydrocarbons from methane, especially the oxidative coupling of methane to ethylene and ethane[1-2]. Although several promising catalysts have been discovered, it is difficult to obtain high activity and high C_2-hydrocarbons selectivity simultaneously as the activation of methane molecules and the deep oxidation of intermediates usually occur at the same reaction conditions. Therefore, the search for more effective catalysts and the investigation of the activation of methane are still great challenges to catalytic science.

More recently, the studies of the activation of methane by noble transition metals has drawn much attention[3-5]. The virtue of the utilization of noble metals is that the conversion of methane can be carried out in the absence of gas-phase oxygen. Therefore, two steps was adopted[4-5]. The first step is the activation of methane by metals, and the second step involves the hydrogenation of adsorbed carbon-containing species. In this way, methane can be converted to higher hydrocarbons at rather low temperatures. However, the problems are that the catalyst deactivates quickly and methane conversion is very low. Except that, the reports on the activation of methane by supported noble metal oxides are very few[5].

The effective paraffins aromatization catalysts (Ga/ZSM-5 and Zn/ZSM-5) are also active for the activation of methane, especially for the aromatization, but the activity is also rather low[6-12]. Methane conversion can reach 4.9% over Ga and Re supported on HZSM-5, and the aromatization selectivity is 51.6%[9]. By using O_2 and N_2O as oxidants, the former mainly turns methane to CO_2 while N_2O gives CH_3OH, $HCHO$, C_2H_4, C_2H_6 and C_6-C_{12} aromatics[10]. Molybdenum supported on HZSM-5 is also an effective catalyst for the aromatization of methane. Over this kind of catalyst, methane conversion of 7.2% and benzene selectivity of 100% at 700°C were obtained under non-oxidizing conditions[12]. To our knowledge, studies on the activation of methane under non-oxidizing conditions by supported transition metals are still very few[12-13].

The present paper reports on preliminary studies of the conversion of methane to ethylene, ethane, benzene and other aromatics over supported transition metal oxide catalysts (Pt, Re, Mo, V and W) under non-oxidizing conditions. The mechanisms of the activation of methane by differently supported transition metal oxides and the formation of aromatics are also discussed.

EXPERIMENTAL

Catalysts preparation

0.5%Re/ZSM-5, 0.5%Re/Al$_2$O$_3$, 0.5%Re/SiO$_2$, 0.5%Re/MgO and Pt-Re/Al$_2$O$_3$ catalysts were prepared by impregnating the corresponding supports with HReO$_4$ and HReO$_4$ + H$_2$PtCl$_6$ aqueous solutions, then dried at 120°C for 4 h and calcined in air at 500 °C for 4 h. NaZSM-5 zeolites were converted into HZSM-5 by repeated ion exchange with NH$_4$NO$_3$ aqueous solution, then dried at 120°C for 6 h, and finally calcined at 500°C for 4 h. The impregnating solutions for the preparation of 5%Mo/ZSM-5, 5%W/ZSM-5 and 5%V/ZSM-5 were ammonium molybdate, ammonium metatungstate and ammonium metavanadate + oxalic acid, respectively. All the catalysts were dried at 120°C for 10 h, then calcined at 750°C in air for 8 h.

Methane conversion tests

Conversion of methane was carried out in a flow type quartz micro-reactor which contained 0.6 g catalyst. The catalyst was first heated at 700°C in air for 1 h, then the flowing gas was switched to helium at 700°C for half an hour to remove remaining oxygen. After that, methane gas(99.95%) was introduced. The GHSV of methane was about 1350 ml/g.cat.h. Reaction products were analyzed by an on line gas chromatography using Porapak QS columns in a temperature-programmed mode and a TC detector. The methane conversion and the products selectivities were calculated on the carbon number basis.

Temperature-Programmed Reaction of methane(TPRm)

Temperature programmed reaction of methane was carried out in an U-type quartz tubular micro-flow reactor. The flow rate ratio of He:CH$_4$ was 10:5 (ml/min) and He:CH$_4$:O$_2$ was 10:5:2 (ml/min). About 0.3 g of sample was used in each run. The reaction products were analyzed by a fast response high sensitive quadrupole mass spectrometer (ANELVA TS-360S) equipped with a computer acquisition and controlling system. The reaction effluent gases were introduced into the analyzing chamber of MS immediately after it left the catalyst bed by using a capillary with a molecule leak at one end, so that we can analyze the reaction products in situ. The mass to charge ratios were calibrated by pure gases or known mixtures and the following ratios of *m/e* were used: H$_2$:2, CH$_4$:16, C$_2$H$_4$:26, CO:28, C$_2$H$_6$:30, O$_2$:32, CO$_2$:44, C$_4$:55, C$_6$H$_6$:78.

RESULTS

Reaction performance of Re/Al$_2$O$_3$ and Pt-Re/Al$_2$O$_3$

The results of the conversion of methane over alumina supported Re and Pt-Re catalysts under non-oxidizing conditions are listed in Table 1. It can be seen that the activities of the two platinum-containing catalysts are lower than that of Re/Al$_2$O$_3$. The selectivity of carbon monoxide decreased and the selectivities of C$_2$ and benzene increased when the on stream time varied from 15 min to 60 min. At the initial stage of the reaction (time on stream is 15 min), a significant amount of CO was produced. The product selectivities as a function of methane on stream time over a 0.5%Pt-0.3%Re/Al$_2$O$_3$ catalyst are given in Fig.1. The CO selectivity decreased significantly with the increase of reaction time. As the reaction was carried out under non-oxidizing conditions, we can conclude that CO was produced by the oxidation of methane by lattice oxygen of the supported oxides. The continuous production of CO probably resulted from the migration of the bulk oxygen to the surface. After about 280 min of reaction, the active lattice oxygen species were almost completely consumed, then the selectivity of C$_2$-hydrocarbons increases with the proceeding of the reaction. However, the benzene selectivity decreased quickly. After about 2 hours of reaction, it dropped to zero.

Table 1. Methane conversion at 700°C over Re/Al$_2$O$_3$ and Pt-Re/Al$_2$O$_3$ catalysts.

catalyst	on stream time (min)	CH$_4$ conversion (%)	selectivity(%)			
			CO	C$_2$H$_4$	C$_2$H$_6$	C$_6$H$_6$
0.3%Re	15	3.8	52.8	2.5	1.6	43.1
	60	2.3	14.0	8.0	0	78.0
0.15%Pt-0.3%Re	15	1.4	26.6	17.7	6.3	49.4
	60	0.5	28.6	26.5	9.6	35.3
0.5%Pt-0.3%Re	15	3.0	27.4	6.7	19.0	64.0
	60	0.6	24.2	21.1	3.0	51.7

Effect of the support

ZSM-5, Al$_2$O$_3$, SiO$_2$ and MgO are typical catalyst supports. ZSM-5 is a strong acid while MgO is a strong base. The acidity order for these supports are known as: ZSM-5 > Al$_2$O$_3$ > SiO$_2$ > MgO[14]. The purpose for choosing these four supports is to determine the influence of acidity of the support on methane conversion and product distribution. Fig. 2 gives the conversions of methane over rhenium catalysts supported on the above four supports. One can see that the influence of acidity on methane conversion is very significant. After methane on stream of 1 h, methane conversion over Re/ZSM-5 was 5.9% while it was only 0.3% over Re/MgO.

At the initial stage of the reaction over these catalysts, the predominant product was CO, while CO$_2$ was not detected in the whole run. Benzene was formed only over the Re/ZSM-5 and Re/Al$_2$O$_3$ catalysts(Fig.3). With the less acidic SiO$_2$ and the basic MgO supported rhenium, benzene was not produced. As the on stream time increased, the product distribution has changed significantly, namely, the CO selectivity decreased and the selectivities of C$_2$-hydrocarbons and benzene increased. Thus we can see from the above results that the more acidic the support, the higher the methane conversion and the higher the benzene selectivity.

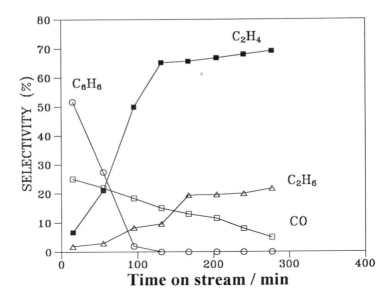

Fig.1. Product selectivities of methane reaction over a 0.2%Pt-0.3%Re/Al$_2$O$_3$ catalyst as a function of methane on stream time.

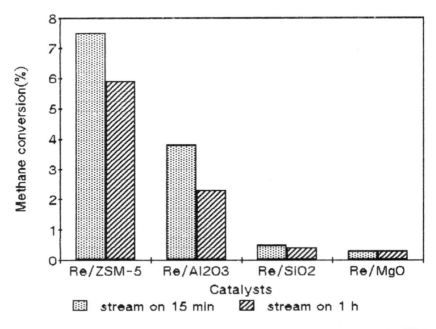

Fig.2. Methane conversion at 700°C over rhenium supported on different supports(2wt%Re).

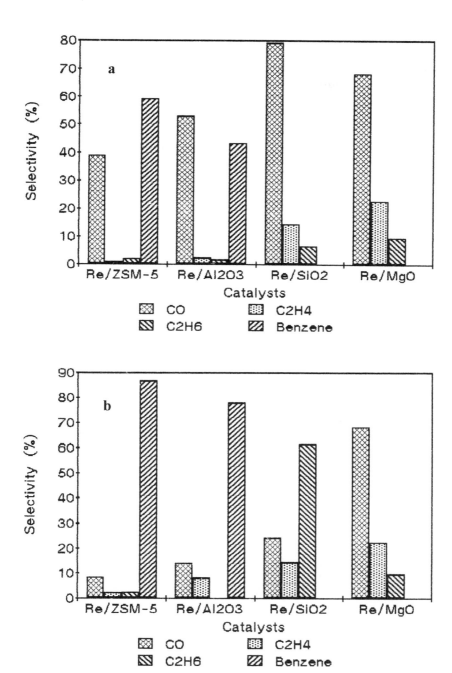

Fig.3. Products distribution of methane conversion at 700°C over various supported rhenium oxides (a) on stream for 15 min, (b) on stream for 1 h.

Methane conversion over ZSM-5 supported transition metals

Reaction performances of Re, Pt, Mo, W and V supported on HZSM-5 were tested under the same reaction conditions as described in the experimental section. Methane conversions and product selectivities after on stream time of 1 hour over these catalysts are given in Table 2.

Table 2. Methane conversion at 700°C over various HZSM-5 supported transition metal oxides. After methane on stream for 60 min.

catalyst	CH_4 conversion (%)	selectivity (%)			
		CO	C_2H_4	C_2H_6	C_6H_6
5%Re/ZSM-5	5.9	8.4	2.4	2.5	86.7
5%Pt/ZSM-5	2.7	11.2	5.6	1.2	82.0
5%Mo/ZSM-5	6.0	9.2	7.1	3.5	80.2
5%W/ZSM-5	3.7	22.3	11.4	2.7	63.6
5%V/ZSM-5	2.4	18.1	7.8	3.8	70.3

One can see that the most active catalysts are 5%Mo/ZSM-5 and 5%Re/ZSM-5, while the 5%Re/ZSM-5 is most selective for benzene formation. There was still a significant amount of CO in the products even after methane has reacted for 1 hour. It is important to state that there was significant amount of hydrogen in the reaction products, although part of which reacted with the active oxygen species to produce water. Fig.4 shows the methane conversions over various catalysts as a function of time on stream. It can be seen that for all the catalysts investigated, the conversions decreased sharply at first, then gradually. It is interesting that the stability of the Pt-Re/Al$_2$O$_3$ catalyst is very different from those of HZSM-5 supported transition metals, that is, Pt-Re/Al$_2$O$_3$ deactivated quickly while HZSM-5 supported catalysts were relatively stable under the reaction time investigated.

The product selectivities as a function of methane on stream time over 5%W/ZSM-5 and 5%V/ZSM-5 catalysts were given in Fig.5. We can see that the product selectivities changed in quite a similar way over these two catalysts. With the increment of methane on stream time, the selectivity for CO decreased, while the selectivities for benzene and ethylene increased. Selectivity for C_2H_6 did not vary significantly. The conversion of methane and the product selectivities as a function of on stream time over Pt/ZSM-5, Mo/ZSM-5 and Re/ZSM-5 catalysts behaved in a similar way as mentioned above(the figures are not shown here).

The activation of methane by lattice oxygen

In order to further investigate the mechanism of methane activation, temperature-programmed reaction of methane was performed. Fig. 6 gives the TPRm spectra of methane on a 0.3 wt % Re/Al$_2$O$_3$ catalyst. It shows that methane can be activated at about 600°C. The initial activation temperature of methane and the temperature corresponding to the maximum production rate of the products are in accordance with the reduction temperatures of rhenium oxide[15]. This consistency reveals that methane is activated by the lattice oxygen of the metal oxides. The formation of ethylene and ethane was prior to that of benzene and C$_4$-hydrocarbon. It is interesting to note that no gas-phase carbon dioxide was detected. In

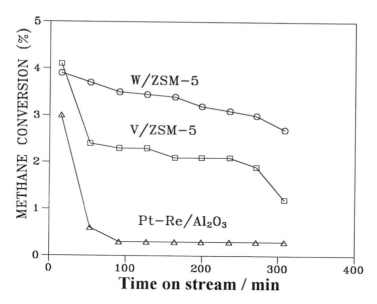

Fig.4. Methane conversion as a function of methane on stream time over various supported transition metal oxides.

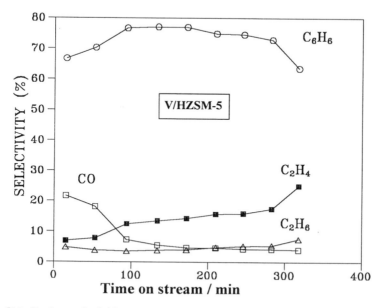

Fig.5(a). Products selectivities as a function of methane on stream time over 5%V/ZSM-5.

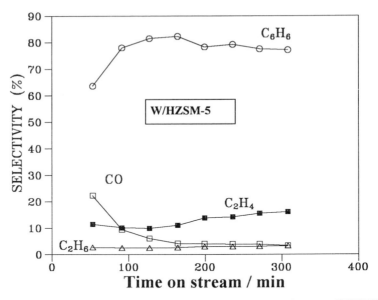

Fig.5(b). Products selectivities as a function of methane on stream time over 5%W/ZSM-5.

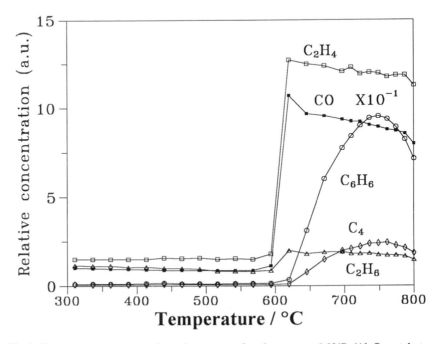

Fig.6. Temperature-programmed reaction spectra of methane over a 0.3%Re/Al$_2$O$_3$ catalyst.

order to exclude the possibility that product CO_2 might adsorb on the catalyst surface, a temperature programmed desorption of the remaining species on the catalyst after methane reaction was carried out. The formation of CO_2 was not observed.

The reaction of methane on a $0.3\%Pt/Al_2O_3$ catalyst took place at a rather low temperature, which was approximately $350^{\circ}C$. However, CO and CO_2 were the main products, and only a trace of benzene and C_4-hydrocarbons were observed over this catalyst. Methane reaction over a hydrogen reduced Pt/Al_2O_3 catalyst mainly gave hydrogen and C_2-hydrocarbons which is similar to that reported by Belgued[3] and Koerts[4]. The activation of methane by a $Pt-Re/Al_2O_3$ catalyst occurred at much lower temperature as compared to that over a Re/Al_2O_3 catalyst. The Re/Pt ratio also exhibited significant influence on the initial methane activation temperature and the product distribution. The TPRm spectra of methane over two $Pt-Re/Al_2O_3$ catalysts with different Re/Pt ratios are given in Fig.7. It can be seen that ethane appeared at much lower temperatures over the catalyst with a lower Re/Pt ratio. Another effect of the Re/Pt ratio was the difference of C_2H_4/C_2H_6 ratio, and it was higher over the catalyst with a higher Re/Pt ratio.

By plotting the onset hydrogen reduction temperatures of the catalysts with the initiate methane activation temperatures over these catalysts, two parallel lines can be obtained(Fig.8). This strongly suggests that methane was activated initially by the lattice oxygen of the transition metal oxides. However, methane molecules are more stable, and are difficult to be oxidized as compared with hydrogen. As a summary of the activation of methane by alumina supported rhenium and platinum oxides, it is important to note that methane can also serve as a reduction agent like H_2, and reacts with the lattice oxygen of the oxides.

Influence of gas-phase oxygen on the reaction performance of Pt-Re/Al₂O₃

In the presence of gas-phase oxygen, the temperature programmed reactions of methane and oxygen over all the aforementioned catalysts mainly gave CO_2, CO and H_2O. These results are in agreement with those reported by Bhasin[16] and Claridge[17]. The role of gas-phase oxygen on the formation of higher hydrocarbons was studied by a transient response method. Fig.9(a-c) shows the transient results over a $Pt-Re/Al_2O_3$ catalyst (Re/Pt = 1.5). When the $CH_4 + O_2$ reaction reached a steady-state at $750^{\circ}C$ (about 2 hours after the start up), the flow of the gas-phase oxygen was abruptly stopped. We can see that the concentrations of CO and CO_2 decreased sharply(Fig.9 a), and then those of benzene and C_4-hydrocarbon increased significantly (Fig.9 c). When the concentrations of benzene and C_4 approached zero, a pulse of oxygen was injected into the reactor. The concentrations of CO and CO_2 then raised suddenly. Increments of concentrations of benzene and C_4 were observed only after the decreasing of CO and CO_2. The time lag between the formation of CO_2 and CO and the formation of benzene and C_4 indicates that gas-phase oxygen or weakly adsorbed oxygen can oxidize the reaction intermediates C_2H_4 and C_2H_6 to CO_x and H_2O. Only when these oxygen species were completely consumed, then the intermediates could be further converted to C_4 and C_6 compounds. It can also be seen that the active sites which were responsible for the production of benzene and higher hydrocarbons could be regenerated by pulse injection of gas-phase oxygen.

DISCUSSION

The activation of methane by lattice oxygen

The supported transition metal catalysts, such as Pt/SiO_2, for the conversion of methane to higher hydrocarbons have been studied extensively[3-5,18]. However, in this paper, we have

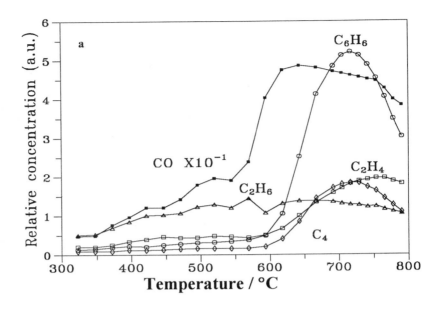

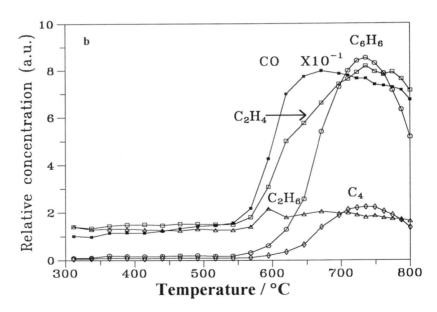

Fig.7. Temperature-programmed reaction spectra of methane over (a)0.5%Pt-0.3%Re/Al₂O₃, (b) 0.2%Pt-0.3%Re/Al₂O₃ catalysts.

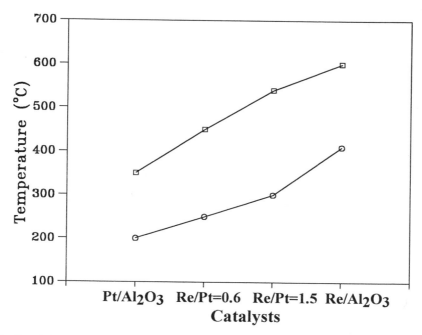

Fig.8. Correlation of the onset reduction temperatures(H_2-TPR)(-o-) with the initial methane activation temperatures(-□-) over Pt/Al_2O_3, Re/Al_2O_3 and Pt-Re/Al_2O_3 catalysts.

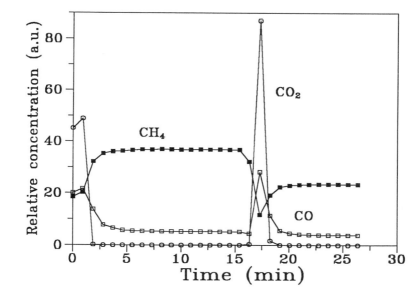

Fig.9(a). Transient experiment results of reaction of $CH_4 + O_2$ over a 0.2%Pt-0.3%Re/Al_2O_3 sample. Reaction temperature was 700°C. The oxygen flow was cut off at 1 min and a pulse of oxygen was injected at 16 min.

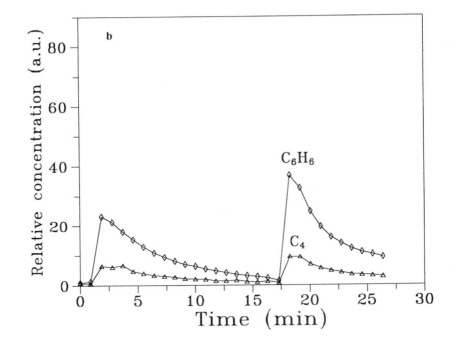

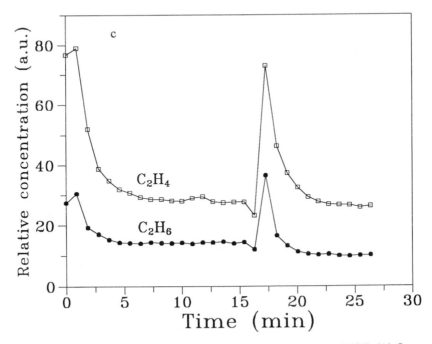

Fig.9(b-c). Transient experiment results of reaction of $CH_4 + O_2$ over a 0.2%Pt-0.3%Re/Al_2O_3 sample.
Reaction temperature was 700°C. The oxygen flow was cut off at 1 min and a pulse of oxygen
was injected at 16 min.

studied the activation of methane by these catalysts both in their oxidized forms and reduced forms, instead of in their reduced forms only, as were the cases in the above mentioned literature.

As the reaction tests were carried out under non-oxidizing conditions, we can conclude that methane is activated by lattice oxygen of the catalysts from the evidence of the formation of CO and H_2O. Fig. 8 shows that the initiating methane activation temperatures correlated well with the onset temperatures of H_2-TPR of the catalysts. This result strongly suggests that methane is activated by the lattice oxygen of the catalysts at the initial stage of the reaction. Because methane is very stable, the onset reaction temperatures were higher than the onset temperatures of hydrogen reduction.

The activation of methane by lattice oxygen was first reported by Keller and Bhasin[16]. Sofranko[19] also carried out methane coupling reactions using lattice oxygen of the catalysts. These authors proposed that methane was activated by lattice oxygen, and meanwhile the catalyst was reduced. Recently, Claridge et. al.[20] have tested molten salt mixtures in a redox mode for the activation of methane at 750°C. They observed by ESR measurements that the transition metal ions were reduced by methane. Therefore, we also proposed that the possible mechanisms for the activation of methane by supported transition metal oxides are as follows:

$$CH_4 + [O] \longrightarrow [CH_3OH] \longrightarrow [HCHO] + H_2 \longrightarrow CO + H_2 \quad (1)$$
$$CH_4 + [O] \longrightarrow [CH_3\cdot] + [HO] \longrightarrow [CH_2\cdot] + H_2O \quad (2)$$
$$CH_4 + [M] \longrightarrow [CH_3\cdot] + [H] \longrightarrow [CH_2\cdot] + H_2 \quad (3)$$

where [O] represents the lattice oxygen of the oxides, and [M] refers to the metals.

At the initial stage of the reaction, scheme (1) is the predominant reaction path. The reaction intermediates CH_3OH and/or HCHO are very unstable under reaction conditions, they give out CO and H_2 quickly. It is interesting to note that no gas-phase CO_2 was detected over these catalysts except on Pt/Al_2O_3 at low temperatures. Most probably CO_2 was not formed under our reaction conditions. CO cannot be oxidized to CO_2 by lattice oxygen. The formation of C_2H_6 probably resulted from the reaction of methane with lattice oxygen, via the homolytic mechanism by the coupling of methyl radicals as in reaction (2), or via a heterolytic mechanism as suggested by Sokolovskii[21], that is,

$$CH_4 + Me^{2+}O^{2-} \longrightarrow CH_3^-Me^{2+} + HO^- \quad (4)$$
$$CH_3^-Me^{2+} \longrightarrow CH_3\cdot + Me^{2+} \quad (5)$$

As the on stream time increased, the active lattice oxygen species(both surface and the bulk) were completely consumed. At that time, since the transition metal oxides were reduced, the active species were the metals, that is, scheme (3) would be predominant. The formation of carbene intermediate was supported by the formation of toluene and naphthalene which were formed via the carbene free radicals. Benzene probably was resulted from the aromatization of C_2H_4 over the acidic sites of the catalysts. This suggestion was proven by the evidence that C_2H_4 and C_2H_6 were formed before C_4-hydrocarbons and benzene (Fig.6).

Effect of acidity on the activation of methane

Results of Fig.2 show that the acidity of the catalyst support exerted significant influence both on methane conversion and product distribution. Methane conversion is higher over the more acidic catalysts. The basic catalyst Re/MgO showed very low methane conversion. Only over the more acidic Re/ZSM-5 and Re/Al_2O_3 catalysts that benzene can be formed. This indicates that the formation of benzene is caused by the synergistic effects between the activity of the lattice oxygen and the acidity of the support.

It is interesting to see from Fig. 4 that the stability of Pt-Re/Al_2O_3 catalyst is different from those of W/ZSM-5 and V/ZSM-5 catalysts. Pt-Re/Al_2O_3 catalyst deactivated quickly while W/ZSM-5 and V/ZSM-5 deactivated slowly at first, then approached a constant level, and finally deactivated. The trend of the product selectivity changes of the Pt-Re/Al$_2$O$_3$ catalyst (Fig.1) was also different from those of the ZSM-5 supported catalysts(Fig.5). For example, at the initial stage of the reaction the selectivity of benzene over the Pt-Re/Al$_2$O$_3$ decreased with methane on stream time, while it increased over ZSM-5 supported catalysts. The above facts demonstrate that the activation of methane by HZSM-5 supported transition metal oxides not only occurred via the action of lattice oxygen, but the acidity of the supports also played a significant role. Scheme (1) also took place in this case. The intermediate CH$_3$OH was either oxidized to CO and H$_2$, or produced C$_2$ over the acidic HZSM-5 surface. C$_2$-hydrocarbons can aromatize to aromatics(mainly benzene and a small amount of toluene and naphthalene compounds) as suggested by Shepelev[6-8] and Anderson[10]. Therefore, besides the activation by lattice oxygen, another possible mechanism for methane activation over HZSM-5 supported transition metal oxides is the heterolytic cleavage of methane molecules[22-23]:

$$CH_4 + [MHZSM-5] \longrightarrow CH_3^+ + [H-MHZSM-5] \qquad (6)$$

M refers to the metal ions in ZSM-5 channels.

The carbenium ion CH$_3^+$ is extremely reactive, it can react with excess methane to start a growth reaction[24]. That is,

$$CH_3^+ + CH_4 \xrightarrow{-H_2} [C_2H_7^+] \xrightarrow{} [C_2H_5^+] \underset{+H^+}{\overset{-H^+}{=\!=\!=}} C_2H_4 \qquad (7)$$

It is interesting to note that the benzene selectivity increased with the increment of methane on stream time, and attained a constant level when CO selectivity decreased to its minimum. This result shows that the best aromatization selectivity over the ZSM-5 supported transition metal catalysts can be obtained only after the consumption of removable lattice oxygen while the reverse is true for a Pt-Re/Al$_2$O$_3$ catalyst. As we can see from Fig.4 that ZSM-5 supported metal catalysts were quite stable even after the consumption of active lattice oxygen. This suggests that the lattice oxygen is not decisive for the activation of methane in this case. However, the metal ions and the strong acid centers should play predominant roles in methane activation. With the proceeding of the reaction, coking substances deposited on both the metallic and the acidic sites, and the catalyst was deactivated. After regeneration of the catalysts by burning of the coke, the activity of the catalysts can be revived.

It is important to note that methane probably was activated via different mechanisms when different supports were used. However, it is sure that higher hydrocarbons can be synthesized by alternately switching methane and air in a fix bed reactor, or by using a riser reactor in a fluidize bed system as used in the conversion of butane to maleic anhydride, or by optimizing the reaction conditions by using air or oxygen as oxidants over HZSM-5 supported transition metals.

CONCLUSIONS

The results of this preliminary report suggest that in the reaction of methane over supported transition metal oxides under non-oxidizing conditions, methane was activated via different mechanisms. Catalysts showed a support effect when HZSM-5, Al$_2$O$_3$, SiO$_2$ and MgO were used as supports. The acidity of the supports played significant roles on the conversion of methane and the product distribution. When a superacidic support such as HZSM-5 was used, heterolytic cleavage of methane was predominant in the activation of methane. When alumina was used as the support, methane was initially activated by the lattice oxygen of the oxides. With the increase of on stream time, the oxides were reduced. Then methane was activated by transition metals. Higher hydrocarbons could not be formed in the presence of gas-phase oxygen. However, they could be synthesized from methane over transition metals when using acidic supports under non-oxidizing conditions. Re/HZSM-5 and Mo/HZSM-5 were believed to be the most effective methane aromatization catalysts. The catalyst could be regenerated by the burning of coke after its deactivation.

REFERENCES

1. T. Ito, J.X.Wang, C.H. Lin and J.H. Lunsford, *J. Am. Chem. Soc.*, 107:5062(1985)
2. J.S. Lee and S.T. Oyama, *Catal.Rev.-Sci.Eng.*, 30:249(1988)
3. M. Belgued, J. Saint-juste, P. Pareja. and H. Amariglio, *Nature*, 352:789(1991)
4. T. Koerts, M.J.A.G. Deelem and R.A. Van Santen *J.Catal.*, 138:101(1992)
5. K. Ravindranathan Thampi, J. Kiwi and M. Gratzel, *Catal. Lett.*, 4:49(1990)
6. S.S.Shepelev and K.G.Ione, *React. Kinet. Catal. Lett.*, 23(3-4):319(1983)
7. S.S.Shepelev and K.G. Ione, *React. Kinet. Catal. Lett.*, 23(3-4):323(1983)
8. S.S.Shepelev and K.G. Ione, *J.Catal.*, 117: 362(1989)
9. EP 228267
10. Anderson, J.R. and Tsai, P., *Appl. Catal.*, 19:141(1985)
11. T.Inui, Y. Ishihara, K.Kamachi and H.Matsuda, *in* "Zeolites: Facts, Figures, Future", Jacobs, P.A. and Van Santen R.A.(Editors), (Studies in Surface Science and Catalysis, Vol49), Elsevier, Amsterdam, 1989, p.1183
12. L.Wang, L. Tao,M. Xie,G. Xu, J.Huang and Y.Xu, *Catal. Lett.*, 21: 35(1993)
13. E.G.Ismailov, S.M.Aliev, S.S.Suleimanov and V.D.Sokolovskii, *React. Kinet. Catal. Lett.*, 45(2):185(1991)
14. K.Tanabe, M.Misono, Y.Ono and H.Hattori, *in* " New Solid Acids and Bases. Their Catalytic Properties." (Studies in Surface Science and Catalysis, Vol 51), Elsevier, Amsterdam, 1989.
15. L.Y.Chen, M.S. thesis, Dalian Institute of Chemical Physics, Chinese Academy of Sciences, Dalian, 1992.
16. G.E.Heller and M.M.Bhasin, *J. Catal.*, 73:9(1982)
17. J.B.Claridge, M.L.H.Green,S.C. Tsang and A.P.E.York, *Appl. Catal.*, A: 89: 103(1992)
18. E.Mielczarski, S.Monteverdi, A.Amariglio and H.Amariglio, *Appl. Catal.*, A: 104: 215(1993)
19. J.A.Sofranko, J.J. Leonard and C.A.Jones, *J.Catal.*, 103:302(1987)
20. J.B.Claridge, M.L.H.Green, R.M.Lago, S.C.Tsang and A.P.E.York, *Catal. Lett.*, 21(1-2):123(1993)
21. V.D.Sokolovskii, G.M.Aliev, O.V.Buevskaya, A.A.Davydov, *Catal. Today*, 4:293(1989); V.D.Sokolovskii and E.A.Mamedov, *Catal. Today*, 14:331(1992)
22. Iglesia and Baumgartner, *Catal. Lett.*, 21(1-2): 55(1993)
23. S.Kowalak and J.B.Moffat, *Appl. Catal.*, 36:139(1988)
24. G.A. Olah and R.H.Schlosberg, *J.Am. Chem. Soc.*, 2726(1968)

MECHANISM AND MODELING OF METHANE-RICH OXIDATION

EFFECT OF DIFFUSION LIMITATIONS OF SURFACE PRODUCED RADICALS ON THE C₂ SELECTIVITY IN THE OXIDATIVE COUPLING OF METHANE

P.M. Couwenberg, Q. Chen, G.B. Marin[1]

Eindhoven University of Technology
Laboratorium voor Chemische Technologie
P.O. Box 513, 5600 MB Eindhoven
The Netherlands

INTRODUCTION

The oxidative coupling of methane occurs through a mechanism in which heterogeneous catalytic reactions interact with gas-phase chain reactions. The existence of such a mechanism was first reported by Polyakov et al. (1) for the total oxidation of methane. Fang and Yeh (2) studied methane pyrolysis over ThO_2/SiO_2 surfaces and proposed a route towards C_2 products consisting of a heterogeneous production of methyl radicals followed by coupling in the gas phase.

Evidence that methyl radicals are formed on the catalyst surface during the oxidative coupling of methane was provided by matrix isolation electron spin resonance which allows to identify the produced radicals (3,4,5). Further confirmation about catalytic radical production and subsequent coupling in the gas phase was provided by experiments using Temporal Analysis of Products (TAP) (6,7). At pressures sufficiently low to inhibit gas-phase reactions, methane was converted into methyl radicals but no ethane was detected.

Internal diffusion limitations can play an important role in the oxidative coupling of methane. Follmer et al. (8) reported that internal diffusion limitations of oxygen have a beneficial effect on the selectivity towards C_2 products. McCarty (9) calculated concentration profiles of both oxygen and surface-produced methyl radicals in the catalyst pores and in the film surrounding a single catalyst pellet. Reyes et al. (10) presented a reactor model taking, at least in principle, the existence of intraparticle concentration gradients into account. The present work reports on the potential effects of both interstitial and intraparticle concentration gradients of reactive gas-phase

---

[1] to whom correspondence should be addressed

Methane and Alkane Conversion Chemistry
Edited by M. M. Bhasin and D. W. Slocum, Plenum Press, New York, 1995

intermediates on the selectivity to C_2 products.

MODEL EQUATIONS

In order to investigate the effects of the pellet scale concentration profiles of the surface-produced radicals during fixed bed operation, a computer program was developed that integrates the model equations 1 and 2. As the reactor was considered to be isothermal these consist of the mass balances of all components in the kinetic model in the interstitial and the intraparticle phase. In the axial direction the flow occurs by convection. On pellet scale transport takes place through diffusion from the interstitial phase to the intraparticle phase. The corresponding boundary conditions are given by equations 3 to 7.

Interstitial phase:

$$\frac{1}{\epsilon_b A} \frac{\partial F_v C_i}{\partial z} = \frac{1}{r} \frac{\partial}{\partial r}\left(r \, D_i \, \frac{\partial C_i}{\partial r}\right) + R_{i,g}(C) \tag{1}$$

Intraparticle phase:

$$-\frac{D_{e,i}}{\xi^2} \frac{\partial}{\partial \xi}\left(\xi^2 \, \frac{\partial C_i}{\partial \xi}\right) = R_{i,c}(C) + \epsilon_c R_{i,g}(C) \tag{2}$$

Boundary conditions:

$$z=0 \quad \wedge \quad 0<r<R : \qquad C_i = C_{i,0} \tag{3}$$

$$z > 0 \quad \wedge \quad r = 0 : \qquad \frac{\partial C_i}{\partial r} = 0 \tag{4}$$

$$z > 0 \quad \wedge \quad r = R : \qquad -a_g D_m \frac{\partial C_{i,g}}{\partial r} = a_c D_e \frac{\partial C_{i,c}}{\partial \xi} \tag{5}$$

$$\xi = 0 : \qquad \frac{\partial C_i}{\partial \xi} = 0 \tag{6}$$

$$\xi = \frac{d_p}{2} : \qquad C_{i,g} = C_{i,c} \tag{7}$$

114

The kinetic model that was used firstly consists of 38 gas-phase reactions between 13 molecules and 10 radicals. One of the essential features of this model is that the radicals are mainly produced through a branching step (11). The reactions and the rate coefficients were taken from Chen et al. (12).

These gas-phase reactions are combined with the following catalytic reactions:

$$O_2 + 2* \rightleftharpoons 2\,O* \tag{8}$$

$$CH_4 + O* \rightarrow CH_3\bullet + OH* \tag{9}$$

$$C_2H_6 + O* \rightarrow C_2H_5\bullet + OH* \tag{10}$$

$$C_2H_4 + O* \rightarrow C_2H_3\bullet + OH* \tag{11}$$

$$CH_3\bullet + O* \rightleftharpoons CH_3O* \tag{12}$$

$$CH_3O* \rightarrow ... \rightarrow CO_2 + H_2O \tag{13}$$

$$OH* + OH* \rightleftharpoons H_2O + O* + * \tag{14}$$

$$CO + O* \rightarrow CO_2* \tag{15}$$

$$CO_2* \rightleftharpoons CO_2 + * \tag{16}$$

Reactions 9, 10, and 11 are an extra source of radicals, i.e. they are heterogeneous initiations. Under coupling conditions these reactions are the main source of radicals. Using the above catalytic reactions combined with the gas-phase reactions did not allow an adequate description of the experimental data obtained for a tin promoted lithium on magnesium oxide catalyst. Especially the calculated selectivities at high space times were too low. This can be attributed to an overestimation of the importance of the gas-phase reactions. Tulenin et al. (13) suggested that MgO does not only produce radicals but also quenches radicals. This radical quenching has a inhibiting effect on the reactions occurring in the gas phase. Therefore the heterogeneous termination reaction 17, suggested by Cheaney et al.(14), was added to the reaction network. The hydrogen-peroxy radical is one of the most important chain carriers in the gas-phase reactions.

$$HO_2\bullet \rightarrow (HO_2)_{ads} \rightarrow stable\ products \tag{17}$$

The rate coefficients of the catalytic reactions and the heterogeneous termination were chosen, within physico-chemical reasonable limits (15), in such a way that the conversions and selectivities observed over a broad range of conditions were adequately described.

SIMULATION RESULTS

The above model equations were used to calculate the concentration profiles in a fixed bed reactor operated at the conditions summarized in Table I. Strong concentration gradients for reactive intermediates, e.g. methyl radicals and hydrogen-

Table 1. Conditions used during the simulation.

T	/ K	998
$CH_4/O_{2,0}$		4.0
p_t	/ kPa	135
$F_{mol,0}$	/ mol s^{-1}	1.5 10^{-4}
W_c	/ kg	0.375 10^{-4}
d_p	/ m	2.0 10^{-4}

peroxy radicals, are developed but no concentration gradient exists for the stable molecules, such as methane, oxygen, ethane, and ethene. This is shown in figure 1 and figure 2 where the intraparticle and interstitial concentration of the methyl and the hydrogen-peroxy radicals are plotted versus the axial reactor position and the pellet coordinate. In these figures the zero on the pellet-coordinate axis represents the centre of the catalyst pellet, the gas-solid interface is located at 1.0 10^{-4} m, and the space between 1.0 10^{-4} m and 1.5 10^{-4} m corresponds to the interstitial gas phase.

The strong concentration gradient for the radicals is caused by the fact that the

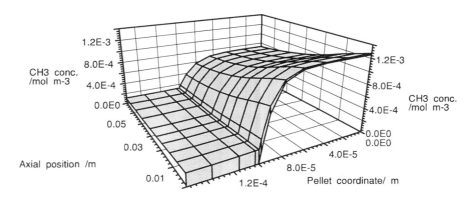

Figure 1. Calculated concentration profile of methyl radicals on pellet scale and on reactor scale.

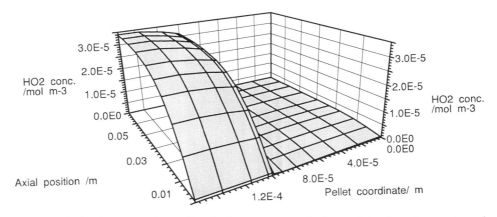

Figure 2. Calculated concentration profile of hydrogen-peroxy radicals on pellet scale and on reactor scale.

time scale for reaction of the radicals is much lower than the time scale for transport by diffusion. This is in contrast to oxygen of which the time scale for diffusion is lower than the time scale for reaction and, hence, no concentration gradient is developed.

In the flat part of the concentration profile in the centre of the pellet, the methyl-radical concentration is equal to the steady-state concentration, i.e. it follows from the balance between radical production and radical disappearance through reaction. Near the interface the concentration decreases strongly, due to diffusion into the interstitial gas phase. In the gas phase the profile is again almost flat because the diffusivity is much higher, but the concentration is approximately one order of magnitude lower than inside the catalyst pellet.

The intraparticle concentration of the surface-terminated hydrogen-peroxy radical is much lower than the interstitial concentration because of the high reactivity of this species. The radicals that are produced through the branched chains far away from the catalyst surface, i.e. at a distance comparable to the diffusion length (9), are hardly influenced by the high rate for surface termination because their lifetime is too short.

The effects of the diffusion limitations of the reactive intermediates on the selectivity were investigated by changing the pellet diameters. Based on the calculated concentration profile of the methyl radical an increase in selectivity is expected, because the volume where the high steady-state methyl radical concentration exists is larger when bigger pellets are applied. This would favour their coupling towards ethane, since this is a second order reaction, all other steps being first order with respect to the methyl radical concentration. The results of these calculations presented in figure 3, however, show a decrease in the C_{2+} selectivity with increasing pellet sizes.

This can be explained by the occurrence of the heterogeneous termination reaction 17. When larger catalyst pellets are applied the distance between two catalyst pellets automatically increases. The lifetime of HO_2 is so short that species far away from the catalyst surface are hardly influenced by the catalyst surface. The heterogeneous termination of the hydrogen-peroxy radical becomes thus less important when larger pellets are applied. This leads to a higher hydrogen-peroxy concentration in the interstitial phase, and thus to a higher contribution of the non-selective gas-phase reactions.

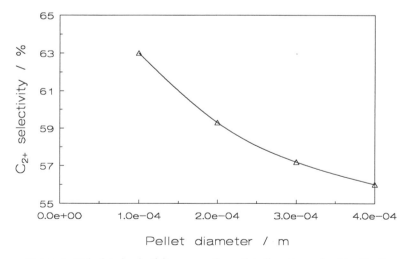

Figure 3. Calculated selectivity versus the pellet diameter, using the kinetic network with a heterogeneous termination. Condition see Table 1.

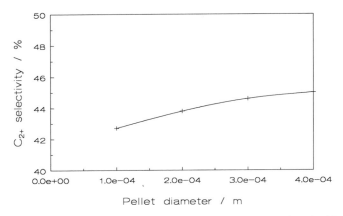

Figure 4. Calculated selectivity versus the pellet diameter, using the kinetic network without a heterogeneous termination.

This was verified by removing the heterogeneous termination, 17, from the kinetic scheme. The results of the calculations without this reaction are shown in figure 4. It can be seen that the selectivity indeed increases with increasing pellet size. Another significant effect of neglecting the heterogeneous termination is the much lower selectivity, as a result of the increased importance of the non-selective gas-phase reactions.

It should be noted that in all the calculations no concentration gradients for the stable components were observed.

EXPERIMENTAL RESULTS

Experiments were performed in a continuous-flow α-Al_2O_3 tubular reactor using catalyst pellets of different diameters in order to verify the simulation results. The experimental set-up was described in detail elsewhere (12). The catalyst used was tin promoted lithium on magnesium oxide that was prepared according to the procedure

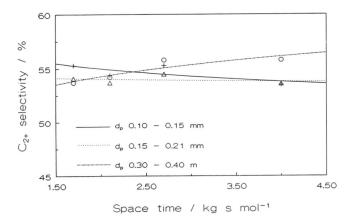

Figure 5. Experimental selectivity vs. space time. + d_p = 0.10-0.15 mm, ▲ d_p = 0.15-0.21 mm, ○ d_p = 0.30-0.40 mm. Conditions see Table 1.

reported by Korf et al. (16). It contains 3.5 wt% Li and 2.6 wt% Sn after 50 ks on stream at 1023 K. The data reported here were all taken after 50 ks on stream. The catalyst bed was diluted at a 10-to-1 weight ratio using α-alumina pellets of the same size as the catalyst particles. Three different pellet sizes were used, 0.10-0.15 mm, 0.15-0.21 mm, and 0.30-0.40 mm. The space time was varied between 1.7 and 4.0 kg s mol^{-1}, keeping all other conditions constant at the values given in Table I. The results of these experiments are shown in figure 5. No significant differences in selectivity are observed. This contradiction with the calculated results can be caused by experimental difficulties. For instance during all experiments a range of pellet diameters was used whereas in the calculation a fixed diameter was used. Furthermore the specific surface area, and the porosity of the pellets of different diameters are not exactly identical. Also the use of smaller pellets results in a higher pressure drop over the catalyst bed.

CONCLUSIONS

Pellet scale concentration gradients for the reactive intermediates, e.g. methyl and hydrogen peroxy radicals, can develop even under conditions where no concentration gradients exist for the stable components. Diffusion limitations of the methyl radicals can have a beneficial effect on the selectivity towards the C_{2+} products because the corresponding high internal methyl-radical concentration favours their coupling, since this is the only second-order reaction with respect to the methyl-radical concentration. The strong concentration gradients are caused by the high reactivity of the methyl radicals.

This potential effect of diffusion limitations of the surface-produced radicals on the C_2 selectivity was not found by both experimental and mathematical-simulation means using a tin promoted lithium on magnesium oxide catalyst.

The computer simulations using a model that adequately describes the experimental data obtained over a broad range of conditions, even showed a decrease in selectivity when the pellet diameter, and thus the importance of the diffusion limitations, increases. This is caused by the fact that the catalyst does not only generate methyl radicals but also quenches, for instance, HO_2 radicals. Larger catalyst pellets also lead to a larger distance between two catalyst pellets. Because of the high reactivity of the surface terminated HO_2 radicals, species that are far away from the catalyst surface, i.e. at a distance comparable to the diffusion length, are not, or hardly, influenced by the surface termination. This leads to a higher concentration of these species in the interstitial phase, and thus to a higher contribution of the non-selective gas-phase reactions. This effect can compensate the positive effect of diffusion limitations of surface-produced methyl radicals.

Experimentally no effect on the selectivity could be observed with increasing pellet diameter. This implies that the positive effect of the concentration gradient of the methyl radicals is compensated by the negative effect that larger pellets have on the heterogeneous termination at the investigated conditions.

NOTATION

Roman symbols:

A	cross sectional area	/ m^2_r
a_g	gas-solid interfacial area	/ $m^2_g m^{-3}_r$
a_c	external catalyst surface area	/ $m^2_c m^{-3}_r$
C_i	concentration of component i	/ mol m^{-3}_g

D	molecular diffusivity	/ $m^2 s^{-1}$
D_e	effective diffusivity	/ $m^3{}_g m_c s^{-1}$
d_p	pellet diameter	/ m
F_{mol}	molar flow rate	/ $mol\ s^{-1}$
F_v	volumetric flow rate	/ $m^3{}_g s^{-1}$
p_t	total pressure	/ kPa
r	distance from the catalyst surface	/ m
R	net volumetric production rate	/ $mol\ m^{-3} s^{-1}$
T	temperature	/ K
W_c	catalyst weight	/ kg

Greek symbols:

ϵ_b	bed porosity	/ $m^3{}_g m^{-3}{}_b$
ϵ_c	catalyst porosity	/ $m^3{}_g m^{-3}{}_c$
ξ	pellet coordinate	/ m

Subscripts:

b	bed
c	catalyst
g	gas
i	component
p	pellet
r	reactor
t	total
v	volume
0	inlet conditions

ACKNOWLEDGMENT

This work was done in the framework of the EC-Joule programme under contract no. JOUF-0044-C. The financial support is gratefully acknowledged.

REFERENCES

1. M.V. Polyakov, P.M. Stadnik, I.E. Neimark, Heterogeneous-homogeneous catalysis of CH_4 + O_2 mixtures, *Zh.Fiz.Khim.*, 8(4);584, (1936).
2. T. Fang, C.T. Yeh, Interactions of methane with ThO_2/SiO_2 surface at 1073 K, *J.Catal.*, 9;227, (1981).
3. D.J. Driscoll, W. Martir, J.X. Wang, J.H Lunsford, Formation of gas-phase methyl radicals over MgO. *J.Am.Chem.Soc.*, 107;58, (1985).
4. K.D. Campbell, E. Morales, J.H. Lunsford, Gas-phase coupling of methyl radicals during catalytic partial oxidation of methane, *J.Am.Chem.Soc.*, 109;7900, (1987).
5. M.Y. Sinev, V.N. Korchak, O.V. Krylov, Mechanisms of Oxidative Condensation of Methane into C_2 Hydrocarbons over Oxide Catalysts, In: Proceedings of the 9th International Congress on Catalysis, 2, eds M.J. Phillips and M. Ternan, 968, Ottowa: The Chemical Institute of Canada, (1988).
6. O.V. Buyevskaya, M. Rothaemel, H. Zanthoff, M. Baerns, Transient studies on reaction steps in the oxidative coupling of CH_4 on catalytic surfaces of MgO and Sm_2O_3, *J.Catal.*, 146;346, (1994).
7. E.P.J. Mallens, J.H.B.J. Hoebink, G.B. Marin, The Oxidative Coupling of Methane

over Tin Promoted Lithium Magnesium Oxide: a TAP Investigation. In: Studies in Surface Science Catalysis, 81; 205, eds Curry-Hyde H.E. and Howe R.F., Elsevier Science B.V., Amsterdam, (1994).

8. G. Follmer, L. Lehmann, M. Baerns, Effect of transport limitations on C_{2+} selectivity in the oxidative coupling reaction using a NaOH/CaO catalyst, *Catal.Tod.*, 4;323, (1989).

9. J. McCarty, Mechanism of Cooxidative Methane Dimerization Catalysis: Kinetic and Thermodynamic Aspects, In: Methane Conversion by Oxidative Processes, ed. E.E. Wolf, van Nostrand Reinhold, New York, (1992).

10. S.C. Reyes, E. Iglesia, C.P. Kelkar, Kinetic transport models of bimodal reaction sequences- I. Homogeneous and heterogeneous reaction pathways in the oxidative coupling of methane, *Chem.Eng.Sc.*, 48(14);2643, (1993).

11. Q. Chen, J.H.B.J. Hoebink, G.B. Marin, Kinetics of the oxidative coupling of methane at atmospheric pressure in the absence of catalyst. *Ind.Eng.Chem.Res.*, 30;2088, (1991).

12. Q. Chen, P.M. Couwenberg, G.B. Marin, Effect of pressure on the oxidative coupling of methane in the absence of catalyst, *A.I.Ch.E.J.*, 40(3);521, (1993).

13. Y. Tulenin, A. Kadushin, V. Seleznev, A. Shestakov, V. Korchak, Effect of pressure on the process of methane oxidative dimerization. Part 1. The mechanism of heterogeneous inhibition of the gas-phase reactions, *Catal.Tod.*, 13;329, (1992).

14. D.E. Cheaney, D.A. Davies, A. Davis, D.E. Hoare, J. Protheroe, A.D. Walsh, Effects of surfaces on combustion of methane and mode of interaction of antiknocks containing metals, 7th Symposium on Combustion, Butterworths Scientific Publications, London, 183, (1959).

15. L.M. Aparicio, S.A. Rossini, D.G. Sanfilippo, J.E. Rekoske, A.A. Trevino, J.A. Dumesic, Microkinetic analysis of methane dimerization reaction. *Ind.Eng.Chem.Res.*, 30;2114, (1991).

16. S.J. Korf, J.A. Roos, L.J. Veltman, J.G. van Ommen, J.R.H. Ross, Effect of additives on lithium doped magnesium oxide used in the oxidative coupling of methane. *Appl.Cat.*, 56;119, (1989).

EFFECTS OF PRODUCT SEPARATION ON THE KINETICS AND SELECTIVITY OF OXIDATIVE COUPLING

R. B. Hall* and G. R. Myers

Corporate Research Laboratories, Exxon Research and
Engineering Company Annandale, New Jersey 08801
* to whom inquiries should be addressed

INTRODUCTION:

Over the past decade, the catalytic conversion of methane plus oxygen to form ethane and ethylene via oxidative coupling over metal oxide catalysts has been widely investigated. Several reviews of this area have been published (1, 2, 3). A broad range of oxide materials has been studied, including: transition metal oxides (4, 5), alkali-doped, basic oxides such as MgO (6, 7), and rare-earth oxides (8, 9). With all of these materials, deep oxidation of methane, or of the reaction products, to form CO_2 competes with oxidative coupling. It is estimated that for oxidative coupling of methane to be economic, C_2+ yields in excess of 40% are necessary (10, 11). Here, yield is defined as the product of methane conversion times the selectivity to C_2+ hydrocarbons. Despite considerable research, yields above 25% have not been achieved in a single pass.

It is widely accepted that a significant fraction of the CO_2 comes from secondary reactions of ethylene, and that these reactions limit the yields that can be achieved. It seems obvious that a means of separating ethylene or other desired products before they are oxidized might overcome this limitation. Recently, several novel engineering approaches for achieving higher yields by means of product separation have been reported for oxidative coupling (12), and for partial oxidation of methane to methanol (13, 14). We have used a lab-scale recycle reactor to determine the extent to which yields can be increased via product separation under a variety of conditions, and to investigate the effects that the removal of specific products have on the complex series of reactions that occur following the initial methane activation step.

Two methods of product separation have been employed. We first used cryogenic trapping of ethane, ethylene, and CO_2 during the course of methane conversion. We also employed a membrane separation technique in which only ethylene is removed. In this case, ethane and CO_2 are recycled back to the reactor along with methane. In both cases, product removal enables a dramatic improvement in yields. It also alters the global kinetics of methane conversion. In this paper, we report on the C_2+ yields achieved by removing several specific products. We show that there is a significant increase in catalyst activity upon removing CO_2. The rate of methane conversion increases, and the activation energy for methane conversion decreases from roughly 50 kcal/mol in the presence of CO_2 to about 25 kcal/mol when CO_2 is removed. A simple mechanism for this is proposed, and a kinetic model is described that accounts for the observations in a quantitative way. Finally, we report on the effectiveness of recycling ethane to the oxidative coupling reactor when only ethylene is removed. A detailed analysis of how product removal affects the network of gas-phase reactions that occur in oxidative coupling will be published elsewhere.

EXPERIMENTAL

The reactor used in these experiments was a recycle loop reactor. The system consisted of a 4 mm i.d, tubular glass loop, with viton o-ring sealed glass joints and stopcocks, and a 4 mm i.d, tubular quartz reactor section. The total volume of the loop was approximately 150 cc. The reactor volume was 4.5 cc. Only the reactor section was heated, the loop was maintained at room temperature. The output side of the quartz reactor was fitted with capillary tubing to minimize the time reactants and products spent in the post-bed hot zone. Reactant gases in the system were circulated at a flow rate of 150 cc/min by a Micropump®, magnetically-coupled gear pump, model 7002-20. Hence, the recycle period was about 1 min.

At the start of each experiment, the total pressure in the system was about 115 kPa. This pressure decreased during the course of the experiment to about 100 kPa due to pressure changes associated with the conversion chemistry, and to the periodic removal of an aliquot of gas (about 0.2 kPa per sampling) for analysis by GC. The initial gas composition for the high-yield investigations was $He:CH_4:O_2 = 5:2:0.2$. Additional oxygen was injected during the course of reaction to maintain a CH_4/O_2 ratio of 10. The initial gas composition for the investigations of the effects of CO_2 removal on the kinetics of methane conversion was $He:CH_4:O_2 = 5:2:1$. No additional oxygen was added in these experiments. Methane (99.97%), oxygen (99.6%), and He (99.95%) were obtained from Matheson and used without further purification.

Roughly 300 mg of catalyst was held in place in the reactor with quartz wool. The catalyst used was lithium-magnesia, first reported by Lunsford (15), and now arguably a benchmark for comparison of any oxidative coupling catalyst. The catalyst was prepared according to the method described by Lunsford. The fresh catalyst had a Li/Mg ratio of 0.29. Exposure to reaction conditions resulted in a gradual (multiple days) loss of selectivity, Li, and surface area (to less than 1 m^2/g). The rate of methane conversion over this catalyst was roughly 50 $\mu mol/s/g$ at 750 °C and a gas composition of $He:CH_4:O_2 = 5:2:1$. With a recycle flow rate of 150 cc/min, the conversion per pass over the catalyst was approximately 1%.

For cryogenic trapping of products, the gas stream was passed through a 30 cm section of coiled, 1/4" o.d. copper tubing maintained at -165± 2 °C. At this temperature, C_2H_4, C_2H_6 and CO_2 partial pressures were kept below 0.3, 0.05 and 1.3×10^{-4} kPa, respectively. Partial pressures of methane and oxygen used in the current experiments were not affected. The content of the trap was sampled at (typically) 4 min. intervals and analyzed by GC to obtain a measure of product yields as a function of time.

In other experiments, a permeation-selective membrane was used to selectively remove ethylene from the gas stream. The membrane was a cross-linked, polyvinyl alcohol membrane, impregnated with silver nitrate. The silver cations/silver nitrate complexes selectively with the ethylene and provide a means for facilitating the transport of ethylene across the membrane. Negligible ethane and methane, and only minimal amounts of CO_2 diffused across the membrane, ethane, carbon dioxide and methane were recycled to the reactor. The membrane, and its transport characteristics, are described in more detail elsewhere. (16, 17, 18, 19). The general application of membranes for product separation in methane conversion processes has also been discussed recently. (13, 14)

Gas composition was analyzed using a Micro Gas Analyzer, model M200, from MTI Inc. This system was capable of providing a complete analysis (through C4 hydrocarbons) in 60 s.

RESULTS AND DISCUSSION:

A summary of the dependence of the C_2^+ selectivity on methane conversion under various conditions is shown in figure 1. The triangles show the C_2^+ selectivity observed at

750 °C and an initial ratio of $CH_4/O_2 = 10$. Oxygen was added during the course of reaction to maintain the oxygen partial pressure and allow higher methane conversions to be reached. All products were recycled to the reactor in these experiments. These data are typical of those reported by others for this catalyst, and for many of the better performing catalyst materials. (1, 2, 3, 7, 8). A rapid decline in selectivity is observed at conversions above roughly 20% due to oxidation of the hydrocarbon products in the gas-phase and on the catalyst. At low conversion, the dominant product is ethane. At conversions above 40%, ethylene becomes the most abundant product, although ethane yields remain significant.

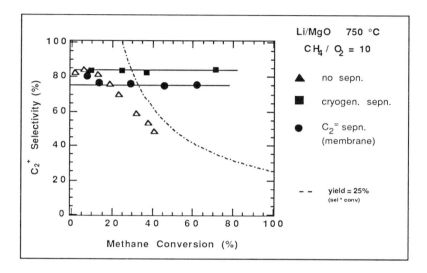

Figure 1: Comparison of C_2^+ Selectivity as a function of methane conversion for several different product separation conditions. In all cases, operating conditions are: 300 mg Li/Mgo catalyst, 750 °C, $PCH_4=0.25$ atm., $PO_2=0.025$ atm., $PHc=0.75$ atm.. PO_2 is held approximately constant by adding oxygen as it is consumed. Triangles - no products removed; squares - ethane, ethylene, carbon dioxide and water removed by cryogenic trapping; circles - only ethylene removed using permeation-selective membrane.

The square symbols in fig. 1 show the selectivity observed when ethane, ethylene, carbon dioxide and water were removed in each pass through the loop by cryogenic trapping. These data are obtained at 750 °C and an initial ratio of $CH_4/O_2 = 10$. The partial pressure of oxygen was held roughly constant during these experiments by adding oxygen as it was consumed. Not too surprisingly, the data in fig. 1 show that the maximum selectivity achievable in single pass operation can be extended to as high a conversion as is desired, depending on the extent to which one is willing to continue to recycle unreacted methane. In these experiments, ethane was trapped out before it had a chance to be converted to ethylene. Ethane made up about 95% of the hydrocarbon products recovered, the remainder was ethylene. There were some important changes in the kinetics of methane conversion resulting from the removal of these products from the gas stream that are not reflected in fig. 1. One of the important effects on the catalytic activity is discussed below. Effects on the gas-phase network of reactions will be discussed elsewhere.

Finally in fig. 1, the circles indicate the selectivity observed when only ethylene was removed using a facilitated-transport membrane. In these experiments, ethane and carbon dioxide were recycled along with the unreacted methane. The temperature and feed-gas conditions were otherwise similar to those described above. Here the C_{2^+} selectivity is about 70% at high conversion. This is somewhat below the maximum achievable in single pass operation. This reflects the selectivity of recycling ethane back to the catalyst and converting it to ethylene. In separate experiments in which ethane is co-fed with methane, the selectivity for converting ethane to ethylene was found to be about 75 to 80%. The total C_{2^+} selectivity obtained when ethylene is removed (circles) is about 80% of that obtained when ethane is removed (squares). Although there is a loss in overall C_{2^+} selectivity, the product distribution now is dominantly ethylene. The ratio of ethylene to ethane recovered in these experiments was about 50:1. Again, the selectivity can be maintained to as high a conversion as desired.

The effects on the kinetics of conversion of the reactant gases due to product removal by cryogenic trapping are shown in fig. 2. The data were obtained at 750°C and an initial ratio of $CH_4/O_2 = 2$. No additional oxygen was added. The fraction of methane converted as a

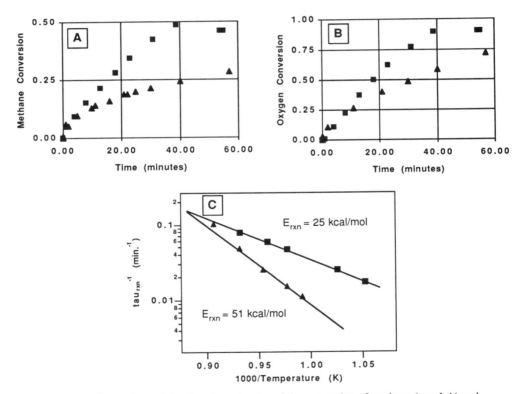

Figure 2: Comparison of the time dependencies of the conversion of methane (panal A) and oxygen (panel B) without product separation (triangles) and with cryogenic separation of ethane, ethylene and CO_2 (squares). Panal C - temperature dependence of the rate of methane conversion with and without product separation. The characteristic reaction time, trxn, is defined in the text. In all cases, operating conditions are: 300 mg Li/Mgo catalyst, 750 °C, PCH4=0.25 atm., PO2=0.125 atm., PHe=0.75 atm.

function of time is shown in fig. 2a. At short times (less than about 8 min.) there is little difference between the runs with product separation, and those without. This is expected because product concentrations are quite low at early times. At later times, however, there is a significant difference. The rate of methane conversion with product separation is roughly 3 times the rate without. Qualitatively, there is a similar effect on the rate of oxygen consumption (fig. 2b). Because the selectivity is substantially higher in the product-separated runs, the efficiency of oxygen utilization is also higher. The temperature dependence of the methane conversion rate is shown in fig. 2c. An approximate half-life, $\tau_{1/2}$, for methane conversion is obtained from the concentration versus time data (e.g. fig. 2a) at a time sufficiently long that there is enough buildup of products to influence the kinetics. Effective activation energies for methane conversion with and without product separation obtained from the plot of $1/\tau_{1/2}$ versus 1/Temperature shown in fig. 2c are 25 and 50±2 kcal/mol respectively.

These activation energies are in the same range as those obtained by Lunsford and coworkers in recent experiments.[20] These authors reported that the rate of methane conversion over Li/MgO catalysts declined as they added increasing partial pressures of carbon dioxide to the feed. At relatively high CO_2 pressures, the observed activation energy for methane conversion was about 52 kcal/mol. On the other hand, at low total pressure, where CO_2 from reaction was negligible, they estimate an activation energy of 26 kcal/mol in the absence of added CO_2. We found in our experiments that if we add small amounts of CO_2 to the recycle loop, the methane conversion rate declines immediately. Addition of small amounts of ethane, ethylene or water do not produce the same effect. We thus attribute the increase in conversion rates and the change in the temperature dependence shown in fig. 2 to removal of CO_2.

It is possible to explain quantitatively the change in rates observed in both our results and those of Lunsford et al. with a very simple model. We start by assuming that CO_2 interacts with the same site on the catalyst at which methane is converted. It is likely that this is a surface oxygen site, and that CO_2 interacts with it by forming a surface carbonate. We than calculate a Langmuir adsorption-isotherm for CO_2 binding at this site, with a heat of adsorption of 50 kcal/mol. The fraction of sites at equilibrium not occupied by CO_2 is shown in fig. 3a and b. (This fraction does not depend on the actual concentration of sites for an adsorption rate that is first order in the concentration of available (open) binding sites, and a desorption rate that is first order in the concentration of occupied sites. Thus, it is not possible to determine the absolute concentration of sites from these results). In fig. 3a, the fraction of available sites, at a partial pressure of CO_2 of 0.01 atm, is plotted versus temperature. This pressure is typical of that encountered under catalytic conversion conditions for methane feed pressures above 0.1 atm, and conversions above 10%. The exponential increase in the fraction of open sites with increasing temperature is clear. In fig. 3b, the fraction of sites available at 1000 K as a function of CO_2 pressure is shown. These data show that the surface can be largely depopulated of surface-bound carbonate if the CO_2 partial pressure can be decreased from anything above several percent of an atmosphere to something less than 1%.

Finally, the methane conversion rate is assumed to be first order the concentration of available sites (i.e., the total concentration of sites, ρ_{sites}, times the fraction of open sites, $\frac{\theta_{open}}{\theta_{total}}$). Thus, the methane conversion rate will vary as indicated in eqn. 1.

$$\text{conversion rate} \quad \alpha \quad (\frac{\theta_{open}}{\theta_{total}})(\, e^{-S_{rxn}/R}\, e^{-E_{rxn}/RT}) \qquad 1)$$

where S_{rxn} is the entropy of reaction, and E_{rxn} is the activation energy. The log of the calculated conversion rate as a function of reciprocal temperature, at several different partial pressures of CO_2 is shown in fig. 3c. The parameters assumed are $E_{rxn} = 25$ kcal/mol, and $S_{rxn} = 0$. The ratio, $\frac{\theta_{open}}{\theta_{total}}$, is taken from the from the isotherm calculation above. To facilitate comparison to experimental data, the rate is scaled to units of $\mu mol\ g^{-1}\ s^{-1}$, assuming a surface area of 1 m^2/g, and a site density of $4 \times 10^{14}\ cm^{-2}$. However, it should be noted that we are only comparing relative rates here, and in most experimental data, the absolute reaction rate is seldom accurately known because of significant uncertainty in the surface area of the catalyst under operating conditions. The relative rates are determined solely by E_{rxn}, and by $\frac{\theta_{open}}{\theta_{total}}$.

Site blocking by methane or its reaction products is assumed to be negligible, thus the latter is a function only of P_{CO2}, $\Delta H_{CO2\ adsorption}$, and temperature.

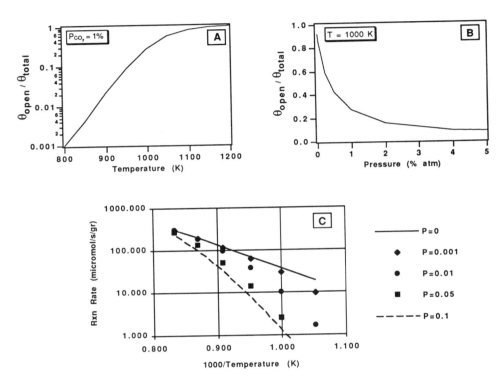

Figure 3: Calculation of the fraction of surface sites not occupied by CO_2 at equilibrium as a function of temperature at PCO_2=0.01 atm (panal A), and as a function of CO_2 pressure at T=1000 K (panal B). Panal C - calculation of the effective rate of methane conversion as a function of temperature at several different CO_2 pressures, as indicated in the legend at the right hand side of panal C.

From a comparison of fig. 3c and fig. 2c, it is clear that this simple model reproduces the experimentally observed change in the apparent activation energy for reaction when CO_2 is removed from the system. It explains why the methane conversion rate exhibits an activation energy of 50 kcal/mol in the presence of more than several tenths of a percent of an atmosphere of CO_2 (when it is determined in the temperature range typically investigated, 675 to 800 °C). It explains the decrease in the apparent activation energy to 25 kcal/mol when the CO_2 partial pressure is kept below 0.1% atm. It predicts the factor of roughly 3 increase in rate observed under the conditions of fig. 2a. Finally, in accounting for the results obtained in the earlier work (20), the model predicts quantitatively the CO_2 pressure dependence of the methyl radical production rate observed at lower pressures, and qualitatively fits the trends observed at higher total pressures. (A quantitative fit to the latter would require a more sophisticated kinetic analysis to take into account the variation of reactant and product concentrations across the catalyst bed.)

The ability of this simple model to reproduce the experimental data suggests that there is a one-to-one correspondence between the site to which CO_2 binds and the site at which methane is converted. Although a bulk carbonate will form upon prolonged exposure of Li/MgO to CO_2, it is not likely that formation of the bulk carbonate is responsible for the changes in the reaction rates observed here and in the earlier work (20). The apparent binding energy of the surface carbonate is significantly below that of the sublimation temperature of bulk lithium carbonate. It would require hundreds of degrees higher temperatures to remove CO_2 from bulk carbonate. A lower binding energy for the surface carbonate relative to the bulk carbonate is expected due to the isolated nature of the surface species.

SUMMARY:

We have used several separation techniques to remove selected products during the catalytic conversion of methane and oxygen to form ethane and ethylene over Li/MgO. Ethane yields in excess of 70% have be achieved by removing ethane as it is formed. Selective removal of ethylene by means of a permeation-selective membrane, and recycling ethane to allow it to be oxidatively dehydrogenated, allows methane to be converted almost exclusively to ethylene, with yields in excess of 50% . Time-resolved measurements of the evolution of the reactants and stable products allow us to determine the influence of the removal of specific products on the global reaction kinetics. We find that the activation energy for methane conversion is significantly reduced by the removal of CO_2 from the reactor, and that the rate of methane conversion is enhanced. The change in activation energy is attributed to the formation of a surface carbonate, with a binding energy of 50 ± 2 kcal/mol. The activation energy for methane conversion in the absence of CO_2 is 25 ± 2 kcal/mol.

Literature Cited:

1. Y. Amenomiya, V. I. Birss, M. Goledzinowski, J. Galuszka, A. R. Sanger, *Catal. Rev. - Sci. Eng.* **32**, 163 (1990).
2. G. J. Hutchings, M. S. Scurrell, J. R. Woodhouse, *Chem. Soc. Rev.* **18**, 251 (1989).
3. M. S. Scurrell, *Appl. Catal*. **32**, 1 (1987).
4. G. E. Keller, M. M. Bhasin, *J. Catal.* **73**, 9 (1982).
5. C. A. Jones, J. S. Leonard, J. A. Sofranko, *Energy and Fuels* **1**, 12 (1987).
6. D. J. Driscol, W. Martir, J.-X. Wang, J. H. Lunsford, *J. Am. Chem. Soc.* **107**, 58 (1985).
7. J. H. Lunsford, *Catal. Today* **6**, 235 (1990).
8. K. Otsuka, K. Jinno, A. Morikawa, *J. Catal.* **100**, 353 (1986).
9. A. Ekstrom, A. Lapszewicz, *J. Am. Chem. Soc.* **110**, 5256 (1988).
10. S. Field, S. C. Nirula, J. G. McCarty, Report of SRI International Project No. 2352, (1987),
11. J.M. Fox, T.-P. Chen, B. D. Degen, *Chem. Eng. Prog.* **42**, (1990).
12. A. L. Tonkovich, R. Aris, R. W. Carr, *Science* **262**, 221 (1993).
13. Y. V. Gokhale, Q. Liu, J. Rogut, J. L. Falconer, R. D. Noble, in *International Congress on Membrane and Membrane Processes* R. Rautenbach, Eds. Heidelberg, Bermany, 1993),
14. R. D. Noble, in *ACS Symposium - Natural Gas Upgrading II* G. A. Huff, D. A. Scarpiello, Eds. (ACS Preprints, Petroleum Division, San Francisco, 1992), pp. 11-15.
15. T. Ito, J. X. Wang, C. H. Lin, J. H. Lunsford, *J. Amer. Chem. Soc.* **107**, 5062 (1985).
16. W. S. Ho, D. Dalrymple, *J. Membrane Sci.* **to be published**, (1994).
17. W. S. Ho, (U.S. Patent, 1991),
18. R. D. Noble, in *New Directions in Separation Technology* E. N. Lightfoot, P. Kerkhof, Eds. (Engineering Foundation, New York, NY, Leeuwenhorst Congress Center, Holland, 1993), pp. 26.
19. R. D. Noble, in *International Congress on Membrane and Membrane Processes* R. Rautenbach, Eds. Heidelberg, Bermany, 1993),
20. C. Shi, M. Xu, J. H. Lunsford, PREPRINTS, Div. of Petrol. Chem., ACS **37** 1180 (1992).

REACTIVE VS. ADSORBED OXYGEN IN THE HETEROGENEOUS OXIDATION OF METHANE OVER Li / MgO

A . J . Colussi and V . T . Amorebieta

Department of Chemistry, University of Mar del Plata,
7600 Mar del Plata, Argentina

Key words: Methane; Oxidation; Catalysis; Oxygen; Adsorption.

INTRODUCTION

The kinetics of $O_2(g)$ adsorption, desorption, and $CH_4(g)$ or $H_2(g)$ oxidation on 7% Li-doped MgO were investigated by on-line, modulated beam mass spectrometry in a low-pressure, continuous flow reactor between 930-1130 K. Under these conditions, which suppress gas-phase reactions, only purely heterogenous processes take place in this system. Dosage of the amounts of bound oxygen removed at several stages of evacuation, or consumed by reaction with methane or hydrogen, indicates that oxygen adsorption is a reversible process leading to at least two major types of chemisorbed species. The kinetics of $O_2(g)$ exchange is a slow process in the time scale of the experiments, that can not be described by Langmuir-type equilibrium isotherms involving either dissociative or non-dissociative adsorption on one or two types of active sites. Fast adsorption on a precursor state, followed by slow conversion into a more strongly bound species seem essential to model oxygen take up by the solid. On an oxygen saturated oxide, but in the absence of $O_2(g)$, methane reacts rapidly with about 14% of the pool of exchangeable oxygen, after which a slower oxidation reaction sets in. Hydrogen behaves similarly. During the slow oxidation stage methane is oxidized into carbon oxides and ethane, but not water, and partially decomposes into unvolatile carbon-containing species without releasing $H_2(g)$.

EXPERIMENTAL

The flow reactor (fused silica, cylindrical, 90 cm^3) was connected to the gas reservoirs through a 3-way switching valve, also leading to a bypass, and to the mass spectrometer via an adjustable needle valve for variable residence times (between 3 and 180 s, depending on the species) (1). Flow rates were varied between 4-300 nmoles/s, corresponding to concentrations 10-9000 nM (0.1-560 mTorr). The fast mass spectrometric detection system (EI, 50 eV) for the continuous monitoring of reactants and products, has been described elsewhere (1). Transient signals for the different species in the absence of catalyst were measured following gas flow admission or interruption, and were stored in a personal

Methane and Alkane Conversion Chemistry
Edited by M. M. Bhasin and D. W. Slocum, Plenum Press, New York, 1995

computer for numerical analysis. Under effusive flow conditions, the residence time t_i of a species of molecular mass M_i is related to the residence time t_N of $N_2(g)$, a inert species throughout, according to: $t_i = t_N (M_i/28)^{1/2}$ (M_i in daltons). The catalyst was prepared by mixing Li_2CO_3 with MgO in a Li/Mg ratio of 0.07. The powder was moistened with distilled water into a paste and then treated with HNO_3; the solution was evaporated under air at 250-270 C during 24 h, and the resulting solid calcined at 680 C for 16 h (2). The catalyst (ca. 1 g) was inserted in the reactor by means of a sliding holder, and activated under oxygen at reaction temperatures prior use.

OXYGEN ADSORPTION AND DESORPTION

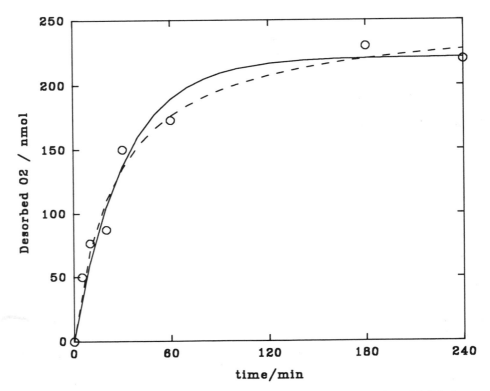

Figure 1 — Amounts of O2 desorbed from 1 g Li/MgO
previously kept under 19 mTorr O2 at 1130 K for 1 h

Figure 1 shows the cumulative amounts of oxygen desorbed from the oxide as function of evacuation time, after being saturated under 19 mTorr of $O_2(g)$ for 1 hour at 1130 K. The curves correspond to first (solid line), and second (broken line) order kinetics. They are indistinguishable within experimental error. We estimate that the total amount removed, ca. 220 nmoles O_2, corresponds to about 0.13% (5%) of the oxygen associated with superficial lithium atoms. We verified that undoped magnesium oxide does not release oxygen, and that inert gases (nitrogen, argon) do not adsorb on 7%Li/MgO under the same conditions.

In Figure 2 we present the reduced $O_2(g)$ partial pressure: $f = I(t')/I_{ss}$ ($I(t')$ and I_{ss} are the mass spectral signal intensities of mass 32 at reduced time t', and at steady state, respectively), vs the reduced time: $t' = t/t_O$ (t_O is the residence time of O_2), measured upon switching on steady flows of $O_2(g)$ on oxide samples that had been evacuated for 1 hour.

Curve 1 corresponds to the filling of the reactor with $O_2(g)$ in the absence of catalyst. Curves 2-4 show the results for various combinations of residence times and O_2 flow rates.

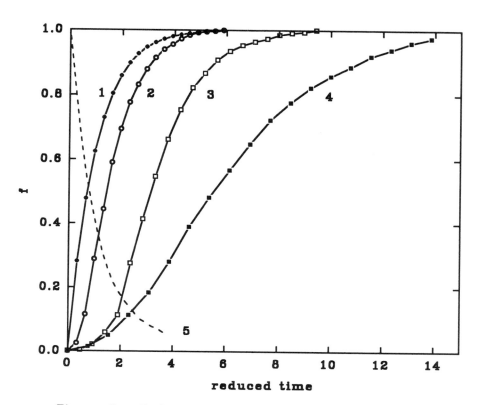

Figure 2 — Reduced O2 partial pressures as function
of reduced time under various conditions (see text)

The difference between the areas under curve 1 and curve i, D_i, and the stationary state concentration of $O_2(g)$ yield the amount of oxygen required to replenish the oxide: $n(O_2) = D_i\, C_{ss}(O_2) = (167 \pm 4)$ nmoles. Curve 5 in Figure 2 corresponds to oxygen desorption at the end of any of the experiments described above. It deviates only slightly from the evacuation curve for an inert gas, such as argon or nitrogen, implying negligible release of $O_2(g)$.

The last observation excludes fast equilibration between the gas and solid phases in the time scale of these experiments. In Figure 3 we show the type of behavior that would be expected if the instantaneous $O_2(g)$ concentration were only determined by flow rates and an adsorption isotherm (3).

Thus, although experimental data for oxygen adsorption (open circles) could be satisfactorily reproduced by means of a dissociative-type Langmuir isotherm (which yields a zero initial slope, as observed, in contradistinction with non-dissociative adsorption), the calculated desorption curve is considerably slower than measured (filled squares). We are currently developing mechanisms consistent with these experimental findings.

METHANE OXIDATION

In Figure 4, the solid curve and symbols correspond to the transient signal of mass 16 (CH_4^+) observed after switching on a steady flow of 4.3 nmoles/s (0.11 mTorr) of methane, over an oxide sample that had been maintained (but no longer) under 0.2 mTorr of $O_2(g)$ at 1130 K for one hour.

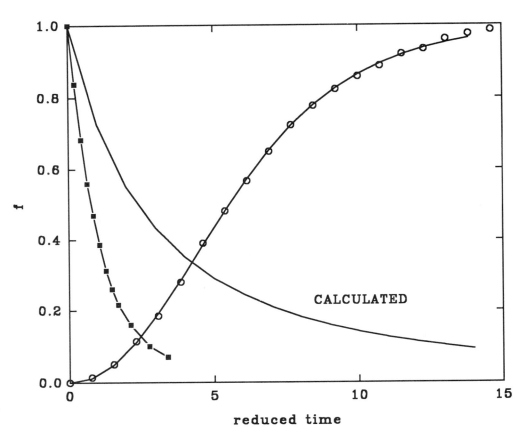

Figure 3 — Experimental and calculated data for O2
adsorption and desorption (see text)

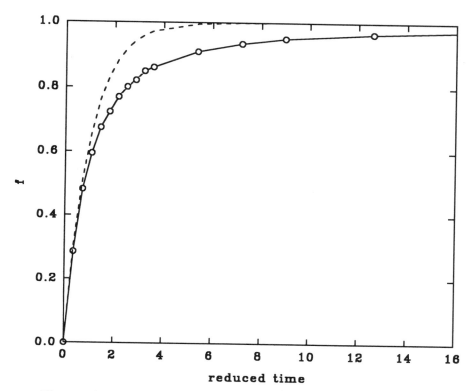

Figure 4 — Reduced CH4 pressures in the absence and presence of catalyst under anoxic conditions (see text)

The broken curve corresponds to the filling of inert gas. Clearly, methane is still reacting on the oxide after 16 residence times. Curve integration up to reduced time $t' = 25$ yields the amount of methane lost (13 nmoles) by heterogeneous reactions. Dosage of oxygen depletion during this experiment (40 nmoles) includes both reaction and evacuation losses. From Figure 1, we obtain 8.7 nmoles for the latter, and therefore the amount of oxygen consumed during reaction roughly corresponds to the (1:2) stoichiometry of full oxidation: $CH_4 + 2\,O_2 = CO_2 + 2\,H_2O$.

Thus, methane only consumes ca. 14% of the pool of exchangeable oxygen available, even by prolonged exposure to the oxide. We verified that the reactive oxygen species had been removed after $t' = 25$, by repeating the experiment of Figure 4, on the already partially depleted oxide. One obtains a filling curve almost overlapping that for an inert gas, i.e. the oxide is left barely reactive. Finally, we found that by coadding 4.3 nmoles/s of $O_2(g)$ to the reactor at $t' = 16$ in Figure 4, f drops to about 0.8, indicating that the presence of even minute concentrations of $O_2(g)$ dramatically accelerate heterogenous oxidation of methane (1,4). We have previously argued that a very reactive species, perhaps different from those presently investigated in anoxic conditions, is formed on the oxide under $O_2(g)$ (4).

The products of methane oxidation in the absence of gaseous oxygen are carbon oxides (m/z = 28 and 44), and minor amounts of ethane or formaldehyde (m/z = 30); H_2O or H_2 were not detected, however. These species could not even be eluted by nitrogen after reaction. Only during oxide regeneration with $O_2(g)$ we could identify water in the effluent gases. Further work is underway.

REFERENCES

1. Amorebieta, V. T. and Colussi, A. J., Gas Phase Free Radicals in the Catalytic Decomposition of Hexane over Tungsten, J. Phys. Chem. 86: 2760 (1982).
2. Chang, Y. F., Somorjai, G. A. and Heinemann, H., Oxidative Coupling of Methane over Mg-Li Oxide Catalysts at Relatively Low Temperature - The Effect of Steam, J. Catal . 141: 713 (1993)
3. a) Renken, A., Instationäre Reaktionsführung zur Modellierung heterogen katalytischer Reaktionen, Chem. Ing. Tech. 62: 724 (1990). b) Weller , S. W., Kinetics of Heterogeneous Catalyzed Reactions, Catal. Rev. Sci. Eng .34 (3): 227 (1992).
4. a) Amorebieta, V. T., and Colussi, A. J. Kinetics and Mechanism of the Catalytic Oxidation of Methane over Lithium-Promoted Magnesium Oxide, J. Phys. Chem. 92: 4576 (1988). b) Amorebieta, V. T., and Colussi, A. J. Mass Spectrometric Studies of the Low Pressure Oxidation of Methane over Samarium Sesquioxide, J. Phys. Chem. 93: 5155 (1989).

THE ROLE OF FREE METHYL RADICALS IN THE SELECTIVE OXIDATION OF METHANE OVER SILICA-SUPPORTED MOLYBDENA

Roy S. Mauti and Charles A. Mims[*]

Department of Chemical Engineering and Applied Chemistry
University of Toronto
Toronto, Canada M5S 1A4

INTRODUCTION

Surface activation of methane and formation of a methyl intermediate is widely accepted as the first step in the reaction mechanism of methane selective oxidation to formaldehyde[1,2,3]. The fate of the methyl intermediate, particularly whether it is desorbed into the gas phase and the surface involvement in the subsequent steps is as of yet unresolved. Various scenarios with differing amounts of surface involvement have been proposed. Dowden *et. al.*[4] have proposed a mechanism for methane partial oxidation to methanol and formaldehyde over mixed oxides (based on molybdena) which involves only surface intermediates. The proposed reaction mechanism involves methane activation to form surface methyl intermediates: this surface methyl reacts with a surface oxygen species to form surface methoxide which in turn reacts to form oxygenated products, methanol and formaldehyde.

Recently greater emphasis has been placed on gas phase intermediates. It has been demonstrated that metal oxide catalysts used in methane oxidative coupling are capable of producing gas phase methyl radicals, detected by MIESR, matrix isolation electron spin resonance technique[5]. The reactivity of methyl radicals with the oxide catalyst determines the product slate. Good oxidative coupling catalysts were shown to be unreactive towards methyl radicals and allow gas phase coupling reactions to predominate[6]. Free methyl radicals can also be produced on multi-valent transition metal oxide catalysts such as supported molybdena. Liu et al.[7] provide spectral data showing formation of methyl radicals (at -196°C) and the formation of surface methoxy (at 25°C) upon exposure of silica-supported molybdena to methane. Thus the overall reactions could proceed by formation of methyl radicals which rapidly react on subsequent surface visits to form surface methoxide:

$$CH_3^{\bullet} \; + \; M^{(n+1)}O^{2-} \; \rightarrow \; M^n(OCH_3)^- \tag{1}$$

Methane and Alkane Conversion Chemistry
Edited by M. M. Bhasin and D. W. Slocum, Plenum Press, New York, 1995

This surface methoxy species reacts to produce formaldehyde or carbon oxides depending on the reaction conditions. Reaction (1) must be rapid enough to compete with the methyl coupling reaction in order to achieve high selectivities to C_1 oxygenates.

Gas phase reactions of free methyl radicals can also lead to methanol and formaldehyde production and if such steps were selective, the role of the catalyst could be limited to methyl radical generation. The following equilibrium has been proposed to control the relative selectivity of oxidation products versus coupled products [3]:

$$CH_3^{\cdot} + O_2 \rightleftharpoons CH_3O_2^{\cdot} \tag{2}$$

The right-hand side of the above equilibrium and thus formation of oxidation products, is favoured by lower temperatures. Conversely, higher temperatures inhibit the oxidation of methyl radicals and favour coupled products. At the temperatures used for methane oxidation 600-650°C, the equilibrium above lies to the left, and all gas phase oxidation channels must compete with the methyl coupling reaction in order to exhibit high selectivity to C_1 oxygenates. This surface-initiated gas phase mechanism is not supported by computer modelling studies [8,9] which show ethane production to dominate under selective oxidation conditions (500C<T<650C). The experimentally observed yields of formaldehyde and oxides of carbon are only possible if a rapid surface-catalysed methyl oxidation step is added to the model.

A test of the mechanism wherein formaldehyde is formed by methyl revisiting the silica-supported molybdena surface is focus of this paper. A previous test of this mechanism was reported by Sun et al. [10] using a dual bed configuration. This study involved allowing methyl radicals generated in a bed of Sr/La_2O_3 coupling catalyst to travel unreacted to a silica-supported molybdena catalyst downstream. They concluded that some formaldehyde was formed by reactions of free methyls to the molybdena catalysts. However, under the reaction conditions used, we estimate a methyl radical half-life of approximately 20 μs, such that the overwhelming majority of methyl radicals generated in the Sr/La_2O_3 bed will have already reacted near their generation site to form ethane. In order to test this mechanism under conditions where a majority of the methyl radicals contact the molybdena catalysts, we generated CH_3 by thermal pyrolysis of azomethane at low pressures:

$$CH_3\text{-}N\text{-}N\text{-}CH_3 \rightarrow 2CH_3^{\cdot} + N_2 \tag{3}$$

Recently thermal pyrolysis of azomethane has been used to study free methyl radical interactions on a variety of surfaces (11).

EXPERIMENTAL

The micro-reactor apparatus described previously [12] was modified for the azomethane pyrolysis experiments by providing a 1 cm^3 pyrolysis zone (for azomethane decomposition to gas phase methyl radicals) followed by the catalyst bed. In order to achieve adequate penetration of methyl radicals into the catalyst bed, the partial pressure of azomethane in the reactor was lowered to a range of 0.001 to 0.8 torr. This was achieved by operating the reactor and a large GC sample loop at low pressure (0.2 to 8 torr). A material balance on the sample loop contents assured that non-steady hold-up of products in the reactor and catalyst bed had not occurred. A stream containing comparable amounts of CH_2O was also analyzed to assure adequate analysis (and absence of wall holdup) for the anticipated low levels formaldehyde.

The silica-supported molybdena catalyst (SSMo) was prepared by the method described in reference [13]. Fumed silica (Cabosil M5, Cabot Corp.) was wetted with a solution of ammonium heptamolybdate (Baker) of appropriate strength. After drying and calcining in air at 600°C, the catalyst was pressed, crushed, and sieved to produce particles of 80 to 120 μm diameter. The catalyst used in this study contained 1.8 wt% Mo, and the total surface area was 175 m²/g.

The synthesis of azomethane was based on the procedure in reference (14). The purity of the material was checked by MS analysis and the azomethane stream to the reactor was produced by metering gas flow from a container under autogenous pressure. The products of slow decomposition in the azomethane container were removed periodically by freeze-pump-thaw cycles.

RESULTS AND DISCUSSION

Azomethane pyrolysis experiments and a simple model were used to estimate how efficiently our experiment provides contacting of the catalyst by methyl radicals. The methyl concentration in the gas is governed by the balance of azomethane pyrolysis and the loss of methyl radicals (largely by coupling and by oxidation). For these calculations, the rate constant for azomethane decomposition was based on our azomethane decomposition experiments ($k_3 = 12,000$ s^{-1} at 650 °C) and was found to fall in a range predicted by rate expressions determined by previous azomethane shock tube experiments [15]. The rate constant for methyl coupling at 650°C was estimated using the recommended expression by Tsang and Hampson (16) including the adjustments necessary in the low pressure fall-off region. The coupling reaction was calculated to dominate any gas phase methyl oxidation reactions. Assuming that the pyrolysis zone of the reactor is completely backmixed (based on very rapid diffusion rates at 0.2 torr) the fraction of azomethane carbon which reaches the catalyst bed as methyl radicals was estimated to be 0.79. The remaining 0.13 and 0.08 of azomethane carbon reach the catalyst bed as C_2 products and unreacted azomethane.

Table 1 contains the results of azomethane pyrolysis experiments at a reactor pressure of 0.2 torr and temperature of 650°C. Mixtures of azomethane, helium, and oxygen were fed to an empty reactor and as well as to the SSMo catalyst. Total azomethane conversion was observed in all experiments. Even without the catalyst present, carbon oxides form a major portion of the product.

Table 1: Product selectivities from azomethane decomposition at 0.2 torr

	Selectivities			
Expt	CO_x	C_2's	CH_4	HCHO
a	0.67	0.16	0.17	<0.05
b	0.91	0.06	0.03	<0.05
c	0.83	0.02	0.15	<0.05

Conditions: (a) empty reactor, (b) and (c) 0.05g of SSMo. For (c) no gas phase O_2 was used. Reaction conditions for all - P=25Pa, T=650°C, reactant mixture: azomethane (0.006)/O_2(0.054)/He(0.94); for (c) O_2 is replaced with He. Reactor residence time =1.9ms.

The reactant gas should exit the high temperature zone largely as methyl radicals and many of the oxidation products are probably formed in the downstream volume, especially on the walls of the sample loop. However, 16% of the methyls recombine to form C_2 hydrocarbons and a similar number abstract hydrogen atoms from other species to form methane.

The effect of the catalyst can be seen by comparing the blank reactor results in Table 1 (experiment a) with experiment b where catalyst and gas phase O_2 are present and with experiment c, where catalyst is present but oxygen is not. The yield of C_2 hydrocarbons is decreased substantially in both cases while the yield of carbon oxides is increased. The silica-supported molybdena as either an oxidant or as a catalyst is capable of oxidizing methyl radicals. However the products of this increased oxidation are CO and CO_2. In all runs neither methanol nor formaldehyde were observed, thus contradicting the hypothesis that gas phase methyl radicals react rapidly with this catalyst to produce formaldehyde. Product selectivity for methane reactions over 0.12g of SSMo, (d) with O_2, and (e) without O_2. Reaction conditions for (d) T=650°C, total flowrate 45ml(STP)/min, $CH_4(0.9)/O_2(0.1)$. For (e) O_2 is replaced by He.

The absence of formaldehyde or methanol cannot be explained by efficient secondary oxidation of these products in the reactor. Mixtures of these materials with helium and oxygen were fed to the reactor at the conditions of the azomethane experiment. In all cases the conversions were very small at the contact times employed in these low pressure experiments, 1.9 ms. Therefore secondary oxidation of formaldehyde is not the source of the increased CO_x yields. Taken at face value, these results contradicts the methyl readsorption mechanism of formaldehyde formation.

The conclusion above is clouded by the fact that these experiments take place at conditions quite different from ambient pressure catalytic experiments. The differences could be explained by the fact that the catalytic surface is more selective when higher pressures of gas phase oxygen are present. The data in Table 2 are included to examine this hypothesis. Methane conversions and product selectivities for methane oxidation over SSMo at ambient pressure both with and without molecular oxygen are shown. The data without oxygen were taken during the first minute after the addition of methane to a previously oxidized catalyst. ith oxygen in the gas the selectivity to formaldehyde is 0.48. As shown by previous studies [5,6], almost all of the carbon monoxide results from secondary oxidation of formaldehyde during the 0.15 s catalyst contact time. Thus approximately 90% of the reaction pathways involve primary formation of formaldehyde. . ithout molecular oxygen, a much smaller methane conversion is achieved. The selectivity to formaldehyde also drops in the absence of oxygen but significant quantities are still formed from methane under these conditions. Thus higher oxygen partial pressures are not necessary for formaldehyde production; and if free methyls were responsible for this product, then some formaldehyde should

Table 2: Product selectivity for methane partial oxidation

Expt	Methane Conversion%	Selectivities				
		CO	CO_2	C_2's	CH_4	HCHO
d	3.5	0.40	0.10	0.02	-	0.48
e	0.2	0.70	0.15	0.05	-	0.10

Product selectivity for methane reactions over 0.12g of SSMo, (7d) with O_2, and (7e) without O_2. Reaction conditions for (7d) T=650°C, total flowrate 45ml(STP)/min, $CH_4(0.9)/O_2(0.1)$. For (7e) O_2 is replaced by He.

have been seen in the azomethane experiment.

A closer comparison of the methyl reaction rates in the azomethane experiments with the one atmosphere results is provided by the following simple analysis. Free methyl intermediates are assumed to be responsible for both the C_2 products (via methyl coupling) and formaldehyde products

Table 3: Comparison of methyl oxidation rates

Expt	Pressure (kPa)	τ (s)	$[CH_3]$ (mol/cc) $(\times 10^{11})$	k_O (s^{-1})
d	105	0.15	2.2	42,000
e	105	0.15	0.8	24,000

(via surface oxidation). This assumption allows the average methyl radical concentration to be calculated from the C_2 productivity and the known methyl radical coupling constant, k_C, by the following expression:

$$[CH_3] = \sqrt{\frac{[C_2H_6] + [C_2H_4]}{k_{C^\bullet}\, \tau}} \qquad (4)$$

This methyl concentration is then used to calculate the value of the heterogeneous methyl oxidation rate constant which would be required to achieve the observed formaldehyde yields in Table 2. These calculations are summarized in Table 3 and represent lower limits to the oxidation rate constant, k_O, since the formaldehyde yields only represent the fraction which survived secondary oxidation. Comparision of these rate constants to the residence times in the azomethane experiments predicts that over 90% of the methyl radicals should have been oxidized to formaldehyde.

In summary, these results indicate that the formaldehyde production pathway in the steady state methane oxidation over silica-supported molybdena does not involve gas phase methyl radicals. If surface methyl is formed upon methane activation, the resulting surface intermediate is immediately oxidized to formaldehyde at the same catalytic site. One possibility is that the methane selective oxidation site represents a very small fraction of the surface and consequently gas phase methyl radicals predominantly react with non-selective sites on the surface. Other surface modifications as a result of other species (such as H_2O) may be critical to maintaining a selective surface. In any case, the surface would have to be both much more active and selective than the results show.

ACKNOWLEDGMENTS
We wish to thank the Natural Sciences and Engineering Research Council of Canada (NSERC) and Exxon Research and Engineering Company for support of this work and J. L. Robbins for helpful discussions.

REFERENCES

1.   Pitchai, R. and Klier, K. *Catal. Rev.- Sci. Eng. 28,* 13 (1986).
2.   Sinev, M.Yu., Korshak, V.N., Krylov, O.V. *Russ. Chem. Rev. 58,* 22 (1989).
3.   Brown, J.J. and Parkyns, N.D. *Catalysis Today, 8* 305 (1991).
4.   Dowden, D.A., Schnell, C.R. and Walker, G.T. Reprints of papers for IVth International Congress on Catalysis, Moscow 1988, ed. J. Hightower, The Catalysis Society, Houston, p. 1120.
5.   Driscoll, D.J., Martir, W., Wang, J.-X. and Lunsford, J.H. *J. Am. Chem. Soc. 107,* 58 (1985).
6.   Tong, Y., Rosynek, P. and Lunsford, J.H. *J. Phys. Chem. 93,* 2896 (1989).
7.   Liu, H.-F., Liu, R.-S., Liew, K.Y., Johnson, R.E., and Lunsford, J.H., *J. Am. Chem. Soc. 106,* 4117 (1984).
8.   Mauti, R. and Mims, C.A. manuscript in preparation.
9.   Dean, A.M., Mims, C.A. *J. Phys. Chem.98,* 13357 (1994).
10. Sun Q., Di Cosimo J.I., Herman, R.G., Klier, K. and Bhasin, M.M. *Catalysis Letters, 15,* 371 (1992)
11. Smudde, G.H. Jr., Yu, M. and Stair, P.C. *J. Am. Chem. Soc. 115,* 1988 (1993).
12. Mauti, R and Mims, C.A. *Catalysis Letters 21,* 201, (1993).
13. Spencer, N.D., Pereira, C.J., *A.I.Ch.E. J. 33,* 1808 (1987).
14. Renaud, R. and Leitch, L.G. *Can J. Chem. 32,* 545 (1954).
15. Chiltz, G., Aten, C.F., and Bauer, S.H., *J. Phys. Chem., 66,* 1426 (1962).
16. Tsang, W., and Hampson, R.F., *J. Phys. Chem. Ref. Data 15(3),* 1087 (1986).

TRANSIENT STUDIES OF METHANE OXIDATIVE COUPLING OVER Li PROMOTED LaNiO3 CATALYSTS

Zbigniew Kalenik and Eduardo E. Wolf*

Department of Chemical Engineering
University of Notre Dame
Notre Dame, IN 46556

ABSTRACT

The oxidative coupling of methane was studied in steady state and transient modes over lithium promoted lanthanum-nickel oxide catalysts. Lithium promotion improves ethane and ethylene selectivity on otherwise completely unselective catalysts. Transient and oxygen isotopic exchange experiments indicate involvement of lattice oxygen in the methane activation process. It is also shown that the presence of lithium significantly lowers a degree to which lattice oxygen can interact with gas phase isotope by blocking oxygen exchange sites.

INTRODUCTION

The most active catalysts for the formation of ethane and ethylene during oxidative coupling of methane are reducible metal oxides, rare-earth metal oxides, or the oxides of alkali and alkaline earth metals. The main issues which have to be better understood in order to improve higher hydrocarbons yields are the nature of the methane surface interactions and the oxygen species responsible for methane activation, the role of unselective oxidation reactions, and the influence of different promoters on the activity of the catalysts. A book relating state of the art in the field up to 1992 summarizes the main issues and challenges in this area (1). In general it has been found that reducible metal oxides are involved in a redox cycle similar to the Mars-van Krevelen mechanism (2), with lattice oxygen involved in the process. In the case of alkali promoted transition metal oxides and rare-earth metal oxides different forms of oxygen on the surface are the sites responsible for hydrogen abstraction from methane to form a free methyl radical that dimerizes in the gas phase (3, 4, 5, 6). The pathway for the formation of CO_2 can involve direct oxidation of methane, or the combustion of C_2 products.

In this study the oxidative coupling of methane was investigated on a series of lithium promoted nickel lanthana oxide catalysts using steady state and transient experiments. Lithium, an effective promoter in alkaline oxides such as TiO_2 and La_2O_3 catalysts (7, 8, 9) has been chosen to promote $LaNiO_3$ oxide. This double oxide was selected as a catalyst since Ni has proven to be able to activate methane at much lower temperatures than those usually used in the oxidative coupling reaction (10, 11). Furthermore, lanthana provides the appropriate reaction pathway to lattice oxygen interaction with surface species.

The objective of this study is to report the effect of lithium promotion on $LaNiO_3$ metallo oxide, and to determine the role of lattice oxygen in the formation of C_2 products. Catalyst characterization techniques (BET, XRD) were also applied in order to correlate catalytic activity with the influence of lithium promotion.

EXPERIMENTAL

The reactor used during steady state studies was a 9 mm i.d. quartz tube especially designed to decrease dead volumes, particularly in transient pulse and isotopic exchange experiments. Minimization of pre and post reactor volumes was achieved by the use of two separate quartz cylinders (fused at the ends) placed between the catalyst bed. This modification allowed for a fast response to changes in the reactant flow rates (12, 13). A typical set of operating conditions during standard activity experiments was as follows: atmospheric pressure, 50 mg of catalyst, total flow rate of 100 cc/min (STP), and methane/oxygen mole ratio of 4. In some cases, experiments were repeated twice to check for the reproducibility of the results. Reactants were diluted in helium, so the reactants partial pressure, $P^*=(P_{CH4}+P_{O2})/P_{tot}$, was equal to 0.4 to minimize gas phase reactions (14). Effluent gases were analyzed by an on-line gas chromatography (GC), equipped with a TCD and FID detectors. Two chromatography columns operated in parallel allowed for the sufficient separation of effluent gases: a HayeSep Q polymer-packed column for the separation of CO_2, C_2H_4, C_2H_6, C_3H_6, and C_3H_8, and a Carbosphere-packed column for the separation of H_2, O_2, CO, CH_4 and CO_2. An ice trap was placed at the reactor exit in order to remove water from the reaction products. Throughout this paper, conversion is based on the amount of carbon detected in the products, selectivity is defined as the ratio of the amount of carbon converted to a given product to the total amount of carbon converted, and yield is defined as the product of conversion and selectivity. In most of the experiments closures on carbon mass balances were reached within 2-4 %.

During transient pulse experiments involving methane and oxygen both gases were diluted in helium before entering a loop of the sampling valve. The ratio of methane to helium in the pulse was constant and was equal to 1:10. Isotopic $^{18}O_2/^{16}O_2$ exchange experiments were conducted between 600°C to 750°C temperature range and with 20 mg of catalyst loading. Oxygen flow rate was adjusted to 2 cc/min. Gases were diluted in helium, so the total flow rate was 60 cc/min. During transient isotopic experiments effluent gases were analyzed by the UTI 100C quadrupole mass spectrometer after appropriate calibration.

Temperature programmed isotopic exchange (TPIE), a technique introduced in our laboratory, has proven to be an effective tool in the investigation of the lattice oxygen pathways during oxidation reactions (15). The experimental procedure consists of exchanging first the $^{18}O_2$ isotope with the $^{16}O_2$ of the oxide lattice at 750°C by sending a step of $^{18}O_2$/He mixture over the catalyst over a period of about 7 minutes. Monitoring of the signals for $^{16}O_2$ (referred hereafter as O_2), $^{18}O_2$ isotope, and $^{16}O^{18}O$ scrambled oxygen permits to determine the amount of isotope exchanged as well as its mobility in the lattice (16, 17). After the initial exchange, the reactor is brought to the room temperature, and the flow of the oxygen and helium is introduced over a sample while the temperature program is started. In this study, $^{16}O_2$ flow rate was fixed at 2 cc/min while the total flow was maintained at 60 cc/min. Of interest is to follow the signals of $^{18}O_2$ and $^{16}O^{18}O$ species as a function of temperature, over a period of predetermined time. The initial exchange of the $^{18}O_2$ with the lattice oxygen allows the TPIE experiment to be performed with significant economy in the isotope use. The oxygen flow rate can be varied, mixed with other gases, or substituted by another oxygen containing species such as CO, CO_2 or N_2O.

Nickel lanthana oxide was prepared by combining stoichiometric amounts of nitrates of lanthanum and nickel into de-ionized water. After dissolving the nitrates citric acid was added to the mixture in such amount that the molar ratio of the sum of the nitrates to that of citric acid was equal 1. After evaporating water, the remaining powder was heated to 450°C and later calcined at 700°C in oxygen flow for two hours.

Lithium promoted catalysts were prepared by the same method described above, however a small amount of lithium carbonate corresponding to a particular weight percentage of Li was added to the solution of Ni and La nitrates. Prior to each experiment, all catalyst samples were pretreated in an oxygen flow for 2 hours at 700°C.

The BET total surface area measurements of the catalysts were conducted using Quantachrome QS-8 unit with nitrogen as the adsorbing gas and helium as a carrier. X-ray Diffraction patterns were measured with a Diano diffractometer using Cu Ka radiation.

RESULTS   AND   DISCUSSION

CATALYST CHARACTERIZATION:

Results of the X-ray diffraction experiments for unpromoted and 1% Li promoted catalyst are presented in Fig. 1. XRD patterns indicate that there are significant changes in the major XRD lines of $LaNiO_3$ with the addition of small amount of Li. While the unpromoted catalyst has XRD lines characteristic only for the perovskite structure of $LaNiO_3$, the Li promoted catalyst exhibits additional lines characteristic of Li_2CO_3. This behavior indicates that the presence of lithium plays a significant role during preparation of the Li promoted catalyst. When sodium was used as an alkaline dopant in $LaNiO_3$ (results not presented in this paper), no changes in the XRD structure were observed. Hence, it can be concluded that in the case of lanthanum nickel oxide the presence of lithium but, not sodium, can modify the final composition of the catalyst.

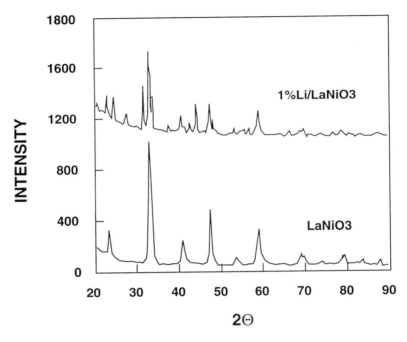

Figure 1. XRD patterns of unpromoted and Li promoted $LaNiO_3$.

BET, surface areas were measured in a Quantachrome QS-8 unit using the flow adsorption method with nitrogen as the adsorbing gas and ultra high purity helium as a carrier. The surface area for the $LaNiO_3$ phase was found to be around 4.2 m^2/g, and after lithium promotion, the total surface area decreased to 2.9 m^2/g for the 1% Li promoted $LaNiO_3$. Measurements were taken on fresh samples at room temperature and prior to each experiment, the catalyst was degassed for 2 hours in a helium flow at 120°C. The decrease in the surface area after Li promotion is a typical behavior observed for the majority of oxidative coupling catalysts that can be explained by the formation of surface species such as Li carbonates and mixed oxides which block pores of the precursor. Due to this behavior it is also likely that the addition of Li is capable of changing the nature of the interaction between CH_4 and the surface of the catalyst during oxidative coupling reaction.

ACTIVITY AND SELECTIVITY

In the presence of the unpromoted LaNiO₃ catalyst CH₄ and O₂ start to react at about 450°C, and complete oxygen conversion is achieved around 550°C. These temperatures are much lower than those usually encountered in oxidative coupling indicating a significant interaction of methane with the catalyst surface. This interaction however, is too strong since it leads to a complete oxidation of methane to yield CO_2.

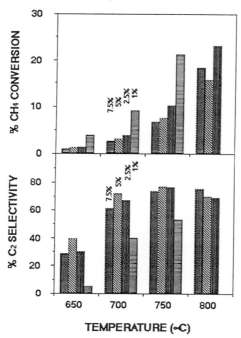

Figure 2. CH4 conversion and C2 selectivities vs reaction temperature during OC reaction at different Li loadings.

The effect of lithium loading on conversion, and product selectivity as a function of reaction temperature is presented in Figure 2. Addition of Li suppresses significantly the high activity for total oxidation displayed by the unpromoted catalysts and shifts the reaction temperature to a range typical for the methane oxidative coupling. At a given temperature as lithium loading increases, the conversion decreases except at 800°C. These results indicate that Li promotion eliminates some of the sites for significant methane-surface interaction present in the unpromoted catalyst, and shifts the reaction to the free radical mechanism which predominates at high temperature (18).

The bottom half of Figure 2 shows the effect of Li loading on C₂ selectivity. In this case, except for the 1% loading, the selectivity is about the same in each case and increases with temperature levelling off at around 70% in the temperature range of 700-800°C. The catalyst promoted with 1% Li has a slightly lower selectivity but exhibits the highest conversion, hence it has the highest yield and it was selected for further study. These results indicate that the suppression of activity coincides with the elimination of the combustion sites which allow for the formation of the free radicals which recombine in the gas phase to form C_2 hydrocarbons.

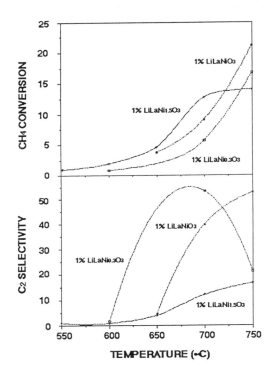

Figure 3. Conversion of CH4 and C2 selectivity on LaNiO3 catalyst with different Ni/La ratio.

Previous work on methane steam reforming has shown that methane interaction with a nickel based catalyst (nickel aluminates) was affected by the ratio between the metal oxide and the alkaline oxide (11). Figure 3 shows the conversion-selectivity results as a function of temperature for two other catalysts with excess and deficiency of Ni in relation to the La oxide. The stoichiometric material exhibits the highest yield whereas the selectivity of the nickel deficient catalyst goes through a maximum and it is rather low for the catalyst with excess Ni. The latter material yields mainly CO_2. Thus, the presence of Li modifies the sites associated with surface Ni since when they exceed the Li loading such sites are still capable of complete combustion of methane or the C_2 products.

Time on stream experiments with the 1% Li loading catalysts reveal that during a five hour period, conversion remains fairly constant, but selectivity decreases with time. This decrease is likely due to the loss of Li leading to the exposure of the combustion sites present in the unpromoted catalysts. To further understand the role of Li promotion a series of transient experiments was conducted.

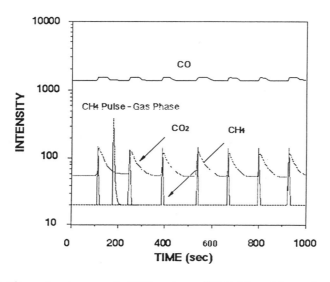

Figure 4. Pulses of methane over LaNiO3 catalyst at 700°C. Mass of the catalyst 20 mg.

TRANSIENT   STUDIES.

Pulses of pure methane diluted in helium with no oxygen were passed over 20 mg of the unpromoted LaNiO3 at 700°C. The results presented Fig. 4, show the appearance of a mass signal corresponding to CO_2 and no C_2 signal. The CO_2 signal has a significant tail indicating

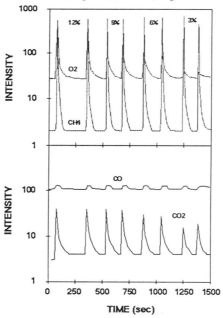

Figure 5. CH4 pulses with different O2 concentration on LaNiO3 at 550°C.

a strong interaction of CO_2. Also displayed is the signal obtained in the absence of the catalyst showing no CO_2 formation in the gas phase. These results indicate that either there is a participation of lattice oxygen in the formation of CO_2 since no oxygen was present in the pulses, or that some reduction of the nickel oxide is taking place.

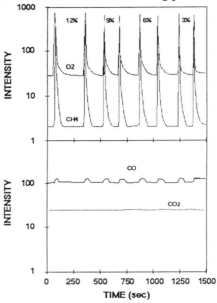

Figure 6. CH4 pulses with different O2 concentration on 2.5% Li/LaNiO3 at 700°C.

Results of pulses of methane containing 12, 9, 6 and 3% oxygen (vol), passed over the unpromoted Li catalyst at 550°C are shown in Figure 5. The reaction temperature was lowered due to the higher reactivity of LaNiO3 found in the presence of oxygen. As in the previous case with pure methane, only CO_2 is observed as a reaction product with no traces of C_2 formation and only a minor trace of CO. In both cases with and without oxygen, the only product detected is CO_2 showing that fast combustion of methane or the C_2 products is occurring on the combustion sites.

Similar experiments consisting of pulses of methane containing oxygen at various oxygen concentration over a 2.5% Li/LaNiO3 at 700°C show that there is no C_2 product evolving in this case, in fact not even CO_2 is detected (Fig. 6). This rather interesting finding indicates that whatever product is formed during pulses over Li promoted catalyst, it is captured on the surface of the catalyst. Since the catalyst used in this experiment has been exposed to reactants, the experiments were repeated at lower temperatures but on a fresh catalyst and on a CO_2 treated catalyst. The results, shown in Figure 7, indicate that indeed some CO_2 formation is observed in the presence of the fresh catalyst, but the signal decreases with the number of pulses used. In the case of the CO_2 treated catalyst (1 hr at 500°C), no measurable CO_2 signal appears in agreement with the experiments presented in Fig. 6 for an aged catalysts. These results show that during transient experiments some surface compound is formed from the reaction products and becomes trapped on the surface. The most likely product formed is a surface carbonate. For the fresh catalysts, the results indicate that such reaction is initially less significant but it becomes more important as time on stream (or number of pulses) increases. If the surface intermediates were carbonates it would imply that they are involved in the formation of CO_2.

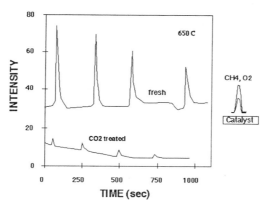

Figure 7. CO_2 signal during CH_4 and O_2 pulses on 2.5% $Li/LaNiO3$, fresh and CO_2 treated at 650°C.

A lack of the product detection during the pulse experiments, suggested that the pulses used were too small, and whatever C_2 products were formed, they were converted to CO_2 due to the large residence time of the pulse in the reactor. During steady state experiments, when

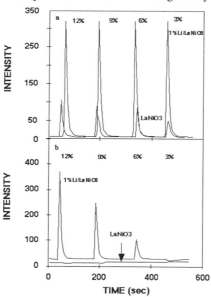

Figure 8. C_2H_6 and O_2 pulses with different O_2 concentration at 650°C. a- CO_2 signal, b- O_2 signal.

C_2 products are detected, the amount of C_2 formed is large compared with the amount of catalysts, hence they are detected at the reactor outlet before they are converted to CO_2. To test this possibility, a step input of methane and oxygen was sent into the reactor at 700°C for two catalysts. One was a 2% Sr/La$_2$O$_3$ catalyst previously studied in our group (16) and the other was the 2.5% Li/LaNiO$_3$ catalyst. The results of such experiments (not shown in this paper) using a step input show that both C$_2$ and CO$_2$ are detected by MS for both catalysts although the signals for the C$_2$ product (and methane conversion) is about six times smaller for the Li/LaNiO$_3$ catalyst than for the Sr/La$_2$O$_3$ catalyst. It follows that whatever C$_2$ products were produced on the Li promoted catalysts during the step experiments, they become converted to CO$_2$ that it is trapped on the catalyst surface.

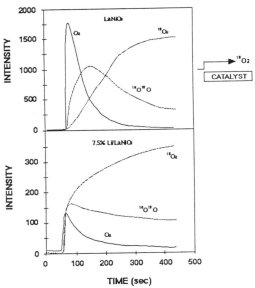

Figure 9. $^{18}O_2$ isotopic exchange at 700°C on LaNiO$_3$ and 7.5% Li/LaNiO$_3$.

To further verify this possibility, pulses of ethane and oxygen were passed over the Li (1%) promoted and unpromoted LaNiO$_3$ catalysts. The CO$_2$ signal for ethane pulses with different oxygen concentration (Fig. 8) is significantly larger for the unpromoted catalyst than for the promoted catalyst. The oxygen consumption in the unpromoted catalysts is 100% percent, indicating complete conversion of the C$_2$ products. In the 1% Li/LaNiO$_3$ catalysts, Li promotion significantly reduces the combustion of the reaction products.

Isotopic oxygen tracing experiments were conducted in order to determine the effect of Li on the catalyst ability to exchange its lattice oxygen with the oxygen available in the gas-phase. Figure 9 shows the intensity signals at m/e 32, 34, and 36 when a step of $^{18}O_2$ in a He carrier stream was introduced into the reactor at 700°C. The results for the unpromoted catalysts show the appearance of a large peak of $^{16}O_2$ and the disappearance of the $^{18}O_2$ signal and subsequent recovery to the feed level. This demonstrates that isotopic oxygen exchanges rapidly with the oxygen from the lattice. The formation of the scrambled $^{16}O^{18}O$ oxygen goes through a maximum as the amount of ^{16}O is depleted within the solid. The same experiment is presented in the bottom half of Fig. 9 for a heavily loaded Li catalyst to ascertain with better accuracy the effect of Li on the O_2 exchange. In this case the exchange, although qualitatively similar, is significantly reduced. This indicates that only a fraction of oxygen can exchange in the Li promoted catalyst due to the blockage of the sites for oxygen exchange or alternatively due to the diffusion of Li into the catalyst. This effect is probably linked to the suppression of the combustion activity with Li loading.

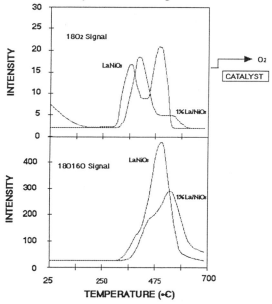

Figure 10. $^{18}O_2$ and $^{18}O^{16}O$ esignals during TPIE experiment with oxygen.

A temperature programmed isotopic exchange experiment (TPIE) permits to ascertain the effect of temperature on the rate of oxygen exchange. In this case isotopic oxygen is exchanged first with catalysts at high temperature and then the reactor is cooled down to room temperature. The oxygen incorporated in the catalyst is the isotopic oxygen. A step of normal oxygen is passed over the catalyst while the temperature is increased. The appearance of the isotopes indicates the temperature at which exchange between catalyst and gas phase oxygen starts occurring. Fig. 10 displays results for the unpromoted and a 1% Li promoted catalysts. Oxygen exchange starts taking place at temperatures as low as about 325-340°C, which agrees with previous results on La_2O_3 (15). However, in the case of the $1\%Li/LaNiO_3$ catalyst an exchange occurs at a slightly higher temperature that in case of the unpromoted catalyst. This effect is different than the results found during the promotion of La_2O_3 with Sr (15). In the later case Sr promotion lowered the temperature for oxygen exchange due to the increase of oxygen mobility and by the formation of oxygen vacancies. Furthermore, the shape of the oxygen exchange curve is different when Li is added to the catalysts with the disappearance of the higher temperature peak for the ^{18}O signal. It follows that there are two types of sites for oxygen exchange in these catalysts and Li promotion suppresses one of them. This finding agrees with the results for oxygen exchange at 700°C on the 7.5%Li promoted catalyst. The

TPIE results show more specifically that Li promotion suppresses the high temperature sites for the oxygen exchange. Since Li promotion results in higher C_2 selectivity due to the decrease in C_2 product combustion, it can be concluded that Li promotion blocks the higher temperature oxygen exchange sites which are presumably associated with C_2 product combustion.

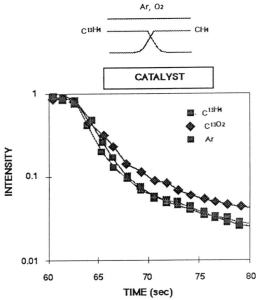

Figure 11. Methane 13 isotopic switch in the presence of Ar and oxygen on LaNiO3 catalyst at 650°C.

To ascertain the extent of the methane surface interaction, isotopic switch experiment were conducted. In these experiments, described in detail elsewhere (10, 19), a feed containing $^{13}CH_4$ and oxygen is switched with a feed containing a similar concentration of normal methane and oxygen. A trace of an inert Ar is also used to measure the transient response of a system in which no surface interaction occurs. The virtue of this method is that the surface remains under steady state conditions while the transient response of the isotopes can be measured. The results of the switch experiments presented in Fig. 11 for the unpromoted catalyst, show that the response of the $^{13}CH_4$ trace cannot be distinguished from that of Ar. Consequently, within the time resolution of this experiments the interaction of methane and the surface is too fast and cannot be distinguished from that of the inert gas such as argon. The $^{13}CO_2$ response observed during the methane switch experiment is also presented. It can be seen that this signal experiences a measurable delay with respect to that of Ar signal indicating that there is surface interaction leading to this delay. This indicates the formation of a surface species which retards the evolution of CO_2 with respect to Ar.

Methane interaction with the surface of the catalyst was the subject of the studies conducted by Ekstrom and Lapszewicz (20), and by Hatano and Otsuka (21). The former group, which measured adsorption of methane on the Sm_2O_3 oxide, suggested that methane acts as a weak acid, while the catalyst acts as a strong base. On the basis of the isotopic experiments, these authors concluded that adsorbed methane does not take part in the oxidative coupling reaction. On the other hand, Hatano and Otsuka presented evidence, that in the case of lithium nickelate oxidative dimerization of methane proceeds through the dissociation of methane on the catalyst surface. Results of the experiment presented in Fig. 11 suggest that for the LaNiO3 catalyst and under reaction conditions applied in this study, the concentration of the long-lived methane species is too low to be differentiated with this technique from the interaction with argon over the surface of the catalyst.

SUMMARY

Results presented in this paper indicate that Li promotion of LaNiO3 metallo oxide significantly changes catalytic properties of the latter compound. Steady state results indicate that unpromoted catalyst oxidizes methane at temperatures significantly lower than those used during oxidative coupling. After Li promotion selectivity towards C_2 products increased from 2% up to 75% at 750°C for the 5% Li promoted catalyst. Transient experiments suggest that lattice oxygen may participate in the process of abstracting a hydrogen atom from the methane molecule. Isotopic exchange experiments conducted on unpromoted and Li promoted catalysts indicate that presence of lithium significantly decreases exchange capabilities of lanthanum nickel oxide by blocking oxygen exchange sites which may also be responsible for the total oxidation of methane. Methane 13 isotopic switch conducted over LaNiO3 showed no detectable interaction with the surface of the catalyst under reaction conditions. The observed delay in the response of the $^{13}CO_2$ signal suggested that this particular product can be considered as a secondary and it is mainly formed from the combustion of C_2 hydrocarbons.

REFERENCES

1. *Direct Methane Conversion by Oxidative Processes.* Edited by E. E. Wolf. Van Nostrand Reinhold Catalysis Series, New York 1991.
2. Mars, P., and van Krevelen, D. W., *Chem. Eng. Sci.,* Suppl. **3**, 4 (1954).
3. Otsuka, K., Jinno, K., *Inorg. Chim. Acta,* **121**, 233 (1986).
4. Otsuka, K., *Inorg. Chim. Acta,* **132**, 123 (1987).
5. Ito, T., Wang, J.-X Lin, C.-H and Lunsford, J. H., *J. Chem. Soc.,* **107**, 5062 (1985).
6. Driscoll, D. J., Campbell, K. D., and Lusford, J. H., *Adv. Catal.,* **35**, 139 (1987).
7. Lane, G. S., Miro, E. E., and Wolf, E. E., *J. Catal.,* **119**, 161   (1988).
8. Lane, G. S., and Wolf, E. E., *Proc. Ninth Int. Congress Catal.,* Calgary, Cnada June 1988.
9. De Boy, J. M., and Hicks, R. F., *J. Chem. Soc., Chem. Commun.* 982, 1988.
10. Miro, E. E., Kalenik, Z., Santamaria, J., and Wolf, E. E., *Catalysis Today,* **6**, 511 (1990).
11. Al-Ubaid, A., and Wolf, E. E., *Appl. Catal.,* **40**, 73 (1985).
12. Kalenik, Z., and Wolf, E. E., In *Direct Methane Conversion by Oxidative Processes.* Edited by E. E. Wolf. Van Nostrand Reinhold Catalysis Series, p. 30, New York 1991.

13. Kalenik, Z., Miro, E. E., Santamaria, J., and E.E. Wolf., *Preprints of 3 Symposium on Methane Acivation,* 1989 Pacifichem Meeting , Honolulu, Hi, 109, (1989).
14. Lane, G. S., and Wolf, E. E., *J. Catal.* **113**, 144 (1988).
15. Kalenik, Z., and Wolf, E. E., *Catal. Let.* **11**, 309 (1991).
16. Kalenik, Z., and Wolf, E. E., *Catal. Let.* **9**, 441 (1991).
17. Peil, K. P., Goodwin, J. G., and Marcelin, G., *J. Catal.,* **131**, 143 1991).
18. Lunsford, J. H., In *Direct Methane Conversion by Oxidative Processes.* Edited by E. E. Wolf. Van Nostrand Reinhold Catalysis Series, p. 3, New York 1991.
19. Peil, K. P., Marcelin, G., and Goodwin, J. G., In *Direct Methane Conversion by Oxidative Processes.* Edited by E. E. Wolf. Van Nostrand Reinhold Catalysis Series, p. 138, New York 1991.
20. Ekstrom, A., Lapszewicz, J. A., *J. Chem. Soc. Chem. Commun.,* 797 (1988).
21. Hatano, M., and Otsuka, K., *J. Chem. Soc., Faraday Trans.* **I, 85** 2, 199 (1989).

SYNTHESIS OF C$_{2+}$ HYDROCARBONS AND SYNGAS BY GAS PHASE METHANE OXIDATION UNDER PRESSURE

K. Fujimoto, K. Omata, T. Nozaki and K. Asami

Department of Applied Chemistry, Faculty of Engineering,
The University of Tokyo, Hongo, Bunkyo-ku, Tokyo 113, Japan

Oxidative coupling of methane was conducted in the non-catalyzed gas phase system under pressurized conditions. Both partial pressures of oxygen and methane showed marked effects on either reaction rate or product selectivity. Low oxygen pressure was essential for high C$_{2+}$ selectivity. Additives such as CHCl$_2$CHCl$_2$, H$_2$S, CH$_3$COCH$_3$, CO$_2$ promoted C$_{2+}$ yield and the promoting mechanism seems different among these promoters. The carbon dioxide seemed to promote the reaction by the third body effects, without lowering the selectivity of C$_{2+}$ hydrocarbons.

INTRODUCTION

The present authors found that the non-catalyzed oxidative coupling of methane proceeds under pressurized conditions at about 650-850°C [1,2] It has been known that the non catalyzed oxidation of methane proceed via the free radical chain reaction. Several groups tried to simulate the overall reaction by using the combination of about 30 free radicals and about 200 elementary reactions. The results were partially successful. Under pressured conditions even the product distribution of the solid catalyzed systems were markedly affected by the contribution of non-catalyzed gas phase reaction. The characteristic features of the products in the non-catalyzed system is that while the selectivity of C_{2+} hydrocarbons is lower than that of the solid catalyzed system, the by-product is not CO_2 but CO and H_2. Also it has been clarified that in this system the initiation (methane activation) is the key step for controlling the rate and the product distribution.   In the present study methods for promoting the initiation were pursed.

EXPERIMENTAL

Reaction was performed with a flow type reaction apparatus under pressurized conditions. A metallic tubular type reactor was used and quartz tubes were inserted into the reactor to prevent direct contact of the reactant gas with the metal surface of the reactor.   All the promoters were injected into the reaction system in the gaseous form.   Products were analyzed by gas chromatographies.

RESULTS AND DISCUSSION

Effect of Reaction Condition

Figure 1 demonstrates the effect of reaction temperature.   The main products of the non-catalyzed reaction at high temperature are C_{2+} hydrocarbons and synthesis gas.   The yield of each product increased with increasing temperature.   The selectivity to ethylene increased as the reaction temperature was raised and become superior to that of ethane above 750 °C.   The formation of H_2 increased with increasing reaction temperature to reach 1.5 times of in mole the amount of CO at 850°C.

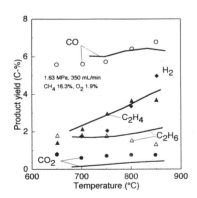

Figure 1.   Effect of reaction temperature

In Fig. 2 to 4 are shown the effect of total pressure and partial pressures of methane and oxygen, respectively.   It is clear that while the oxidation of methane does not proceed, under

158

atmospheric pressure at 750°C, the rate increased acceleratively with increasing total pressure to reach 10% conversion at 1.6 MPa. The increase in the oxygen partial pressure promote the formation of all reaction products but especially that of CO to result in the decrease in the selectivity of C_{2+} hydrocarbons. The high methane partial pressure, on the other hand, promote primarily the formation C_{2+} hydrocarbons below 0.7 MPa but suppress the C_{2+} hydrocarbons about this level. The reason is not clear, yet. It is obvious that a certain pressure is essential for high selectivity of C_{2+} hydrocarbons.

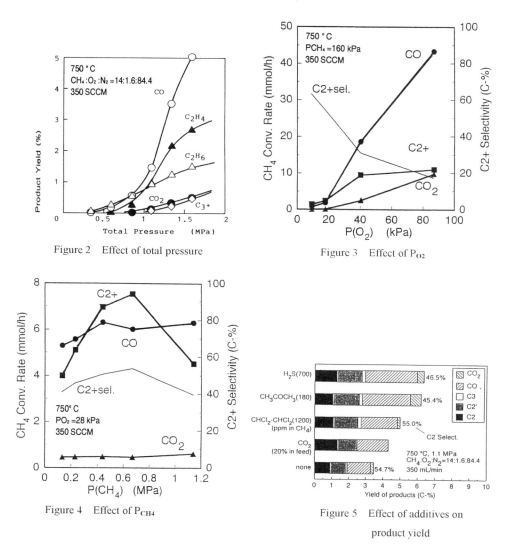

Figure 2   Effect of total pressure

Figure 3   Effect of P_{O2}

Figure 4   Effect of P_{CH4}

Figure 5   Effect of additives on product yield

Reaction Net Work

The solid line in Fig. 1 show the simulated results by the kinetic model [3]. The main reaction path at high O_2 conversion is illustrated in Fig. 5. The simulated values are agreed well with the experimental results, which indicates the validity of this kinetic model. In the reaction network remarkable methyl radial contributes significantly to both the formation and the consumption of C_{2+} hydrocarbons. The maximum yield of C_{2+} was simulated based on these successive reactions and it was clarified that the yield is proportional to the square root of the formation rate of methyl radical.

$$CH_4 + X \cdot \rightarrow CH_3 \cdot + HX \ (X = H \cdot, OH \cdot, HO_2 \cdot) \tag{1}$$

$$2CH_3 \cdot + (M) \rightarrow CH_3CH_3 + (M) \tag{2}$$

$$C_2H_6 + CH_3 \cdot \rightarrow C_2H_5 \cdot + CH_4 \tag{3}$$

Addition of Promoter

From the reaction network the sole reaction which can be modified for higher yield of C_{2+} by the addition of promoters is reaction (1). The yield of reaction products in the presence of promoters are shown in Fig. 5. Hydrogen sulfide and acetone promoted methane conversion with an enhanced CO selectivity. On the other hand, $CHCl_2CHCl_2$ promoted methane conversion without changing C_{2+} selectivity. The difference in the effect of each additive might be attributed to the difference in the initiation mechanism. Activation of H_2S or acetone by oxygen (Eqs. 4 and 5) is accompanied by the formation of $HO_2 \cdot$ radical that of $C_2H_2Cl_4$ needs no oxygen consumption (Eq. 6). The $HO_2 \cdot$ radical seems to act as an oxidant[3] and methyl radical should be transformed to CH_2O and CO successively, through Eq. 8. Thus the addition of H_2S or CH_3COCH_3 promotes both C_{2+} and CO formation.

$$H_2S + O_2 \rightarrow HS \cdot + HO_2 \cdot \tag{4}$$

$$CH_3COCH_3 + O_2 \rightarrow CH_3COCH_2 \cdot + HO_2 \tag{5}$$

$$C_2H_2Cl_4 \rightarrow nCl \cdot + C_2H_2Cl_{4-n} \cdot \tag{6}$$

$$CH_4 + X \rightarrow CH_3 \cdot + HX \ (X = HS \cdot, Cl \cdot, CH_3COCH_2 \cdot) \tag{7}$$

$$CH_3 \cdot + HO_2 \cdot \rightarrow CH_2O + H_2O \tag{8}$$

Addition of large amounts of CO_2 also promoted both C_2 and CO formation. The mechanism is not clear yet. It is probable, however, that the recombination of methyl radical (Eq.2) is accelerated by CO_2 as third body because the mass of CO_2 is lager than that of CH_4 of N_2.

CONCLUSION

Oxidative coupling of methane was conducted under pressure in the non-catalyzed system. The main products were C_2H_4 and syngas. It was found that H_2S, CH_3COCH_3, $C2H_2Cl_4$ and CO_2 promotes the reaction. The promoting mechanism seems to be different among these promoters. The results have proved that some kinds of additives promote the reaction without lowering over even elevating the selectivity of C_{2+} hydrocarbons.

REFERENCES

1. K. Asami, K. Omata, K. Fujimoto, and H. Tominaga, *J. Chem. Soc., Chem., Commun.*, 1287 (1987).
2. K. Asami, K. Omata, K. Fujimoto, and H. Tominaga, *Energy & Fuels*, 2:574 (1988).
3. H. Zanthoff, and M. Baerns, *Ind. Eng. Chem. Research*, 29:2 (1990).

INFRARED SPECTROSCOPIC STUDIES OF ADSORBED
METHANE ON OXIDE SURFACES
AT LOW TEMPERATURES

Can Li, Guoqiang Li, Weihong Yan, and Qin Xin

State Key Laboratory of Catalysis
Dalian Institute of Chemical Physics
Chinese Academy of Sciences
Dalian 116023, China

INTRODUCTION

Methane activation has long been a challenging task in catalysis. There has been much theoretical and experimental work done for methane activation on metal and oxide surfaces,[1-5] and especially significant achievements have been made in oxidative coupling of methane[6-8] since 1982. But at a fundamental level, the mechanism of methane activation including how methane interacts with surface and how the C-H bond cleaves, is still not well understood. It is difficult to obtain experimental information on methane activation at catalyst surfaces because the surface reaction of methane normally takes place at high temperatures where methane collides on the surface and subsequently, surface-generated radical fragments or methane itself leave the surface instantaneously. For example, a general conclusion from methane oxidative coupling is that the reaction involves generation of methyl radicals at the catalyst surface which is then followed by radical coupling in the gas phase at high temperatures.[9,10] From the catalysis principle, the adsorption of methane on catalyst is a necessary step for methane activation, although the residence time of methane at the surface is very short at high temperatures.[11] However, methane is the most inert molecule among hydrocarbons and it is difficult to adsorb on catalyst surface even at room temperature. An expedient measure to approach the activation mechanism is to study the adsorption of methane at low temperatures where methane may be adsorbed on the catalyst surface.

Although the methane adsorption at low temperatures differs from the reaction conditions at high temperatures, it is still an effective way to demonstrate the interaction nature of methane with catalyst surfaces since the interaction potential of methane with a surface should not show a large difference between low and high temperatures. This contribution presents the results of methane adsorption on oxide surfaces at 173 K and highlights the methane adsorption in the activation mechanism. Our results have found that the adsorbed methane was distorted structurally and this distortion becomes significant when methane interacts with active oxygen species. It is proposed that the distortion of methane could be a vital step in governing the methane activation and reaction at catalyst surfaces.

EXPERIMENTAL

In this communication IR spectra of adsorbed methane on CeO_2, MgO and Al_2O_3 are reported. These three samples were chosen because CeO_2 is an active catalyst for methane complete oxidation, MgO is a basic catalyst used for methane oxidative coupling, and Al_2O_3 is an acidic oxide which is nearly inert towards methane. MgO and Al_2O_3 have been obtained from commercial sources. CeO_2 was prepared by calcinating cerium hydroxide gel at 773 K as described in a previous paper.[12] The surface areas of CeO_2, MgO and Al_2O_3 were measured to be 20, 40, and 250 m^2/g respectively. The oxide samples were pressed into self-supporting discs for IR studies.

A quartz IR cell is available for *in-situ* studies over a wide temperature range(100-1000 K). The sample disc in the IR cell can be treated *in situ* in various ways, such as outgassing, reduction and oxidation. A well-outgassed sample was obtained by treating the sample in the cell with O_2 at 873 K and then outgassed at 973 K.

The adsorption of CH_4 was performed at temperatures 173-273 K as reported previously.[13] The sample disc outgassed at various temperatures was cooled down to 173 K in vacuum, and background spectra were recorded during the cooling process. The IR spectra were recorded on a double-beam FT-IR spectrophotometer (Perkin-Elmer 1800) equipped with a liquid-nitrogen cooled mercury-calcium-telluride(MCT) detector. All IR spectra were recorded with 4-cm^{-1} resolution and 4 scans.

RESULTS AND DISCUSSION

Methane Adsorption on Oxide Surfaces

To compare spectra of adsorbed methane with those of gas phase and liquid methane, figure 1 shows IR spectra of gas phase and liquid methane recorded at room temperature and 110 K, respectively. Only the IR bands near 3000 cm^{-1} region are shown here because these IR bands exhibit the most variation for adsorbed methane. The band at 3019 cm^{-1} is the v_3 mode of methane and the fine bands around the two sides of this band are due to the rotational structure of methane.[14] The band at 2828 cm^{-1} is a combination mode of $v_2 + v_4$. When methane was cooled down to its boiling point(~110 K) the 3019 cm^{-1} band shifts slightly to 3014 cm^{-1} and the rotational structure disappears, but no new bands were produced. IR spectra for adsorbed methane were not clearly observed until temperature was lowered to 250 K under a pressure of 5 torr methane. At temperatures near 173 K, methane adsorbs on many metal oxides and the adsorbed states vary with different metal oxides. Most of the adsorbed methane can be removed by evacuation at 173 K or by elevating the temperature to above 250 K, indicating that the adsorption is relatively weak.

Figure 2 shows IR spectra of adsorbed methane at 173 K on CeO_2, MgO, and Al_2O_3. The bands in the 3000 cm^{-1} and 2900 cm^{-1} regions are respectively assigned to v_3 and v_1 modes of methane.[13,15] There is an additional band located at 1305 cm^{-1} which is not shown here because this band does not change significantly for the different oxide samples. The bands in the 3000 cm^{-1} region show red-shifts ranging from 10 to 30 cm^{-1} compared with that of free methane. The bands in the 2900 cm^{-1} region vary more significantly in both relative intensity

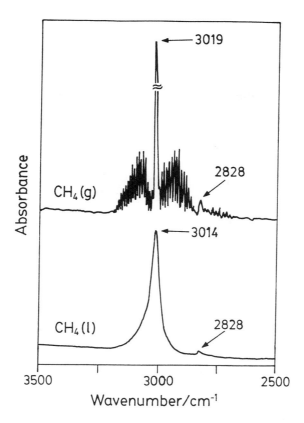

Figure 1. IR spectra of CH_4 in gas phase, $CH_4(g)$, and liquid state, $CH_4(l)$.

and frequency position for the three samples. For example, the red-shifts of v_1 band of methane adsorbed on CeO_2, MgO and Al_2O_3 are 42, 27, and 17 cm^{-1}, respectively. By comparing with figure 1, it can be seen that the rotational structure of methane in the adsorbed and liquid states is not present, while no any IR band can be observed at 2900 cm^{-1} region for liquid methane. These results prove that adsorbed methane on oxide surfaces is naturally different from the condensed methane. The v_1 mode is a symmetric stretching vibration, so this mode is an infrared forbidden vibration for free methane. But the v_1 mode becomes an infrared detectable band for the adsorbed methane. The appearance of the v_1 band reflects the distortion from *Td* symmetry due to the structural distortion of methane and the anisotropic environment at the surface. The relative intensity and the red-shift of the v_1 band largely depend on the degree of interaction of methane with the catalyst surface. A strong interaction between methane and the surface will lead to a more intense band and a large red-shift of the v_1 band. Comparing to figure 1 the red-shift of the v_1 band of adsorbed methane on the oxides shows the degree of interaction of methane with these oxides following the order, CeO_2 > MgO > Al_2O_3, which is consistent with the reactivities of these surfaces.

Interaction Models of Methane with Oxide Surfaces

Separate experiments confirmed that the surface sites available for methane adsorption are different for these three oxides. For CeO_2, two types of adsorbed methane species were

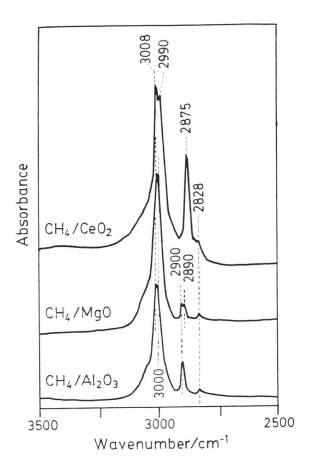

Figure 2. IR spectra of adsorbed CH₄ on CeO₂, MgO and Al₂O₃ at 173 K(the three samples were preoutgassed at 973 K).

observed, one interacts mainly with surface O⁻-like species and the other with surface lattice oxygen anions.[13,15] The adsorbed methane formed *via* the interaction with the surface lattice oxygen seems to be common for many oxides. It was found that the adsorbed methane on MgO could be formed through the interaction of methane with both surface coordinatively unsaturated cations and anions, *i.e.*, the Lewis acid and base sites, $Mg^{2+}_{LC}O^{2-}_{LC}$. The Lewis acid site was demonstrated to be necessary for the formation of adsorbed methane since adsorbed methane can be displaced by adsorbed CO. In particular, this type of adsorbed methane becomes dominant for samples outgassed at high temperatures.[16] It is worthwhile to note that the Lewis acid sites, *i.e.*, surface cations, play an important role in methane activation for some catalysts.[17,18] Adsorbed methane on alumina interacts not only with surface oxygen anions but also with surface hydroxyl groups. The surface hydroxyls are perturbed strongly by the adsorbed methane resulting large red-shifts of the IR bands of the surface hydroxyls.[19] The similar phenomenon has been observed for other acidic catalysts, such as HZSM-5 and mordenite.[20]

The interaction of methane with metal oxides is described schematically in figure 3. There are at least four types of adsorbed methane formed on metal oxides. Species I is formed mainly on the surface having O⁻-like species which can be created by high-temperature outgassing or by oxygen adsorption. This kind of species shows the largest red-shift of the v_1 band indicating the strongest interaction between methane and active oxygen species on the surface. This is reasonable since the O^- species is the most active oxygen species which can abstract hydrogens from methane most effectively at high temperatures. It should be mentioned that the interaction geometry of species I can be depicted a number of ways, $i.e.$, mono-hydrogen, di-hydrogen or tri-hydrogen interacting with O^-, but it is difficult to distinguish the slight differences from the spectra because the heterogeneity of surface makes the IR band broad. Species II can be derived on most of metal oxides and its spectrum is similar to that of liquid methane which only produces a small red-shift of v_3. Species III is formed mainly on Lewis acid-base sites of basic oxides such as MgO. The Lewis acid site may interact with C-H bond or carbon atom of methane while a hydrogen atom is attached to the Lewis base site. Species IV is the adsorbed methane which involves interactions with surface oxygen anions and hydroxyl groups on acidic oxides or zeolites.

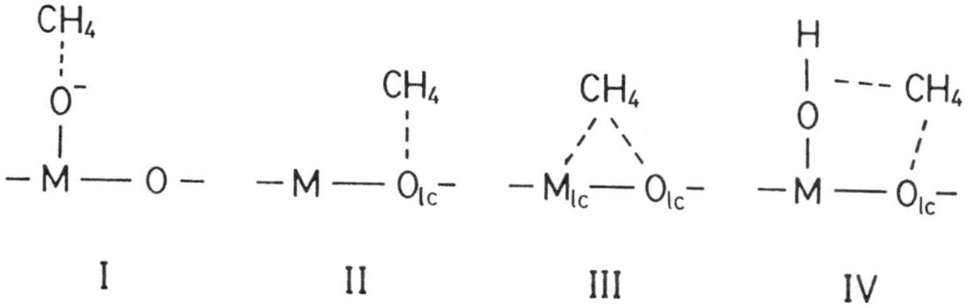

Figure 3. Schematic description of methane interaction models on metal oxides.

Methane Activation at Surface

Methane adsorption is very weak for most metal oxides, and the adsorption is a physically or chemically weak interaction in nature. Though the low-temperature adsorption is weak and far beyond the reaction conditions, the degree of interaction of adsorbed methane with surfaces is found to be correlated well with the methane reaction activity over metal oxides. Table 1 lists the red-shift of the v_1 band of adsorbed methane and the initial reaction temperature of methane on CeO_2, MgO and Al_2O_3. The initial reaction temperature is defined as the temperature at which the reaction of methane is detectable, mostly oxidation or coupling of methane. It has been found that smaller red-shifts correspond to lower initial reaction temperatures. This relationship was found to be true for many other metal oxides also. Therefore the red-shift of the v_1 band may be a good indicator of methane reactivity over a catalyst surface, $i.e.$, the larger the red-shift of the v_1 mode implies the stronger interaction of methane with the catalyst surface, which accordingly lowers the initial reaction temperature.

In a formalism of molecular orbital theory, methane is stable partly because of its high symmetry, Td, which makes the frontier levels highly degenerate. As a result, its HOMO levels are relatively lower and its LUMO levels are relatively higher comparing with other hydrocarbons, making the activation of methane difficult. When methane is adsorbed, the Td symmetry can be distorted. The surface-induced symmetry distortion may change the frontier levels, HOMO and LUMO, of adsorbed methane, and this may be the one reason why the distortion of adsorbed methane is related to the reactivity of methane.

According to Ceyer $et\ al.$,[21] methane seems to be a spherical ball making it difficult to

Table 1. Initial reaction temperature of methane and the observed v_1 vibrational band of adsorbed methane on CeO_2, MgO and Al_2O_3.

Catalyst	Initial reaction temperature(K)	$v_1 (cm^{-1})$	Red-shift(cm^{-1})
CeO_2	473	2875	42
MgO	773	2890	27
Al_2O_3	>973	2900	17

attack the carbon atom. A distortion in the structure of adsorbed methane may make the carbon atom more close to the surface sites, thereby allowing an interaction between the carbon atom and the surface sites which could assist the methane activation.

There is a wide range of oxides which can catalyze the methane conversion at high temperatures, for example, methane oxidative coupling to ethene and ethane. The methane reaction can occur even on relatively inert catalyst at high temperatures. At high temperatures the reaction mechanism is quite complicated so the reaction of methane may involve vibrationally excited states[22] which are thought to be the precursors of methane dissociation. The following scheme is proposed for methane activation and reaction on surface at high temperatures,

$$CH_4(g) \xrightarrow{1} CH_4(a) \xrightarrow{2} CH_4(a)^* \xrightarrow{3} CH_3 + H$$
$$\xrightarrow{4} CH_4(g)^* \xrightarrow{5} CH_4(g)$$

where $CH_4(g)$ and $CH_4(a)$ are the methane in the gas phase and adsorbed state, respectively. The asterisk "*" denotes a transition state indicating a vibrationally excited state. The methane activation at surface may experience adsorption(step 1), vibrational excitation(step 2) and C-H bond dissociation(step 3). Step 2 is suggested as the rate-determining step for the reaction of methane. Steps 4 and 5 are the desorption and quenching of activated methane respectively. At high temperatures these steps occur too quickly to be followed. It is assumed that the adsorbed state, $CH_4(a)$, might influence the following steps(steps 2-4). The $CH_4(a)^*$ could be generated by overcoming an energy barrier by using translational energy[21,23] and/or surface vibrational energy, surface phonons.[24] The former may not strongly depend on the chemical properties of the surfaces but the latter might be intimately related to the adsorbed state, $CH_4(a)$, so the exciting process from $CH_4(a)$ to $CH_4(a)^*$ might be partly dependent on the adsorbed state of $CH_4(a)$. It could be envisioned that a $CH_4(a)$ having a suitable symmetry at the surface and a strong interaction with surface will be favorable for energy transfer from a surface vibrational mode to the adsorbed methane leading to an effective dissociation of C-H bond. Another contribution of adsorption to methane activation is assumed to be that the distortion of the *Td* symmetry makes the four C-H bonds unequal which could be favorable for the formation of a highly excited local-bond vibration in methane, *i.e.*, the energy from translational collisions and/or from surface vibrational modes can be instantly localized on one of the C-H bonds, and as a result, the C-H bond breaking of methane becomes facile at the surface.

ACKNOWLEDGEMENTS

We acknowledge the financial supports from National Natural Science Foundation of China, and Science and Technology Commission of Liaoning Province.

REFERENCES

1. A. B. Anderson, in: "Theoretical Aspects of Heterogeneous Catalysis," J. B. Moffat, ed., Van Nostrand Reinhold, New York (1990).
2. M. C. McMaster, and R. J. Madix, *J. Phys. Chem.* 98:9963(1993).
3. O. Swang, K. Faegri, Jr., and O. Gropen, *J. Phys. Chem.* 98:3006(1994).
4. S. G. Brass, and G. Ehrlich, *Surf. Sci.* 197:21(1987).
5. Q. Y. Yang, A. D. Johnson, K. J. Maynard, S. T. Ceyer, *J. Am. Chem. Soc.* 111:8748(1989).
6. G. E. Keller, and M. M. Bashin, *J. Catal.* 73: 9(1982).
7. J. H. Lunsford, *Studies in Surface Science and Catalysis* 75:103(1992).
8. M. Baerns, and J. R. H. Ross, in: "Perspectives in Catalysis," J. M. Thomas, and K. I. Zamaraev, eds., Oxford Blackwell Scientific Publications, London(1992).
9. D. J. Driscoll, K. D. Campbell, and J. H. Lunsford, *Adv. in Catal.* 35:139(1987).
10. T. A. Garibyan, and L. Ya. Margolis, *Catal. Rev.-Sci. Eng.* 31:355(1989-1990).
11. D. J. Statman, J. T. Gleaves, D. McNamara, P. L. Mills, G. Fornasari, and J. R. H. Ross, *Appl. Catal.* 77:45(1991).
12. C. Li, K. Domen, K. Maruya, and T. Onishi, *J. Am. Chem. Soc.* 111:7683(1989).
13. C. Li, and Q. Xin, *J. Phys. Chem.* 96:7714(1992).
14. N. Sheppard, and D. J. C. Yates, *Proc. R. Soc. London* A238:69(1956).
15. C. Li, and Q. Xin, *J. Chem. Soc., Chem. Commun.* 782(1992).
16. C. Li, G.-Q. Li, and Q. Xin, *J. Phys. Chem.* 98:1933(1994).
17. M. R. A. Blomberg, P. E. M. Siegbahn, and M. Svensson, *J. Phys. Chem.* 98:2062(1994).
18. V. R. Choudhary, and V. H. Rane, *J. Catal.* 130:411(1991).
19. C. Li, W.-H. Yan, and Q. Xin, *Catal. Lett.* 24:249(1994).
20. C. Li, G.-Q. Li, Z.-M. Liu, and Q. Xin, *Progress in Natural Sciences(China)* (1994).
21. S. T. Ceyer, *Langmuir* 6:82(1990).
22. F. F. Crim, *Science* 249:1387(1990).
23. M. B. Lee, Q. Y. Yang, and S. T. Ceyer, *J. Chem. Phys.* 87:2724(1987).
24. G. C. Bond, *Catal. Today* 17:399(1993).

COMPUTER-AIDED ELUCIDATION OF REACTION MECHANISMS:
PARTIAL OXIDATION OF METHANE

Raul E. Valdes-Perez

Computer Science Department and
Center for Light Microscope Imaging and Biotechnology
Carnegie Mellon University
Pittsburgh, PA 15213 - USA

INTRODUCTION

We have applied artificial intelligence principles to the problem of inferring simple reaction mechanisms that are consistent with available experimental evidence. The interactive
computer program MECHEM is based on a combinatorial heuristic search, and has recently
found an interesting new mechanism on a long-studied model reaction in heterogeneous
catalysis. Here, we show that MECHEM is applicable to methane and alkane conversion
chemistry. First, we give a cursory description of the program, and then describe the results
of its application to illustrative cases of the partial oxidation of methane.

OVERVIEW OF MECHEM

MECHEM is a 5000-line Lisp program for elucidation of reaction mechanisms[1,2,3,4,5]
that has been developed over the last six years, and which represents an application of artificial intelligence principles.[6,7] So far, we have focused MECHEM on catalytic industrial
chemistry as a primary application because catalytic pathways are cyclic and complicated,
but the molecules involved are typically small. Hence only one large combinatorial problem
has to be solved, that of multi-step pathways, not that of large-molecule bond breaking and
formation.

The genesis of the program was this question: what is the "space" of possible reaction
mechanisms, and how can an algorithm generate them in order of simplicity? This question
asks not how mechanisms can be predicted from starting materials, nor how a target molecule
can be synthesized. Rather the issue is: what are the simplest mechanisms consistent with
experimental evidence? The different question can lead to very different chemical inference.

A difficult obstacle is that, in practice, the experimental evidence does not include
all reaction intermediates or even products, hence they must be conjectured. Some familiar
recourses are to predict these unseen species using ab initio quantum chemistry or to use
knowledge-based methods. Instead, MECHEM's approach is to invent wild-card species
literally in the form of variables. The program's combinatorial search over the space of
mechanisms draws freely on these variables to generate abstract reaction mechanisms that

$$
\begin{array}{rrcl}
1. & CH_4 + M & \rightarrow & \mathcal{U} + \mathcal{V} \\
2. & \mathcal{U} & \rightarrow & M + {:}CH_3 \\
3. & O_2 + \mathcal{U} & \rightarrow & \mathcal{W} \\
4. & 2\mathcal{V} & \rightarrow & H_2 \\
5. & 2({:}CH_3) & \rightarrow & CH_3CH_3 \\
6. & \mathcal{V} + \mathcal{W} & \rightarrow & H_2O + \mathcal{X} \\
7. & {:}CH_3 + \mathcal{W} & \rightarrow & CH_3OH + \mathcal{X} \\
8. & \mathcal{X} & \rightarrow & M + CH_2O \\
9. & \mathcal{W} & \rightarrow & \mathcal{V} + \mathcal{Y} \\
10. & CH_2O + \mathcal{Y} & \rightarrow & \mathcal{W} + \mathcal{Z} \\
11. & \mathcal{Y} + \mathcal{Z} & \rightarrow & \mathcal{W} + CO \\
12. & CO + \mathcal{Y} & \rightarrow & \mathcal{X} + CO_2 \\
\end{array}
$$

Figure 1: An Abstract Mechanism that Includes Wild-Card Species

contain a mix of starting materials, variables, and experimentally observed intermediates and products. Figure 1 illustrates such an abstract mechanism for a representative mechanism for the partial oxidation of methane (M is a catalyst reaction site, not a variable). The abstract mechanism contains six variables $\mathcal{U}$, $\mathcal{V}$, $\mathcal{W}$, $\mathcal{X}$, $\mathcal{Y}$, and $\mathcal{Z}$; that is, six unseen species are being conjectured.

From these abstract mechanisms, MECHEM arrives at specific mechanisms by using simple chemical laws: steps are balanced, and most elementary steps (except exchange reactions, which can be allowed by setting a parameter) involve three or fewer topological alterations to molecular structure, that is, bond formations or cleavages. We have designed algorithms using these laws to infer molecular formulas and structures for the species wild cards in an abstract mechanism.[2,4] Inferring molecular structure was the more challenging algorithm design, since the formulas can be found by adapting the standard Gaussian elimination method.

In principle, MECHEM's generator will not miss any plausible reaction mechanisms, since it carries out an exhaustive search. (In practice, heavy combinatorial demands may sometimes require simplifying assumptions that will rule out some mechanisms.) Therefore, the space of potential mechanisms is huge, due to the combinatorics implied by forming all possible reactants, products, and steps. The solution depends on combining various factors: one searches for the simplest mechanisms by introducing only one extra wild card at a time, and only after the current set proves inadequate[2]; partial mechanisms are pruned by using the available experimental evidence and laws[5]; a divide-and-conquer heuristic subdivides the original problem into more manageable pieces[5]; mechanisms are generated in canonical order to prevent redundant mechanisms from appearing in multiple branches of the search.[1] By means of such devices, the potential for combinatorial explosion is controlled enough to make a practical program that is effective on a significant class of chemistry. This approach is especially suitable for the high cyclicity and complicated stoichiometries of catalytic pathways.

A recent result achieved with MECHEM was to find a plausible alternative mechanism for the catalytic hydrogenolysis of ethane.[8] This reaction has been studied for over twenty years, and several mechanisms have been proposed. The program found a seemingly novel mechanism of comparable simplicity which involves hydrogen transfer between adsorbed species, rather than the successive hydrogenation which the accepted mechanisms

$$
\begin{array}{rrcl}
1. & CH_4 + M & \rightarrow & \langle MCH_3 \rangle + \langle H \rangle \\
2. & \langle MCH_3 \rangle & \rightarrow & M + :CH_3 \\
3. & O_2 + \langle MCH_3 \rangle & \rightarrow & \langle MCH_2OOH \rangle \\
4. & 2\langle H \rangle & \rightarrow & H_2 \\
5. & 2(:CH_3) & \rightarrow & CH_3CH_3 \\
6. & \langle H \rangle + \langle MCH_2OOH \rangle & \rightarrow & H_2O + \langle MCH_2O \rangle \\
7. & :CH_3 + \langle MCH_2OOH \rangle & \rightarrow & CH_3OH + \langle MCH_2O \rangle \\
8. & \langle MCH_2O \rangle & \rightarrow & M + CH_2O \\
9. & \langle MCH_2OOH \rangle & \rightarrow & \langle H \rangle + \langle MCH_2OO \rangle \\
10. & CH_2O + \langle MCH_2OO \rangle & \rightarrow & \langle MCH_2OOH \rangle + \langle HCO \rangle \\
11. & \langle MCH_2OO \rangle + \langle HCO \rangle & \rightarrow & \langle MCH_2OOH \rangle + CO \\
12. & CO + \langle MCH_2OO \rangle & \rightarrow & \langle MCH_2O \rangle + CO_2 \\
\end{array}
$$

Figure 2: An Illustrative Mechanism for Partial Oxidation of Methane

have featured.

APPLICATION TO THE PARTIAL OXIDATION OF METHANE

We input to the program the following constraints, which are intended as plausible examples relevant to methane partial oxidation. These constraints do not necessarily correspond to any specific experimental setup.

- The starting materials are CH4, O_2, and M (a catalyst reaction site).

- The observed products (including intermediates) are methyl radicals, ethane, methanol, water, CO, CO_2, formaldehyde, and H_2.

- The catalyst reaction sites (M) form one bond.

- The oxidation numbers cannot decrease across a step.

- Every step involves the catalyst, except possibly radical recombination.

- Every step is either uni-molecular, or involves a gas-phase reactant.

- Every step involves at most 3 topological changes to bonding (cleavage or formation).

- Every species contains at most 2 carbon atoms and at most 2 oxygen atoms.

MECHEM found the two simplest mechanisms shown in Figures 2 and 3. The notation '$\langle\rangle$' indicates a conjectured species, i.e., not given as input to the program. An overall stoichiometry that is consistent with both mechanisms is:

$$
5(CH_4) + 2(O_2) \longrightarrow CH_3CH_3 + H_2O + CH_3OH + CH_2O + 3(H_2) + CO
$$

1.	$CH_4 + M$	$\rightarrow$	$:CH_3 + \langle MH \rangle$
2.	$\langle MH \rangle$	$\rightarrow$	$M + \langle H \rangle$
3.	$O_2 + \langle MH \rangle$	$\rightarrow$	$\langle MOOH \rangle$
4.	$2(:CH_3)$	$\rightarrow$	CH_3CH_3
5.	$2\langle H \rangle$	$\rightarrow$	H_2
6.	$:CH_3 + \langle MOOH \rangle$	$\rightarrow$	$CH_3OH + \langle MO \rangle$
7.	$\langle H \rangle + \langle MOOH \rangle$	$\rightarrow$	$H_2O + \langle MO \rangle$
8.	$:CH_3 + \langle MO \rangle$	$\rightarrow$	$\langle H \rangle + \langle MOCH_2 \rangle$
9.	$\langle MOCH_2 \rangle$	$\rightarrow$	$\langle MH \rangle + \langle HCO \rangle$
10.	$\langle MOCH_2 \rangle$	$\rightarrow$	$CH_2O + M$
11.	$\langle HCO \rangle + M$	$\rightarrow$	$\langle MH \rangle + CO$
12.	$\langle MO \rangle + CO$	$\rightarrow$	$CO_2 + M$

Figure 3: A Second Illustrative Mechanism for Partial Oxidation of Methane

The above constraints are only meant to be illustrative, not necessarily correct for any specific catalyst. However, these results show that MECHEM can serve as a useful tool for the experimentalist who studies the mechanisms of methane and alkane conversion, and who may wish to formulate different constraints to match his own experimental case.

CONCLUSION

Due to recent advances, MECHEM is able to handle moderately-complex reactions of industrial or scientific interest. We have targeted catalytic reactions as a first practical area of application, partly because of their industrial importance, but mainly because their complex nature makes the introduction of computer-aided methods especially opportune.

We believe that MECHEM can serve as a complement to other computer-aided methods such as microkinetic analysis.[9] The latter emphasizes the stages subsequent to developing a serviceable reaction mechanism, whereas MECHEM emphasizes the earlier stages. Computer tools can enhance the model building process across all stages.

Our future plans include applying the program to new reactions posed by experimental problems. We also intend to parallelize the program over distributed workstations in order to speed it up by more than an order of magnitude and enlarge further its scope of application. However, the program outputs reported here were obtained on an ordinary workstation, so even the current version displays a promising capability.

Acknowledgment   This work was supported in part by a Science and Technology Center grant from the National Science Foundation, #MCB-8920118.

REFERENCES

1. R.E. Valdes-Perez, A canonical representation of multistep reactions, *J. Chem. Info. Comp. Sci.* 31:554 (1991).

2. R.E. Valdes-Perez, Algorithm to generate reaction pathways for computer-assisted elucidation, *J. Comp. Chem.* 13:1079 (1992).

3. R.E. Valdes-Perez, Algorithm to test the structural plausibility of a proposed elementary reaction, *J. Comp. Chem.* 14:1454 (1993).

4. R.E. Valdes-Perez, Algorithm to infer the structures of molecular formulas within a reaction pathway, *J.Comp.Chem.* 15:1266 (1994).

5. R.E. Valdes-Perez, Heuristics for systematic elucidation of reaction pathways, *J. Chem. Info. Comp. Sci.* 34:976 (1994).

6. R.E. Valdes-Perez, Conjecturing hidden entities via simplicity and conservation laws: Machine discovery in chemistry, *Artif. Intell.* 65:247 (1994).

7. R.E. Valdes-Perez, Machine discovery in chemistry: New results, *Artif. Intell.* 74:191 (1995).

8. R.E. Valdes-Perez, Human/computer interactive elucidation of reaction mechanisms: application to catalyzed hydrogenolysis of ethane, *Catal. Letters*, 28:79 (1994).

9. Dumesic, J.A. et al., 1993, *The Microkinetics of Heterogeneous Catalysis*, American Chemistry Society, Washington, D.C.

METHANE TO OXYGENATES AND CHEMICALS

SELECTIVE PHOTO-OXIDATION OF METHANE
TO FORMALDEHYDE USING SUPPORTED
GROUP VB AND VIB METAL OXIDE CATALYSTS

Kenji Wada [1] and Yoshihisa Watanabe[1]

Division of Energy and Hydrocarbon Chemistry,
Graduate School of Engineering, Kyoto University,
Sakyo-ku, Kyoto 606-01, JAPAN

INTRODUCTION

Catalytic direct conversion of light alkanes, especially methane, has attracted much attention for many years,[1] although the functionalization of unactivated C-H bond in light alkanes is quite difficult. Since Kaliaguine et al.[2] have investigated the reactions of methane and ethane with hole centers generated by γ-irradiation of metal complexes dispersed on silica, the photo-oxidation of methane using highly dispersed supported metal oxides or solid oxide semiconductors has been studied extensively. However, most of these studies, which were usually performed at ambient temperature, resulted in deep oxidation, and only few attempts were reported to be successful.[3-7]. Ward et al.[3] reported the selective photo-oxidation of methane over the silica-supported MoO_3 or $CuMoO_4$ catalysts to give a small amount of methanol without deep oxidation, but the formation of formaldehyde was not confirmed. Photo-oxidation of propane on the supported molybdenum oxide catalysts has been extensively investigated by Kaliaguine et al .[8]

In the present study, the selective photo-oxidation of methane and ethane into corresponding aldehydes on supported group VB and VIB metal oxide catalysts at around 500 K was performed using two kinds of newly designed fluidized bed flow type reaction apparatus.[9,10] Effects of the reaction temperature, effects of preparation of the catalysts, and the possible reaction paths are briefly discussed.

EXPERIMENTS

Typical silica supported catalysts were prepared by the usual pore volume impregnation method using ammonium salts and silica gel (Alfa, surface area; 300 $m^2.g^{-1}$) followed by calcination in air at 823 K for 2 h. Several silica-supported molybdenum catalysts were prepared by the equilibrium adsorption method using an aqueous solution of ammonium heptamolybdate. The pH of the solution was adjusted using dilute HNO_3 and NH_4OH. One of MoO_3/SiO_2 was prepared by the UV irradiation for 0.5 h at room temperature on $Mo(CO)_6$-adsorbed silica gel. All these catalysts mounted in the reaction apparatus were again subjected to *in situ* calcination in a stream of air at 823 K for 2 h. Amounts of vanadium or molybdenum species in the calcined catalysts were measured by the atomic absorption analysis.

An upstream fluidized bed flow reactor made by quartz glass with an window for UV irradiation (10 mm x 20 mm) was used for the reactions. The reactor was equipped with a quartz glass tube (35 mm o.d.) to cover the whole reactor, and the catalyst bed temperature was maintained uniform by passing heated air around the reactor (Apparatus A).[7] For several runs, a reaction gas preheater (maximum temperature; 823 K) was equipped just below the irradiation window (Apparatus B).[9] UV irradiation was performed using three types of high pressure mercury vapor lamps designated as A, B and C. Relative intensity of the light flux was estimated to be 1 : 1 : 3 for lamps A : B : C. Chemical actinometry using ferric ammonium oxalate revealed that number of photon irradiated into the catalyst bed using lamp C was 3.3×10^{-7} einstein.s^{-1}. Detailed reaction conditions were noted in the footnote of each tables and captions to figures. Liquid products were collected in a cold trap, and gaseous products were collected in gas sampling bags. The products were analyzed using gas chromatography.

UV-visible diffuse reflectance spectra were measured with a Shimadzu model MPS-2000 multipurpose spectrophotometer without any moisture contact. XPS spectra were acquired using a Perkin-Elmer 5500 MT system with Mg $K\alpha_{1,2}$-ray radiation (15 kV, 400 W).

RESULTS

The preliminary studies on the activities of various silica-supported group VB and VIB metal oxide catalysts for the photo-oxidation of methane and ethane at 493 K have been performed. They revealed the highest activity of the vanadium oxide catalyst to selectively give formaldehyde from methane. A molybdenum oxide catalyst showed the second highest activity. Other group VB and VIB metal oxide catalysts did not show any significant activities.

Supported Vanadium Oxide Catalysts

Table 1 shows the activities of various supported vanadium oxide catalysts. The highest formaldehyde yield of 34 μmol.h^{-1} was achieved with 0.5 mol% loaded catalysts. Combined yields of carbon oxides were 11 μmol.h^{-1}, indicating the high selectivity for formaldehyde of *ca.* 70%. Effects of the reaction temperature are shown in Figure 1 (a). The yields of the products were very small below 450 K, and the remarkably improved yields were observed at around 500 K. Further increases in the temperature decreased not only the yield of formaldehyde, but also those of carbon oxides.

Photo-oxidation of ethane using the vanadium oxide-loaded silica catalyst gave acetaldehyde together with formaldehyde and small amounts of ethanol, carbon monoxide and carbon dioxide. With 2.0 mol%-loaded catalyst, the selectivity for oxygen-containing chemicals was over 80% on carbon base. Figure 1 (b) shows the differences in effects of the temperature between the photo-oxidation of methane and ethane. The yields of all the

Table 1. Photo-oxidation of methane and ethane at 493 K using silica-supported vanadium oxide catalysts prepared by the impregnation method.[1]

Run	Loading level (mol% as V)	Alkane	Yield (μmol·h^{-1}) CH$_3$CHO	HCHO	CO	CO$_2$
1	0.5	CH$_4$	-	34	6	5
2	1.0	CH$_4$	-	31	8	6
3	2.0[2]	CH$_4$	-	30	6	tr.
4	6.6	CH$_4$	-	5	tr.	tr.
5	0.5	C$_2$H$_6$	39	4	0	5
6	1.0	C$_2$H$_6$	30	7	tr.	7
7	2.0[2]	C$_2$H$_6$	41	9	tr.	7
8	6.6	C$_2$H$_6$	14	4	tr.	2

[1]Using apparatus A and lamp C, amount of catalyst 0.025g, W/F 0.71 g.h.mol^{-1}, alkane feed rate 7.5 mmol·h^{-1}, alkane : O$_2$: He = 3 : 1 : 10.
[2]Catalyst prepared using excess amount of water.

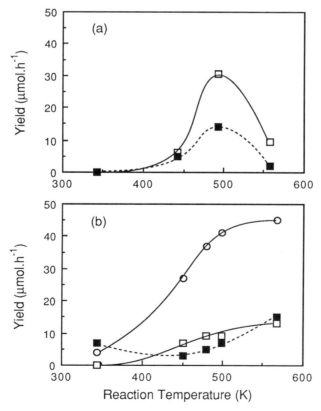

Figure 1. Effects of the reaction temperature on photo-oxidation of (a) methane using the V$_2$O$_5$ (1.0 mol%)/SiO$_2$ catalyst and (b) ethane using V$_2$O$_5$ (2.0 mol%) /SiO$_2$[a]. The yields of (o) formaldehyde, (O) acetaldehyde, and (■) carbon monoxide + carbon dioxide. Amount of catalyst 0.025 g, W/F 0.71 g.h.mol^{-1}, alkane : O$_2$: He = 3 : 1 : 10. [a]Prepared using excess amount of water.

Table 2. Photo-oxidation of methane and ethane using silica-supported
molybdenum oxide catalysts prepared by the impregnation method.[1]

Run	Loading	Apparatus	Lamp	Temp.	Alkane	Yield (μmol·h^{-1})			
	(mol%)			(K)		CH_3CHO	HCHO	CO	CO_2
1	1.0	A	C	493	CH_4	-	13	3	2
2[2]	1.8	B	A	463	CH_4	-	5	0	0
3[2]	1.8	B	C	503	CH_4	-	19	6	0
4[3]	7.9	B	A	463	CH_4	-	3	0	0
5[3,4]	7.9	B	A	463	CH_4	-	tr.	0	0
6	1.0	A	C	493	C_2H_6	42	22	2	7
7	1.0	B	B	463	C_2H_6	14	4	0	tr.
8	1.0	B	C	493	C_2H_6	38	20	tr.	7
9[5]	1.0	B	C	493	C_2H_6	60	22	tr.	5

[1] Amount of catalyst 0.025g, W/F 0.71 g.h.mol^{-1}, alkane feed rate 7.5 mmol·h^{-1},
alkane : O_2 : He = 3 : 1 : 10.
[2] W/F 0.62 g.h.mol^{-1}, CH_4 : O_2 : He = 6 : 2 : 25.
[3] Amount of catalyst 0.050g, W/F 1.2 g.h.mol^{-1}, CH_4 : O_2 : He = 2 : 3 : 17.
[4] With Pyrex filter.
[5] Ethane feed rate 19 mmol·h^{-1}, oxygen feed rate 2.5 mmol·h^{-1}.

products gradually increased with increasing temperature. It should be noted that the yield of
acetaldehyde further increased at higher temperature than 500 K.

Supported Molybdenum Oxide Catalysts

The results of the photo-oxidation of methane using supported molybdenum oxide
catalysts during the first 2 h are shown in Table 2. For apparatus B, blank experiments in
the absence of a catalyst ruled out the possibility of dark oxidation on the wall of the
preheating zone at the temperature below 823 K. The reaction gave formaldehyde and a
small amount of methanol. Formic acid was not detected by GC analysis. The examination
on effects of the reaction temperature using the MoO_3 (1.8 mol%)/SiO_2 catalyst showed that
the yields of the products markedly increased above 440 K, and then leveled off at around
450 ~ 500 K. The reaction below 440 K gave very small amounts of the products. When
incident light flux into the catalyst bed was tripled by the use of lamp C, the higher yield of
formaldehyde (19 μmol·h^{-1}, 2.5 mol·(Mo-mol·h)$^{-1}$) was obtained at 503 K together with 6
μmol·h^{-1} of carbon monoxide, indicating that the yields of the products increases almost in
proportion to the intensity of light flux under the present conditions. The result of the
reaction with UV irradiation through a Pyrex filter (passes >300 nm) suggests that a
wavelength shorter than 300 nm was required to activate the MoO_3/SiO_2 catalyst.
Unsupported MoO_3 and Na_2MoO_4 did not show any catalytic activities.

Figure 2 shows changes in the yield of formaldehyde with the catalysts prepared by
several method with variations in loading levels. The highest yield of formaldehyde was
obtained with the 1.0 mol% loaded catalyst prepared by the impregnation method. Further
increase in the loading of MoO_3 decreased the yield of formaldehyde. Among the catalysts
prepared by the equilibrium adsorption method, the catalyst prepared at pH 1 was an
excellent catalysts to form formaldehyde of 4.4 μmol·h^{-1}. The catalyst prepared using
$Mo(CO)_6$ showed relatively low activity.

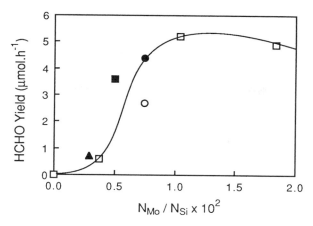

Figure 2. Photo-oxidation of methane at 463 K using the MoO_3/SiO_2 catalysts of various loading levels prepared by the (□) impregnation method, (●) equilibrium adsorption method at pH 1, (■) pH 5.5, and (▲) pH 11, (○) using $Mo(CO)_6$. Apparatus B, amount of catalyst 0.025 g, W/F 0.62 g.h.mol^{-1}, $CH_4 : O_2 : He = 6 : 2 : 25$.

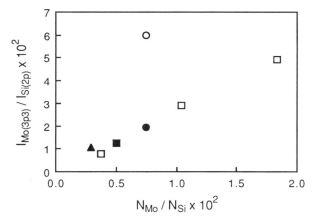

Figure 3. The XPS intensity ratio $I_{Mo(3p3)} / I_{Si(2p)}$ as a function of overall atomic ratio of Mo and Si of the MoO_3/SiO_2 catalysts prepared by the (□) impregnation method, (●) equilibrium adsorption method at pH 1, (■) pH 5.5, and (▲) pH 11, (○) using $Mo(CO)_6$.

The UV diffuse reflectance study revealed that the MoO_3/SiO_2 catalyst showed the absorption bands with maxima around 240 ~ 250 nm and 280 nm, which can be assigned to $3t_2 \rightarrow 2t_1$ and to $3t_2 \rightarrow t_1$ transition, respectively, in the tetrahedrally coordinated molybdenum species.[11]

In Figure 3, the ratio of the XPS intensities of the Mo(3p3) and Si(2s) of the MoO_3/SiO_2 catalysts are plotted as a function of the overall atomic ratio. The catalysts of high activity seem to show rather low observability of Mo, indicating the presence of multilayer species on their surfaces. On the other hand, the catalyst prepared using $Mo(CO)_6$ showed higher observability of Mo than other catalysts, suggesting the presence of more dispersed species. But its activity was low (see above). These results suggest that the less dispersed molybdenum species, but not large crystallite of MoO_3, are responsible for the high activity.[8]

The silica supported MoO_3 (1.0 mol%) catalyst was very excellent for the photo-oxidation of ethane to produce acetaldehyde and formaldehyde. The catalyst bed temperature around 450 ~ 500 K was preferable for the selective formation of acetaldehyde. Further increases in the reaction temperature increased the yields of formaldehyde at the cost of that of acetaldehyde. Optimization of the reaction conditions using lamp C drew the highest yields of acetaldehyde (60 $\mu mol \cdot h^{-1}$) and formaldehyde (22 $\mu mol \cdot h^{-1}$). The combined yield of carbon oxides was 5 $\mu mol \cdot h^{-1}$, implying the high selectivity over 96% for oxygen-containing compounds.

DISCUSSION

The results in our present study, despite of the higher reaction temperature, did agree with the mechanism proposed by Kazansky et al.[2] The first step of the probable mechanism involves the activation of surface species by UV irradiation, resulting in the formation of the charge transfer complex, $V^{4+}O^-$ or $Mo^{5+}O^-$. An alkane is adsorbed on the photoactivated O^- species and subjected to the activation of its C-H bond. Kazansky et al. have suggested that a hydrogen is abstracted by a bridged lattice oxygen.[2] Abstraction of hydrogen by other electron rich oxygen species may be possible. Then molecular oxygen is adsorbed onto a metal ion that can provide an electron to molecular oxygen, resulting in the formation of aldehydes together with the re-oxidation of the surface species. The reaction using the MoO_3 (1.0 mol%) catalyst in the absence of oxygen yielded 2.7 $\mu mol \cdot h^{-1}$ of acetaldehyde during first 1 h, and the yield then gradually decreased. This can be explained by the participation of the lattice oxygen species in place of molecular oxygen. The reduced surface would not be re-oxidized, resulting in the rapid deactivation.

The elevated reaction temperature would promote desorption of the oxygenated products and water from the catalyst. In addition, enhanced rate of re-oxidation of metal oxide at higher temperature would increase the turnover frequency.

CONCLUSION

Silica-supported vanadium and molybdenum oxide catalysts showed excellent activities for the selective photo-oxidation of light alkanes to oxygen-containing chemicals. Above all, the 2.0 mol% vanadium oxide loaded catalyst gave formaldehyde from methane in the highest yield of 34 $\mu mol \cdot h^{-1}$ together with a small amount of carbon oxides at 493 K. Both UV irradiation and the reaction temperature as high as 500 K were indispensable. The supported molybdenum oxide catalysts were especially effective for the photo-oxidation of ethane to give aldehydes in the high yields (~80 $\mu mol \cdot h^{-1}$). The surface terminal tetrahedrally coordinated groups in relatively gathered metal oxide species seems to be responsible for the high activities.

REFERENCES

1. P. Pitchai, and K. Klier, *Catal. Rev.-Sci. Eng.* 28:13 (1986).
2. S.L. Kaliaguine, B.N. Shelimov, and V.B. Kazansky, *J. Catal.* 55:384 (1978).
3. M.D. Ward, J.F. Brazdil, S.P. Mehandru, and A.B. Anderson, *J. Phys. Chem.* 91:6515 (1987).

4. W. Hill, B.N. Shelimov, and V.B. Kazansky, *J. Chem. Soc., Faraday Trans. 1* 83:2381 (1987).
5. N. Djeghri, M. Formenti, F. Juillet, and S.J, Techner, *Faraday Discuss. Chem. Soc.* 58:1850 (1984).
6. K. Wada, K. Yoshida, Y. Watanabe, and T. Suzuki, *J. Chem. Soc., Chem. Commun.* 752 (1991).
7. K. Wada, K. Yoshida, T. Takatani, and Y. Watanabe, *Appl. Catal. A: General* 99:21 (1993).
8. K. Marcinkowska, S.L. Kaliaguine, and P.C. Roberge, *J. Catal.* 90:49 (1984).
9. T. Suzuki, K. Wada, M. Shima, and Y. Watanabe, *J. Chem. Soc., Chem. Commun.* 1059 (1990).
10. K. Wada, K. Yoshida, Y. Watanabe, and T. Suzuki, *Appl. Catal.* 74:L1 (1991).
11. M. Wolfsberg, and L. Helmholz, *J. Chem. Phys.* 20:837 (1952).

A STUDY OF THE IRON/SODALITE CATALYST FOR THE PARTIAL OXIDATION OF METHANE TO METHANOL

Steven Betteridge[1], C. Richard A. Catlow[2*], Robin W. Grimes[2], Justin S. J. Hargreaves[3], Graham J. Hutchings[3*], Richard W. Joyner[3*], Christopher J. Kiely[4], Darren F. Lee[3], Quentin A. Pankhurst[5], Stuart H. Taylor[3] and David Whittle[4]

[1]Department of Physics, University of Liverpool, PO Box 147, Liverpool L69 3BX

[2]Davy Faraday Laboratory, Royal Institution of Great Britain, 21 Albemarle Street, London W1X 4BS

[3]Leverhulme Centre for Innovative Catalysis, Department of Chemistry, University of Liverpool, PO Box 147, Liverpool L69 3BX

[4]Department of Materials Science and Engineering, University of Liverpool, PO Box 147, Liverpool L69 3BX

[5]Department of Physics and Astronomy, University College London, Gower Street, London WC1E 3BT

ABSTRACT

The synthesis, characterisation and reactivity of an iron/sodalite catalyst for the partial oxidation of methane to methanol are described and discussed. In a limited temperature range, ca. 410°C, and at 34 bar total pressure, the selectivity to methanol is found to be similar to that observed for the empty silica glass reactor. Before reaction, powder X-ray diffraction and ^{57}Fe Mössbauer spectroscopy confirm the presence of Fe(III) in the sodalite framework. After reaction the structure is more complex, as revealed by ^{57}Fe Mössbauer spectroscopy and transmission electron microscopy, and both Fe(II) and Fe(III) are present. In particular, small particles of Fe_2O_3, < 1μm in size are also present in the used catalyst. The nature of the active site is discussed in terms of an Fe(II)/Fe(III) couple involving both framework and non-framework iron.

INTRODUCTION

In recent years considerable research interest has been shown in the direct catalytic conversion of methane to methanol. For an industrial process to be established based on a single step process it is necessary that a catalyst is identified that possesses both high activity and selectivity. To date only limited success has been recorded in these studies and generally high selectivities to methanol or formaldehyde are recorded at very low conversion [1]. To some extent this is due to the complex nature of the reactions involved in the oxidation reaction which are dominated by gas phase processes [2]. Recent work by Lyons, Durante and co-workers of the Sun Oil Co. has, however, suggested that iron/sodalite may be a promising catalyst for this reaction [3], although a complex reactor design appears to be required for high methanol selectivity [4]. This material may represent a synergy between iron oxide and sodalite, since individually these materials possess little activity for the synthesis of methanol and hence is worthy of a detailed study. In this paper we present a study of the catalytic performance and characterisation of an iron/sodalite catalyst and discuss the origin of the catalytic activity for this material.

EXPERIMENTAL

The iron sodalite used in this study was prepared from Laboratory Grade reagents, using essentially the method described by Szostak and Thomas [5] and the Si/Fe synthesis ratio was 15. This material was pelleted and sieved without the addition of a binder, to yield particles of diameter 0.6-1.0 mm. The material was then calcined according to the Patent recipe by heating at 540C in flowing argon (38 ml min^{-1}) for one hour, then for two hours at 540C in flowing air (47 ml min^{-1}) and finally at 550C in static air for one hour [3]. The resultant material was light grey in colour.

Methane oxidation experiments were performed using a silica lined stainless steel microreactor of 8 mm internal diameter. Temperature measurement used a thermocouple placed in the reactor furnace. The reaction pressure was maintained using a Tescom back pressure regulator and all reactor lines downstream of the catalytic bed were trace heated to > 120C. A catalyst charge of *ca.* 0.5 g was held in place between silica wool plugs. Gases were supplied by BOC PLC with the following purity: methane, 99.0% minimum; oxygen, 99.5%, and helium, 99.995%. Flowrates were achieved by Brooks 5850 TR mass flow controllers and typical test conditions used methane, oxygen and helium flows respectively of 46, 4 and 11 ml min^{-1}. Product analysis was performed on-line using a Phillips PU4500 gas chromatograph, equipped with thermal conductivity and flame ionisation detectors. Analysis of methane, carbon dioxide, ethane, ethene and methanol was performed via elution from a *Porapak QS* column (1/4" diameter, 80-100 mesh) using a temperature programme regime, while carbon monoxide and oxygen were resolved by a 13X molecular sieve column (1/4" diameter, 65-80 mesh). Methane conversion was calculated on the basis of products formed and oxygen conversion determined by difference. Catalyst performance stabilised after less than one hour and conversion and stability thereafter were stable for at least 3 hours: reported data are the means of three analyses. Methane conversion values are accurate to ± 0.5% and selectivities to ± 3%.

Powder X-ray diffraction patterns were recorded by a Philips 1050W diffractometer modified by Hiltonbrooks Ltd using CuKα radiation and operating at 40 keV and 20 mA. Data were collected in the range $5 < 2\theta < 75^{o}$, with a step size of 0.05^{o}.

^{57}Fe Mössbauer spectra were recorded both at room temperature and at 4.2 K, using a conventional constant acceleration spectrometer and a ^{57}CoRh source. A double-ramp velocity drive signal was used, and the recorded spectra were numerically folded to eliminate any baseline curvature. Calibration was performed with respect to α-iron at room temperature.

Samples for electron microscope examination were prepared by dispersing the powder onto a lacey carbon film supported on a copper mesh grid. Specimens were examined in a JEOL 2000 EX high resolution electron microscope operating at 200 keV.

RESULTS AND DISCUSSION

Characterisation of Unused Iron/Sodalite

Iron/sodalite was characterised by X-ray powder diffraction and was found to be consistent with that previously recorded by Durante *et al* [3]. This powder X-ray diffraction pattern expected for an iron/sodalite, in which iron atoms substitute for silicon atoms in the sodalite framework, has been calculated using **BIOSYM** software. The simulated pattern agrees very well with the experimental results when two iron atoms per unit cell are present (Table 1).

The unused iron sodalite was further characterised by ^{57}Fe Mössbauer spectroscopy at both room temperature and at 4.2K (Figure 1a). At room temperature a doublet was observed with a quadrupole splitting of $\sim$0.94 mm s^{-1} with an isomer shift of $\delta \sim 0.34$ mm s^{-1}.

Table 1. Experimental and calculated X-ray powder diffraction patterns for sodalite catalysts: introduction of iron increases the sodalite lattice constant by *ca.* 2%

Sodalite/2θ°	Iron Sodalite/2θ°	
Experimental	Experimental	Calculated
14.38	14.19	14.03
20.38	20.11	20.00
25.01	24.64	24.58
29.00	28.53	28.33
32.45	31.89	31.67
37.57	35.12	35.42

At 4.2 K these values change slightly to $\Delta \sim 0.96$ mm s^{-1} and $\delta \sim 0.44$ mm s^{-1}. These findings are typical of Fe(III) and the large quadrupole splitting indicating a comparatively disordered atomic environment. This would be expected for random framework substitution of iron in the sodalite lattice. Hence, the characterisation of the unused iron/sodalite indicates that the material consists of a silicon sodalite lattice with random substitution of silica by iron with two iron atoms per unit cell.

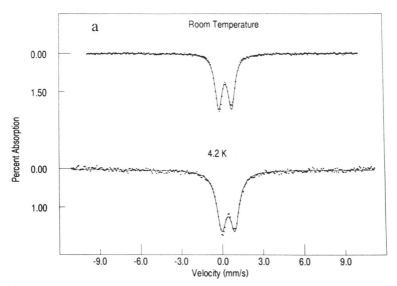

Figure 1a) ^{57}Fe Mössbauer spectra of the iron/sodalite catalyst before reaction : +, experimental data; solid line calculated parameters given in text.

Methane Conversion Over Iron/Sodalite

The performance of the iron/sodalite catalyst for the oxidation of methane has been assessed for a reaction condition that is within the preferred patent range. The results are shown in Table 2 and are contrasted with those of an empty reactor. In our studies the empty reactor gave the better performance which is in agreement with previous studies of the homogeneous methane oxidation which has been found to be reasonably selective to methanol and formaldehyde [6,7]. There is a small region, *ca.* 410°C, when the selectivity to methanol of the iron sodalite catalyst is comparable to that of the empty reactor. At this temperature the conversion of methane is considerably lower in the presence of iron/sodalite when compared to its absence. Further studies which involved the co-feeding of water vapour to the post reactor volume to quench the reaction products did not significantly improve the selectivity to oxygenated products. Although we have been unable to reproduce the high selectivities to methanol (> 60%) indicated by Durante *et al* [3,4] in the original studies, the selectivity of iron/sodalite is interesting, since most other materials would produce only carbon oxides under these conditions. It should be noted that Lyons *et al* [4] obtained high selectivities to methanol only with 'complex' reactor configurations.

Characterisation of Iron/Sodalite After Use

After use as a catalyst the iron sodalite was again examined by X-ray powder diffraction and no iron oxides were determined and little evidence could be found for framework collapse. Examination of the used iron/sodalite by ^{57}Fe Mössbauer spectroscopy (Figure 1b) indicated significant differences for the spectrum of the unused material. At both room temperature and 4.2 K two contributions are evident. At room temperature there are two overlapping doublets. One of these exhibits the parameters typical of Fe(III), namely $\delta \simeq 0.42$ mm s^{-1} and $\Delta \simeq 0.64$ mm s^{-1}. The second doublet has parameters typical of Fe(II), *ie.* $\delta \simeq 0.96$ mm s^{-1} and $\Delta \simeq 1.95$ mm s^{-1}. On cooling to 4.2 K the Fe(II) doublet remains ($\delta \simeq 0.96$ mm s^{-1} and $\Delta \simeq 0.34$ mm s^{-1}) but now the Fe(III) doublet is split into a magnetic sextet with $\delta \simeq 0.45$ mm s^{-1} and a magnetic hyperfine field of $B_{hf} \simeq 50.5$T. Such a hyperfine field is typical of iron oxide, for example hematite (α-Fe$_2$O$_3$) or maghemite (γ-Fe$_2$O$_3$).

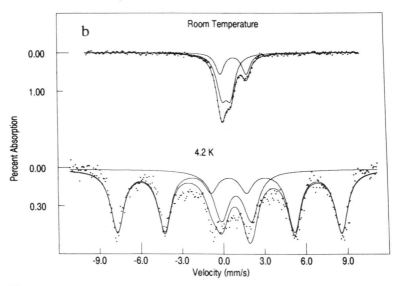

Figure 1 b) ^{57}Fe Mössbauer spectra of the iron/sodalite catalyst after reaction : +, experimental data; solid line calculated parameters given in text indicating presence of Fe(II) and Fe(III).

The observation of a doublet at room temperature, well below the ordering temperature for iron oxides, is indicative of superparamagnetism. In addition, the time averaged field at the iron nuclei is zero, despite the presence of long term magnetic order. These observations are indicative that Fe(III) is present in very small particles, probably < 1μm. The presence of Fe(II) is not surprising given the reducing nature of the reaction conditions, however, the Mössbauer spectra provide no further information on the nature of the Fe(II) species. The presence of both Fe(II) and Fe(III) was further confirmed in the used catalyst by X-ray photoelectron spectroscopy.

A preliminary comparison of the unused and used iron/sodalite catalyst has been attempted using transmission electron microscopy. Profile images, such as those shown in figure 2, seem to suggest that the used catalyst commonly exhibits surface particles (1-3 nm in size), whereas these features are absent in the unused catalyst. The composition of these surface particles is currently being investigated using STEM microanalysis. The origin of the diffuse patches of dark contrast seen within both the iron/sodalite specimens is as yet unclear. We should point out, however, that the TEM results obtained to date should be treated with some caution, since the iron/sodalite sample are highly susceptible to electron beam damage.

Nature of the Active Site of Iron/Sodalite

In their original studies Lyons *et al* [8] have suggested that the iron/sodalite catalyst operates via a mechanism similar to that observed of the non-haem containing enzyme methane mono-oxygenase [9] which involves Fe(IV) and Fe(V) species. These high oxidation states can be stabilised within the enzyme tertiary structure. Calculations to investigate the stability of Fe(IV) within the sodalite framework have been conducted by examining the replacement of framework Fe(III) by Fe(IV) and this has been found to incur a large energy penalty (1190 kJ mol^{-1}). Since it is unlikely that this energy penalty can be compensated by reduction of the remaining iron/sodalite, it must be concluded that Fe(IV) is an unlikely species in this catalyst system. The calculations indicate, as expected, that the Fe(II)/Fe(III) couple is energetically favourable. This finding is consistent with the observation of both Fe(II) and Fe(III) in the used catalyst.

Molecular dynamics calculations performed using the BIOSYM software indicated that methane is unable to diffuse into the sodalite framework. Hence any catalytic activity can only be associated with sites on the surface of the crystallites.

The results of the characterisation studies indicate that the unused iron/sodalite comprises mainly iron substituted into the sodalite lattice. Calcination at 550°C in air, according to the patent procedure [3], does not significantly affect this structure, however, after use in a methane/air atmosphere at temperatures up to 500°C, the structure is significantly changed. Although some iron is still retained in the sodalite framework, the Mössbauer study indicates that a significant amount of Fe_2O_3 is present as small (< 1μm) particles which are possibly occluded within the crystallites. In addition, preliminary transmission electron microscopy studies seem to indicate that some discrete particles (possibly iron oxide) may be present on the surface of the crystallites. In a separate experiment we examined the stability of methanol over Fe_2O_3 over a range of temperatures (Figure 3). It is clear that at temperatures > 400°C almost all of the methanol is converted to CO_2 in the presence of oxygen. Since the oxygen conversion in the methane conversion experiments (Table 2) is < 100% at ≤ 450°C, it is clear that the presence of any significant amounts of surface Fe_2O_3 must be deleterious to catalytic performance. It is most likely that the catalytically active site is an Fe(II)/Fe(III) couple involving both framework substituted iron in the sodalite lattice and non framework iron. It is possible that the catalytic performance of the iron/sodalite could be improved if preparations could be found that did not lead to the formation of separate, discrete Fe_2O_3 particles on exposure to reaction conditions. However, it is also possible that Fe_2O_3 may possess improved catalytic performance when present either as thin films supported on iron/sodalite or as very small particles partially occluded in iron/sodalite crystallites and further studies are required to elucidate this aspect.

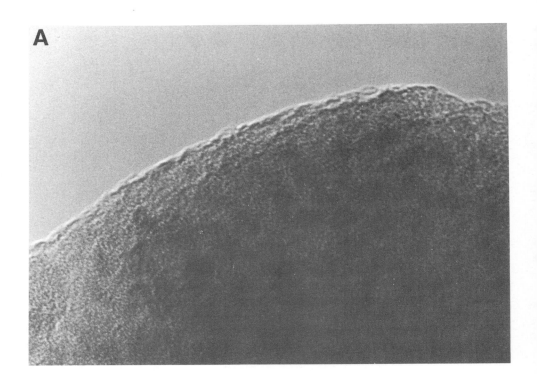

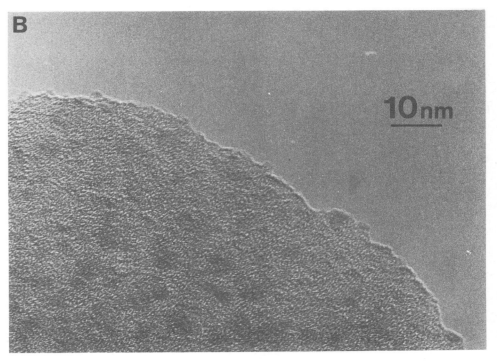

Figure 2 Transmission electron micrographs of the surface iron/sodalite catalyst: A before use, B after use.

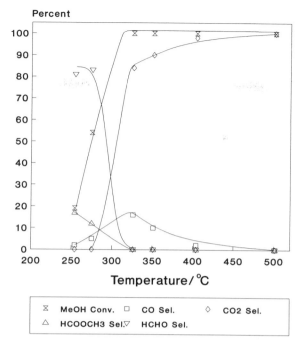

Figure 3 Methanol oxidation over Fe_2O_3, $CH_3OH:O_2:He = 1:4:12$, GHSV = 12000 h^{-1}.

ACKNOWLEDGEMENTS

We thank the Gas Research Institute of Chicago, USA, the SERC and the Industrial Affiliates of the Leverhulme Centre for Innovative Catalysis for financial support.

REFERENCES

1. N.D. Parkyns, C.I. Warburton, J.D. Wilson, *Catal. Today*, 18: 385 (1993).
2. H. Zanthoff and M. Baerns, *Ind. Eng. Chem. Res.*, 29:2 (1990).
3. V.A. Durante, D.W. Walker, S.M. Gussow and J.E. Lyons, US Patent 4918249.
4. V.A. Durante, D.W. Walker, W.H. Seitzer and J.E. Lyons, Preprints, Symposium 3b, Methane Activation, Conversion and Utilisation, Pacifichem '89, (1989).
5. R. Szostak and T.L. Thomas, *J. Chem. Soc., Chem. Commun.*, 113 (1986).
6. N.R. Hunter, H.D. Gesser, L.A. Mostan, P.S. Yarlagadda and D.P.C. Fung, *Appl. Catal.*, 57:45 (1990).
7. T.R. Baldwin, R. Burch, E.D. Squire and S.C. Tsang, *Appl. Catal.*, 74:137 (1991).
8. J.E. Lyons, P.E. Ellis and V.A. Durante, *Stud. Surf. Sci. Catal.*, 67:99 (1990).

PARTIAL OXIDATION OF METHANE TO FORMALDEHYDE OVER VANADIA CATALYSTS: REACTION MECHANISM

B. Kartheuser[1], B.K. Hodnett*[1], H. Zanthoff[2] and M. Baerns[2]

[1]Dept of Chemical and Environmental Sciences,University of Limerick, Limerick, Ireland
[2]Ruhr-University Bochum,Lehrstuhl fur Technische Chemie, PO Box 10 21 48, Bochum 1, Germany

ABSTRACT

A series of vanadium oxide catalysts with loadings in the range 0.4-7.8 wt% vanadium supported on silica was prepared and tested for the selective oxidation of methane into formaldehyde in a fixed bed micro reactor operating at ambient pressure. The catalyst structure was characterised by X-Ray diffraction, X-Ray photoelectron spectroscopy, temperature programmed reduction and by the NO-NH3 rectangular pulse technique. Mechanistic studies were carried out using Transient Analysis of Products (T.A.P.) The best catalysts for formaldehyde production featured 1 wt% vanadium on the silica support. For this catalyst the surface area of the supported phase vanadium oxide was also approximately 1 % of the total B.E.T. surface area. A relationship was observed between the dispersion of the supported vanadium oxide phase and the yield of formaldehyde measured when the methane conversion was held constant. Satisfactory yields of formaldehyde were observed only when the dispersion of the supported phase vanadium oxide, exceeded 25%. Experiments with the T.A.P. technique shows that methane interacts very weakly and oxygen very strongly with the catalyst surface and it is concluded that the initial activation of methane involves a short-lived adsorbed oxygen species. Methyl radicals formed in the first step subsequently extract lattice oxygen to yield formaldehyde.

INTRODUCTION

A variety of transition metal oxide catalysts have been found to be active for partial oxidation of methane to formaldehyde and methanol using N_2O or molecular oxygen as oxidant[1-30]. Formaldehyde is the principal oxygenated product (apart from CO_x) in reactors operating at atmospheric pressure. When gas phase oxidation of methane is carried out at high pressure, or when large amounts of steam are added to the feed significant yields of methanol are obtained [17,31,32]. Generally, optimal formation of the oxygenates occurs at lower temperatures (623-773 K) than those which favour coupling to C_2^+ products (1073 K) [27].

Molybdenum or vanadium oxides supported on silica have been most often employed as catalysts, but a variety of other materials have also been tested [19-21,26,29]. Several studies have outlined the very strong relationship between selectivity to formaldehyde and conversion of methane for this system[22,25]. For the vanadium oxide catalysts in particular it is possible to arrive at 100% selectivity to formaldehyde at very low conversions, but this value falls below 10% even before the conversion reaches 1%. Essentially the selective oxidation product (HCHO) is so unstable by comparison with the reactant (CH_4) that kinetic isolation of the selective oxidation product is very difficult. Large amounts of CO are usually observed in these systems which arise from the decomposition of formaldehyde.

A key feature of the vanadium and molybdenum oxide catalysts used to date for this reaction is that good selectivity to formaldehyde is observed only for very low loadings of the supported phase, usually below 2 wt% [22,25], . In this paper the results of a study of the structure of vanadium oxide catalysts supported on silica is presented and correlations sought between the catalytic performance in methane oxidation to formaldehyde and various structural features [33].

A limited number of mechanistic studies have been carried for this system. In the most significant of these Banares et al. [34] have studied methane oxidation to formaldehyde on MoO_3/SiO_2 catalyst using $^{18}O_2$ as oxidant. They concluded that the selective oxidation of methane proceeded by a Mars-van Krevelen mechanism in which the CH_4 molecule was oxidised by the lattice oxygen of molybdenum oxide while the oxygen consumed was restored by dioxygen from the gas phase. When a mixture of CH_4 and $^{18}O_2$ was contacted with MoO_3/SiO_2 the product HCHO featured almost exclusively ^{16}O, evidently from the lattice. Some ambiguity remained regarding the nature of the species, whether adsorbed or lattice oxygen, which affected the primary activation of the methane and at what stage of the mechanism oxygen was inserted into the hydrocarbon.

Here $^{18}O_2$ as oxidant was used to elucidate the role of the different oxygen species involved in the partial oxidation of methane over silica supported vanadium oxide catalyst using the T.A.P. technique [35].

EXPERIMENTAL

Preparation Of Impregnated Catalysts.

The support used for the preparation of the catalysts was Cab-o-Sil M5, a fumed silica supplied by the Cabot Corporation. The support was made into a paste with water, dried at 523 K and sieved to yield particles in the range of 100-500 μm. The vanadium oxide catalysts were prepared by impregnation of Cab-o-Sil with an aqueous solution of ammonium metavanadate (NH_4VO_3), stirred at 353 K for 5 hours, then dried under vacuum. The vanadium loadings were in a range 0.5 to 10 wt% of vanadium. The catalysts are referred to below by their intended metal loading (wt%) : 0.5 Vcab signifies a Cab-o-sil support treated to give a loading of 0.5 wt% of vanadium metal. Analysed vanadium loadings are also given below.

Catalyst Testing

Testing was carried out in a standard flow quartz micro reactor. Unless otherwise stated the test conditions were Temp = 823 K, p_{CH4} = 81 kPa, p_{air} = 20 kPa, W/F = 0.24 g s ml^{-1}.

Characterisation

Characterisation was by X-Ray diffraction, X-Ray photoelectron spectroscopy, nitrogen adsorption, atomic absorption spectroscopy, temperature programmed reduction and by the NO-NH3 rectangular pulse technique (NARP). The latter was used to measure the dispersion of the vanadium oxide phase on the silica support. The rectangular pulse apparatus used was a modified version of that used by Miyamoto et al. [36]. This technique is based on the reactivity of a mixture of NO and NH3 towards surface V=O species according to the equation:

$$V=O_{(surf)} + NH_3 + NO \quad \text{--------}> V\text{-}OH + N_2 + H_2O$$

In practice the He flow (60 ml/min) over the catalyst was replaced by a rectangular pulse made up of an identical mass flow of a mixture NH3 and NO in He. The NO and NH3 were stored in two different cylinders and mixed together prior to contacting the catalyst. (total flow 60 ml/min : 30ml of 4% NH3/He + 30 ml of 0.2% NO/He , Total pressure 101 kPa, p_{NH3} = 2 kPa, p_{NO} = 0.1 kPa) [33]. Dispersion was calculated by the following equation

$$D(\%) = L/V_2O_5 * 100$$

where dispersion, D(%), is defined as the ratio of the number of moles of surface V=O species, L (measured from the amount of N2 evolved), divided by the number of moles of V_2O_5 on the catalyst.

Temperature programmed reduction (TPR) with 5% H_2 in nitrogen as reducing agent at 20 ml min^{-1} over 100 mg of catalyst was performed by ramping the catalyst temperature at 10 K min^{-1} between 300 and 900K. The NARP and TPR analyses were integrated with the catalyst testing apparatus, so that catalytic work could be interrupted and the working state of the catalyst assessed without exposure to the ambient atmosphere.

Temporal Analysis of Products (T.A.P.)

One hundred mg of the catalyst (1 wt% vanadium supported on silica (Cab-o-sil), refereed to below as 1V/cab), was loaded into the T.A.P. reactor. A mixture of methane and /or oxygen and argon (9:1:1) (pulse size = 3.6 * 10^{15} molecules) was passed through the reactor in the temperature range 300-900K. These conditions correspond to the Knudsen diffusion regime so that gas phase collisions were largely eliminated.

RESULTS AND DISCUSSION

Characterisation

Table 1 presents the measured vanadium loadings and B.E.T. surface areas. There was some lessening of surface area as the vanadium loading was increased, but there was no significant change in porosity (average pore size was 300 A). All the samples were x-Ray amorphous, except 10Vcab, which exhibited very weak lines due to V_2O_5.

Table 1. Measured vanadium loadings, B.E.T. surface areas, concentration of surface V=O species (L) and dispersion (D%) for vanadium oxide catalysts supported on silica.

Sample	V (%wt) measured	S_{BET} m^2/g	L $\mu mol/g$	Disp%
Cab-o-Sil	-	218	-	-
0.5Vcab	0.41	194	11.2	27.8
1Vcab	0.84	189	22.2	26.9
2Vcab	1.83	188	47.0	26.1
4Vcab	3.85	172	91.7	24.2
6Vcab	5.38	159	104.0	19.7
10Vcab	7.67	154	88.8	11.8

Table 1 also shows the concentration of V=O species on the catalyst, L, and the measured dispersion from the NARP technique. The dispersion of V_2O_5 was highest at low loadings and decreased with increasing vanadium content. Further work with this technique indicates that it does not detect isolated vanadyl species on the silica surface 33 , so that the dispersions presented in Table 1 are in all probability smaller than the real values. In fact, the technique appears to measure that fraction of the supported vanadium oxide composed of three or more layers of oxide material. However, the trend observed in consistent with the literature and are supported by the XPS and TPR data in Figures 1-3.

Figure 1 shows the V/Si XPS ratio as function of the V/Si bulk (from chemical analysis) ratio . The measures binding energy for the $V_{2p3/2}$ peak was close to 517 eV for all samples measured and a similar value was recorded for bulk V_2O_5. Figure 1 indicates that at low loadings the vanadium oxide covered the support fairly well (V/Si$_{XPS}$ = V/Si$_{bulk}$). At higher loading (4, 6 and 10Vcab) the V/Si$_{XPS}$ values are smaller than the V/Si$_{bulk}$ values pointing to the formation of larger V_2O_5 particles. This is consistent with the XRD data and with similar observations obtained by other workers for vanadium oxide supported on TiO_2, Al_2O_3 and SiO_2 [9,10, 19, 37-40] . There was no significant difference between the fresh and used catalysts.

The TPR profiles for the fresh and tested (after testing at 823 K) vanadium based catalyst are shown in Figures 2 and 3. For vanadium loadings below 2 wt% the peak maximum in TPR shifted by as much as 80 K to higher temperatures following testing. Up to 2 wt % V, a single sharp peak appears at 813-823 K, which moves to higher temperatures after catalytic work. At a loading equal to 3.8 wt % V a shoulder appears at a higher temperature (890K), which becomes more pronounced and eventually forms a second peak at 7.8 wt % V (Fig 3).

This feature in the TPR profiles was also observed by Roozeboom et al. [41] and Bond et al. [42] . They attributed the single peak to dispersed vanadia monolayer species ; the shoulder and the second peak were attributed to some surface phases and crystalline V_2O_5. Our results are similar to those of Roozeboom et al. [41] . The only difference is the higher peak temperature. This variation can be explained by the difference in the experimental conditions used by Rooseboom et al. (66% H_2) and those used in our conditions (5% H_2).

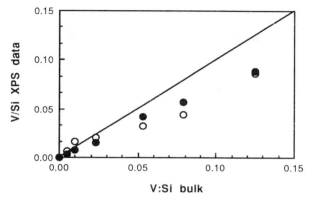

Figure 1. V/Si$_{XPS}$ ratios measured by XPS versus V/Si$_{bulk}$ from chemical analysis (○fresh, ●tested)

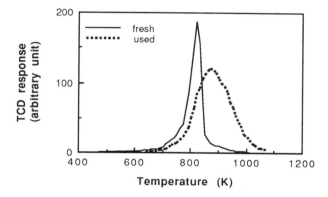

Figure 2. Temperature programmed reduction under H_2 of 1Vcab, fresh and after testing, 100 mg, H_2/N_2 5%, 20 ml/min, heating rate 10 K/min

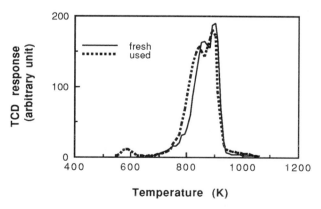

Figure 3. Temperature programmed reduction under H_2 of 10Vcab, fresh and after testing, 100 mg, H_2/N_2 5%, 20 ml/min, heating rate 10 K min^{-1}.

The TPR results tie in nicely with the NARP results (Table 1) and together are in excellent agreement with Bond's observations [42], that a second peak appears in the TPR profile of these catalysts when the number of layers of vanadium oxide per particle is greater than four.

It is now well establish by Raman studies [41, 43, 44] that the vanadium oxide on silica, at low loadings will be present as isolated tetrahedral species as well as octahedral polyvanadates. At low loading, the supported vanadium oxide contains some isolated vanadyl species. When the loading increases, even below the level needed to cover totally the silica surface, particulate V_2O_5 will be present but not as well-defined structures. This is in agreement with our XRD analysis which indicated that no well-defined crystals of V_2O_5 are observed below a loading of about 7.8 wt% vanadium (Table 1). The evolution of the TPR profiles indicated that a poorly defined vanadium oxide phase (which was not detected by XRD) is present as well as isolated vanadyl and polyvanadate species at loadings of 3.85 wt% V and below.

The H_2/V molar ratio in TPR was always between 0.95 and 1.09 for the fresh and used catalysts and did not depend on the vanadium loading. The ratio corresponds to the stoichiometric coefficient for the reduction of V_2O_5 to V_2O_3 by the following reaction :

$$V_2O_5 + 2H_2 \ \text{------->} \ V_2O_3 + 2H_2O$$

and indicates that the vanadium oxide was in an oxidised state throughout catalyst testing in spite of the high $CH_4 : O_2$ ratio used.

Relationship Between Structure Of The Supported Vanadium Oxide And Catalytic Performance.

The catalysts were tested at 823 K in methane rich conditions for a period of 8 to 10 hours. No deactivation was observed during that time. Percentage conversion was approximately constant, in the range 1.3-1.5 % of available methane and greater than 50% of available oxygen, for vanadium loading of 1 wt% and above. Lower loadings were not as effective in converting methane. Figure 4 presents the selectivity to formaldehyde, CO and CO_2 for the full range of catalysts tested.

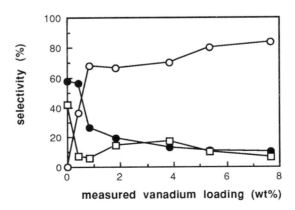

Figure 4 Selectivity to formaldehyde (●), CO (O) and CO_2(□) for the vanadium loadings indicated. Temp = 823 K, p_{CH4} = 81 kPA, p_{air} = 20 kPa, W/F = 0.24 g s ml^{-1}.

Rates of formaldehyde production and percentage yields are presented in Figure 5 for each vanadium loading tested. The highest rate or yield was observed at 1 wt% vanadium on the catalyst.

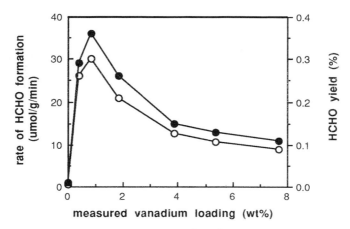

Figure 5. Rate of formaldehyde formation ($\bigcirc$) (mmol g^{-1}min^{-1}) and the percentage formaldehyde yield ($\bullet$) for the vanadium loading indicated. Temp = 823 K, p_{CH4} = 81 kPa, p_{air} = 20 kPa, W/F = 0.24 g s ml^{-1}.

Effectively, the higher selectivity to formaldehyde observed with the bare support and the 0.5Vcab catalyst was primarily a consequence of the lower conversion rather than any intrinsic property of these materials [27, 28, 30, 45, 46, 47]. At vanadium loadings above 1 wt % a significant lowering in selectivity and yield to formaldehyde was observed accompanied by a small increase in the selectivities to CO and CO_2, in spite of the fact that the CH_4 conversion was constant, or falling marginally, as the loading was increased.

Percentage yields of formaldehyde and conversions of methane (always in the range of 1.3 and 1.5 %) are plotted in Figure 6 against the dispersion of the vanadium oxide as measured by the NARP [36] technique. These data clearly show that good yields of formaldehyde, for similar conversions of methane, are achieved only when the dispersion of the vanadium oxide exceeds 25%. Bearing in mind the propensity of this technique to detect only the larger vanadium oxide particles, these data point to a role for larger vanadium oxide particles in the decomposition of formaldehyde.

It is clear from the data presented in Figure 6 that there is a relationship between dispersion, or particle size, of the vanadium oxide and the yield or selectivity to formaldehyde at constant methane conversion. In effect, the selectivity at constant conversion only starts to improve when the NARP measured dispersion exceeds 25%. An alternative explanation that the observed effect was due to some degree of pore constriction by the supported phase is not favoured here because there was no change in pore diameter as the vanadium loading was increased, although there was a reduction in B.E.T. surface area.

It is clear from the data of Table 1 that the surface area of the supported vanadium oxide is just a very small fraction of the B.E.T surface area. This is particularly true of 1Vcab where the vanadium oxide surface area, calculated from the dispersion, is just 1.5% of the B.E.T. surface area. This is an indication that to achieve a reasonable selectivity in formaldehyde the concentration of oxidising sites on the surface of the catalyst must be severely restricted.

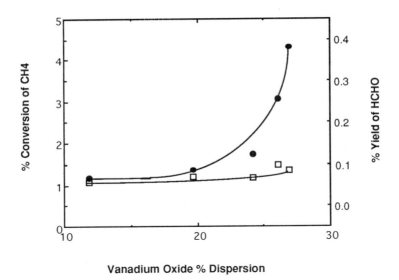

Figure 6. Yield of formaldehyde and (●) percentage methane conversion (□) versus vanadium oxide dispersion (D% = L/V$_2$O$_5$). Temp = 823 K, p$_{CH4}$ = 91 kPa, p$_{air}$ = 20 kPa, W/F = 0.24 g s ml^{-1}

T.A.P. Experiments

Preliminary T.A.P. experiments indicated that methane did not interact in any way with the catalyst surface, but that oxygen was strongly adsorbed. However ^{18}O$_2$ did not exchange with lattice oxygen. On the product side, some carbon monoxide did adsorb strongly on the catalyst surface and could be readily displaced when oxygen was pulsed through the reactor. Carbon dioxide did not interact in any way with the surface of the catalyst, and in fact acted as an inert gas in the system. A further important finding is that methane activation does not occur in this system in the absence of adsorbed oxygen species, whose lifetime was assessed at less than 60 seconds using the T.A.P. apparatus 45 .

Figure 7 shows the T.A.P. curves of the reactants and products when a mixture of CH$_4$ and ^{18}O$_2$ was passed over 1V/cab at 880K. The only reaction products observed were HCH^{16}O, C^{16}O and C^{16}O$_2$. The sequence in which the reactants and products emerged from the T.A.P. reactor was CH$_4$ followed by O$_2$, then the products HCHO, CO and CO$_2$. There is further mechanistic information contained in this sequence, based on the fact that CO$_2$ does not interact strongly with the catalyst. If CO$_2$ were formed directly from methane, corresponding to a parallel route not involving HCHO and CO, we could expect the CO$_2$ peak to emerge from the reactor before the HCHO and CO peaks. Since this was not observed we can argue that CO$_2$ is not produced directly from methane. Methanol was not detected nor were HCH^{18}O, C^{16}O^{18}O or C^{18}O$_2$.

One essential conclusion to be drawn from this work is that there is a very weak interaction between the catalyst surface and methane, whereas interaction with oxygen involved the formation of a surface species of lifetime less than 60 seconds, which were capable of activating the methane, resulting ultimately in the production of formaldehyde. Given the formaldehyde is not produced from methane pulses when 60 seconds is allowed for the flushing of oxygen species from the catalyst surface, it seems reasonable to conclude that, methane is initially activated through interaction with a form of adsorbed oxygen, most likely O_2^- or O^-, with the formation of methyl radicals. Further reaction of this species leading ultimately to the formation of formaldehyde, occurs via a reaction with lattice oxygen leading to its incorporation into the product.

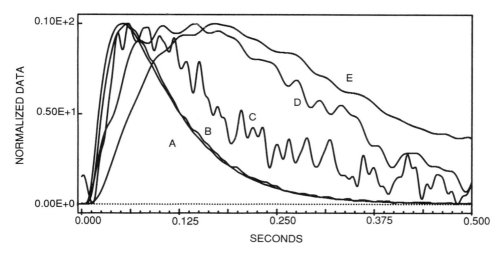

Figure 7. Reactants and products curves. $T = 880$ K, 150 mg 1Vcab, pulse size $3.6*10^{15}$ molecules, average of 100 pulses for the products and 20 for the reactants, mixture $CH_4/^{18}O_2/He$ (9/1/1). $A = CH_4$, $B = {}^{18}O_2$, $C = HCH^{16}O$, $D = C^{16}O$, $E = C^{16}O_2$.

The following sequence can be written to describe the steps involved:

$CH_4 + {}^{18}O(ads)$ ---> $CH_3. + {}^{18}OH$
$CH_3 + {}^{16}O(latt)$ ---> $HCH^{16}O$

CONCLUSIONS

During operation in methane rich conditions, vanadium oxide catalysts supported on silica appear to operate in the 5+ oxidation state.

During catalyst preparation it is important to avoid the creation of large vanadium oxide particles on the support. These species rapidly oxidise formaldehyde and reduce the overall selectivity. The clear implication is that the oxidising system must be extremely diluted on the macro and micro scales.

Primary activation of the methane is by a short lived adsorbed oxygen species.

Methyl radicals formed in the primary activation step extract oxygen from the large lattice oxygen pool with subsequent formation of formaldehyde.

Acknowledgements: This work was funded under the European Community Joule Programme through contract number Contract no: JOUF-0044-c(TT). We also gratefully acknowledge the help of M.Genet of the Universite Catholique de Louvain (Louvain-La-Neuve, Belgium) for the XPS analysis and M.Callant for the BET analysis.

REFERENCES

1. J.Saint-Just, J-M. Basset, J. Bousquet and G.A. Martin, *La Recherche*, 222:730 (1990).
2. Gaz de France, *La Recherche*, 243:230 (1992).
3. P. McGreer and E. Durbin, "Methane: fuel for the future", Plenum Press, New-York and London, (1982).
4. G. Lambert, *La Recherche*, 243:550 (1992) .
5. G.J. Hutchings and R.W. Joyner, *Chem. & Ind.*, p-575 (1991).
6. R. Pitchai and K. Klier, *Catal. Rev.- Sci. Eng.*, 28:13 (1986).
7. C. Shannon, *Chem. and Ind.*, p-154 (1991).
8. J. Haggin, *C&EN*, p-7(1992).
9. J.M. Fox III, *Catal. Rev.- Sci. Eng.*, 35:169 (1993).
10. M.S. Reisch, *C&EN*, p-10(1993).
11. G.J. Hutchings, M.S. Scurrell and J.R. Woodhouse, *Chem. Soc. Rev.*, 18:251 (1989).
12. J.H. Lunsford, *Catal. Today*, 6:235 (1990).
13. Y. Amenomiya, V.I. Birss, A. Goledzinowski, J. Galuska and A.R. Sanger, *Catal. Rev.- Sci. Eng.*, 32:163 (1991).
14. J.C.W. Kuo, C.T. Kresge and R.E. Palermo, *Catal. Today*, 4:463 (1989).
15.. M.J. Brown and N.D. Parkyns, *Catal. Today*, 8:305 (1991).
16. R-S. Liu, M. Iwamoto and J.H. Lunsford, *J.Chem.Soc., Chem. Commun.*, p-78 (1982).
17. H-F. Liu, R.-S. Liu, K.Y. Liew, R.E. Johnson and J.H. Lunsford, *J.Am. Chem Soc*, 106:4117 (1984).
18. M.N. Khan and G.A. Somorjai, *J. Catal*, 91:263 (1985).
19. S. Kasztelan and J.B. Moffat, *J. Chem. Soc., Chem Commun.*, p-1663 (1987).
20. K. Otsuka and M. Hatano, *J. Catal*, 108:252 (1987).
21. Y. Barbaux, A.R. Elamrani, E. Payen, L. Gengembre, J.P. Bonnelle and B. Grzybowska, *Appl.Catal.*, 44:117 (1988)
22. N. Spencer and C.J. Periera, *J. Catal.*, 116:399 (1989).
23. G.N. Kastanas, G.A. Tsigdinos and J. Schwank, *J. Chem. Soc., Chem. Commun.*, p-1298 (1988).
24. T.R. Baldwin, R. Burch, G.D. Squire and S.C. Tsang, *Appl. Catal.*, 74:137 (1991).
25. M.D. Amiridis, J.E. Rekoske, J.A. Dumesic, D. F. Rudd, N.D. Spencer and C.J. Periera, *AIChE Journal*, 1:87 (1991).
26. M. Kennedy, PhD thesis, University of Limerick, Ireland, 1992.
27. M.M. Koranne, J.G. Goodwin, Jr. and G. Marcelin, *J. Phys. Chem.*, 97:673 (1993).
28. Z. Sojika, R.G. Herman and K. Klier, *J. Chem. Soc., Chem. Commun*, p-185 (1991).
29. T. Weng and E.E. Wolf, *Appl. Catal. A : General*, 96:383 (1993).
30. J.C. Mackie, *Catal.Rev.- Sci. Eng.*, 33:169 (1991).
31. P.S.Yarlagadda, L.A. Morton, N.R. Hunter and H.D. Gesser, Ind. Eng. Chem. Res., 27:252 (1988).
32. R.Burch, G.D. Squire and S.C. Tsang, *J. Chem.Soc., Faraday Trans.1*, 85:3561 (1989).
33. B Kartheuser, Ph.D. Thesis, University of Limerick, 1993.
34. M A Banares, I Rodriguez-Ramos, A Guerrero-Ruiz and J L G Fierro, 10th International Congress on Catalysis, Budapest, 1131:B (1992)
35. J.T. Gleaves, J.R. Ebner and T.C. Kuechler, *Catal. Rev.-Sci. Eng.*, 30(1):49 (1988).
36. A. Miyamoto, Y. Yamazaki, M. Inomata and Y. Murakami, *J. Phys. Chem.*, 85:2366 (1981).
37. J. Stringer, *J. Less-Common Metals*, 1:8 (1965).
38. K. Inumaru, T. Okuhara and M. Misono., *J. Phys. Chem.*, 95:4826 (1991).
39 P.Ciambelli, G. Bagnasco, L. Lisi, M. Turco, G. Chiarello, M. Musci, M. Notaro, D. Robba, and P. Ghetti, *Appl. Catal. B : Environmental*, 1:61 (1992).
40. J.C. Bond, J.P. Zurita, and S. Flamerz, *Appl. Catal.* 27:353 (1986), 27, 353.
41. F. Roozeboom, M.C. Mittelmeijer-Hazeleger, J.A. Moulijn, J. Medema,

V.H.J. de Beer and P.J. Gellings, *J. Phys. Chem.*, 84:2783 (1980).

42. G.C. Bond, J. P. Zurita and S. Flamerz, P.J. Gellings, J.G. Van Ommen and B.J. Kip, *Appl. Catal.*, 22:361 (1986).

43. G.T. Went, L.J. Leu and A.T. Bell, *J. Catal.*, 134:479 (1992).

44. B.E. Handy, A. Baiker, M. Schraml-Marth and A. Wokaun, *J. Catal.*, 133:1 (1992).

45. N.D. Spencer and C.J. Pereira, *AIChE Journal*, 33:1808 (1987).

46. N.D. Spencer, *J. Catal.*, 109:187 (1988).

47. B. Kartheuser and B K Hodnett, *J. Chem. Soc., Chem Commun.*, p-1093 (1993)

48. B Kartheuser, B K Hodnett, H Zanthoff and M Baerns, Catal. Letts., 21:209 (1993).

COMPARISON OF THE CONVERSION OF METHANE AND ETHANE ON METAL-OXYGEN CLUSTER COMPOUNDS

S. Hong[1], S. Kasztelan[2], E. Payen[3] and J.B. Moffat[4]

The Guelph-Waterloo Centre for Graduate Work in Chemistry,
Department of Chemistry, University of Waterloo,
Waterloo, Ontario, N2L 3G1, Canada

Present Addresses:
[1]Department of Chemical Engineering, Pusan National University of Technology, 100 Yongdang-dong, Nam-ku, 608-739, Pusan, Korea

[2]Institut Français du Pétrole, Division cinétique et catalyse,
B.P. 311, 92506 Rueil-Malmaison Cedex, France

[3]Laboratoire de Spectroscopie, Infra-Rouge et Raman,
Université des Sciences et Techniques de Lille
59655 Villeneuve d'Ascq, Cedex, France

[4] To Whom Correspondence Should be Addressed

ABSTRACT

Studies of the conversion of methane and ethane on metal-oxygen cluster compounds, in particular silica-supported 12-molybdophosphoric acid ($H_3PMo_{12}O_{40}$, abbreviated to HPMo), show interesting similarities. The partial oxidation products from methane and ethane, formaldehyde and acetaldehyde, respectively, are obtained only in the presence of the catalyst. With both alkanes optimum results are obtained for loadings of approximately 20 wt% of HPMo on the support, corresponding to a molecular surface occupancy of 1000 $Å^2$. The conversions of the alkanes and selectivities to the various products remain constant, for a given reaction temperature, up to a catalyst pretreatment temperature of 500-550°C, a value significantly higher than the usually accepted decomposition temperature of HPMo, thereby suggesting the enhancement of the thermal stability of the active catalyst presumably resulting from a strong interaction between the support and the supported material. With ethane selectivities to acetaldehyde and ethylene of 62 and 30% have been obtained, both of which appear to be primary products. Since the oxidants appear to be primarily functioning as regenerators of the active sites on the catalysts, a mechanism in which the anionic oxygen species act as the active sites is tentatively postulated.

Methane and Alkane Conversion Chemistry
Edited by M. M. Bhasin and D. W. Slocum, Plenum Press, New York, 1995

INTRODUCTION

The conversion of alkanes and, in particular the oxidation of these organic molecules, is a topic of considerable interest to those concerned with heterogeneous catalysis[1]. Although in recent years research efforts have focused on methane less work has been reported on ethane[1-30]. Studies of these alkanes are important not only in providing information on the mechanism of oxidation processes, but also in aiding our understanding of the nature and source of the activity of oxidation catalysts.

Work in this laboratory has been concerned with, among other topics, the surface, bulk and catalytic properties of metal-oxygen cluster compounds (MOCC) (also known as heteropoly oxometalates). Although a wide variety of these materials exists, of particular interest are those with so-called Keggin structure[31]. The MOCC, like the well known faujasitic zeolites, are ionic solids, but, unlike these same zeolites have discrete anions. These are large (approximately 10 Å in diameter) packed structures with a central atom such as phosphorus bonded to four oxygen atoms arranged tetrahedrally[32] (Fig. 1). Twelve octahedra with oxygen atoms at their vertices and a peripheral metal atom such as molybdenum at each of their approximate centres surround and share oxygen atoms with the central tetrahedron as well as each other. Evidently there are three types of oxygen atoms, two bridging and one terminal, the latter of which number twelve in total. The MOCC anions can be charge compensated with a wide variety of cations, although recent evidence suggests that those with oxidation number +2 may be precluded from doing so [33,34].

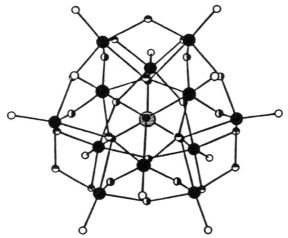

Figure 1. Anion structures in 12-heteropoly acid. Central atom: ◒; peripheral atom: ●; terminal or outer oxygen atom, O_t: ○; inner oxygen atom, O_a: ●; edge-shared, bridging oxygen atom, O_b: ◐; corner-shared, bridging oxygen atom, O_c: ◑.

While earlier photoacoustic FTIR studies in this laboratory provided direct observations of the penetration of polar molecules into the bulk structure of the MOCC, that is, between the cations and anions[35], nonpolar molecules appear to be unable to do so. Consequently, in the earlier work on the oxidation of methane it was found necessary to support the MOCC in order to increase the accessibility of the catalyst[36-38]. Silica was chosen as a suitable support for this purpose although this material is itself catalytically active in oxidation processes[39]. In view of the size of the anions it is not surprising that their stability is less than optimum. Indeed differential thermal analysis shows that the thermal stability is a function of both the composition of the anion and the nature of the cation[40]. However the MOCC supported on silica, for example, appear to have enhanced thermal stability which is of obvious importance for their utilization at the relatively high temperatures normally employed in oxidation processes[41-43].

The present report compares the oxidation of methane and ethane on silica-supported 12-molybdophosphoric acid for various loadings of the catalyst, reaction and pretreatment temperatures and in the presence or absence of the catalyst.

RESULTS

With the HPMo/SiO$_2$ catalyst the products C$_2$H$_4$, CH$_3$CHO, CO, CO$_2$ and water were formed from ethane. In contrast, with unsupported HPMo and with silica alone no acetaldehyde was detected.

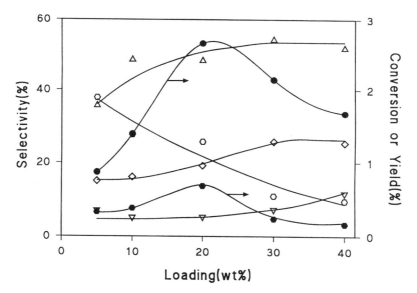

Figure 2. Effect of loading of HPMo on conversion of ethane and selectivity over supported catalyst; T$_R$=540°C, W=0.5 g, F=25 ml/min, C$_2$H$_6$/N$_2$O=4/1, (●) conversion of C$_2$H$_6$, (△) selectivity of C$_2$H$_4$, (○) CH$_3$CHO, (▽) CO$_2$, (◊) CO, (◉) yield of CH$_3$CHO.

At a reaction temperature of 540°C the conversion of ethane and the yield of acetaldehyde attain maximum values at a loading of approximately 20 wt% of HPMo on the silica support (Fig. 2). The rates of production of the various products from the conversion of CH$_4$ also appear to reach maximum values at or near 20 wt% loading (Fig. 3).

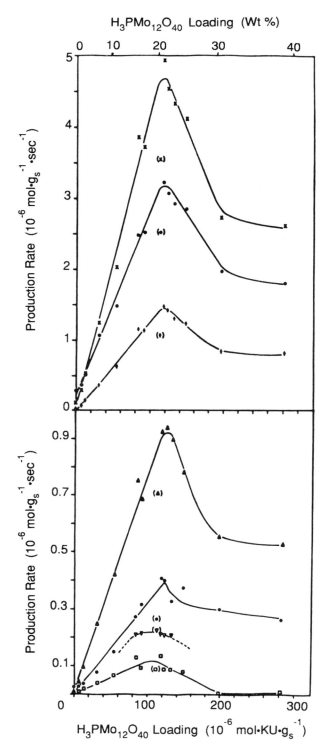

Figure 3. Effect of the HPMo loading of the support on the production rate of the different products of the $CH_4 + N_2O$ reaction at 843 K. Reaction conditions: CH_4 (67%) N_2O (33%), W=0.5 g, F=30 mL/min^{-1}. Symbols: (x) N_2, (+) total carbon detected, (●) H_2O, (∇) CH_3OH, (△)CO, (0) CO_2, (☐) CH_2O.

Not surprisingly the conversion of ethane increases with increasing contact time regardless of the oxidant, N_2O or O_2 (Fig. 4). However with the latter oxidant only small quantities of acetaldehyde are formed at high contact times. With the same catalyst and the latter oxidant no formaldehyde or methanol was formed from methane at the same reaction temperature.

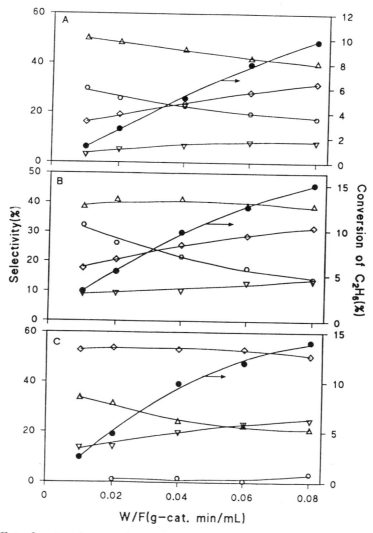

Figure 4. Effect of contact time on ethane conversion and selectivity over 20 wt% HPMo/SiO$_2$: (A) T_R=540°C, C_2H_6/N_2O=4/1; (B) T_R=540°C C_2H_6/N_2O=1/1; (C) T_R=540°C, C_2H_6/O_2=4/1; (●) conversion of C_2H_6, (◊) selectivity of CO, (▽) CO$_2$, (△) C_2H_4, (○) CH$_3$CHO.

With increasing reaction temperature the conversion of ethane and selectivities to carbon monoxide and carbon dioxide increase while that to acetaldehyde decreases (not shown). With oxygen the production of carbon monoxide is favoured at all reaction temperatures. At reaction temperatures less than 540°C no acetaldehyde is formed from oxygen and ethane. In the absence of a catalyst and at 480-510°C ethane is converted entirely to ethylene but the conversions are very low. As the reaction temperature is increased the conversion increases slightly while the selectivity to CO increases significantly.

At a constant reaction temperature and space velocity variation in the partial pressures of ethane and of nitrous oxide permitted orders of reaction to be obtained as 0.8 and 0.6 respectively, to be compared with the corresponding values of 0.5 and 0.5, respectively, for methane and nitrous oxide. The activation energies for ethane and methane conversion were calculated as 20±2 kcal/mole.

Separate experiments in which ethylene and nitrous oxide in various ratios were passed over 20% HPMo/SiO$_2$ at 540°C showed that, with decreasing C$_2$H$_4$/N$_2$O ratios the conversion increased from approximately 2 to 20% while the selectivity to acetaldehyde decreased from 32 to 13%. With oxygen the selectivity to acetaldehyde was very small. The conversions of acetaldehyde with either N$_2$O or O$_2$ were high and the predominant products were CO and CO$_2$. With nitrous oxide ethanol is completely converted, largely to CO, CO$_2$ and ethylene but with a significant selectivity (15%) to acetaldehyde.

With increasing pretreatment temperature the conversion of ethane and the selectivities to the various products remain unchanged up to 500°C (Fig. 5). For higher temperatures the conversion decreases while the selectivities remain approximately constant up to 550°C. With further increases in temperature above 500°C the selectivity to acetaldehyde decreases while those to CO and CO$_2$ increase. The selectivity to ethylene is constant over the entire range of temperatures studied.

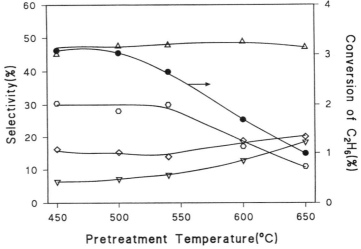

Figure 5. Effect of pretreatment temperature on ethane conversion and selectivity over 20 wt% HPMo/SiO$_2$:T$_R$=540°C, W=0.5 g, F=25 ml/min, C$_2$H$_6$/N$_2$O=4/1, (●) conversion of C$_2$H$_6$, (◊) selectivity of CO, (∇) CO$_2$, (Δ) C$_2$H$_4$, (○) CH$_3$CHO.

With the partial oxidation of methane similar observations of the effect of pretreatment temperature have been made. The conversion of methane is constant up to approximately 500°C while the selectivity to formaldehyde is virtually constant to 550°C. However with methane the selectivities to CO and CO$_2$ remain relatively unchanged up to 600°C (Fig. 6).

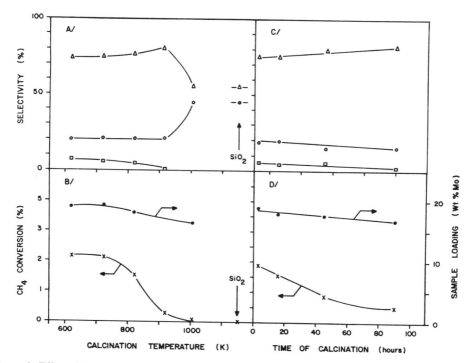

Figure 6. Effect of the temperature of calcination during 16 hours (left) and of the time of calcination at 823 K under air (right) on the CH_4 conversion, selectivity and Mo loading of the 23 HPMo catalyst. Reaction conditions: CH_4 (67%) N_2O (33%) T_R=843K, W=0.5 g, F=30 mL/min^{-1}. Symbols: (Δ) CO, (0) CO_2, (□) CH_2O, (x) CH_4 conversion (●) Mo loading.

From reaction data obtained under a variety of reaction conditions (Fig. 7) the selectivities of C_2H_4 and acetaldehyde decrease with increasing conversion while those to CO and CO_2 increase. Evidently C_2H_4 and acetaldehyde are primary, while CO and CO_2 are secondary products.

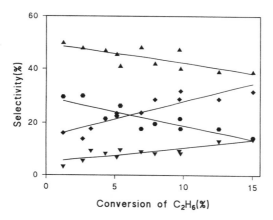

Figure 7. Selectivities to product versus conversion of ethane at different reaction conditions over 20 wt% HPMo/SiO$_2$: T_R=450-570°C, W/F=0.01-0.08 g-cat./ml/min, C_2H_6/N_2O=1/1-4/1, (♦) CO, (▼) CO_2, (▲) C_2H_4 (○) CH_3CHO.

Insight into the contrasting behaviour of the two oxidants with ethane can be found from Fig. 8. In the absence of oxidant the selectivity to CO, CO₂ and acetaldehyde decreases to nearly zero over two hours. Subsequent treatment of the catalyst with nitrous oxide followed by reaction again in the absence of oxidant produces results similar to those obtained in the first stage. Following a treatment with oxygen, however, the results are significantly different from those found in the first and second stages. Importantly no acetaldehyde is observed at any time during the final stage of the process. The conversion of nitrous oxide is significantly higher than that of oxygen in either the presence or absence of the catalyst.

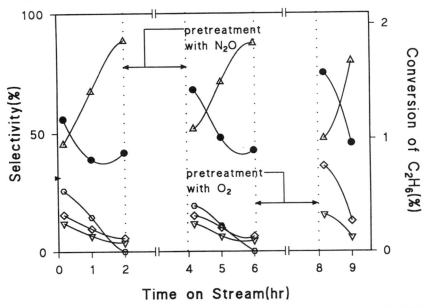

Figure 8. Effect of time on stream on ethane conversion and selectivity over 20 wt% HPMo/SiO₂: T_R=510°C, W=1.00 g, F=5 ml/min, C₂H₆ only and no oxidants, (●) conversion of C₂H₆, (◊) selectivity of CO, (▽) CO₂, (△) C₂H₄, (◯) CH₃CHO.

DISCUSSION

The conversions of both methane and ethane depend on the presence of the HPMo/SiO₂ catalyst for the generation of partial oxidation products. However there is evidently an optimum loading of 20% on the silica support with conversion of both alkanes. Since this loading corresponds to a coverage of approximately 1000Å²/anion considerably larger than the 100Å² estimated as the anionic cross-sectional area, the anions are apparently isolated on the silica support. The reduction in conversion at loadings higher than 20% may be attributed to the inaccessibility of the active sites on the anions to methane and ethane. Although polar molecules are capable of penetrating into the bulk structure of the metal-oxygen cluster compounds, nonpolar molecules cannot do so[35].

With both methane and ethane selectivities and conversions show changes occurring after pretreatment of HPMo/SiO₂ at 500-550°C. Temperature-programmed desorption studies of unsupported HPMo show high temperature peaks beginning at approximately 350°C which are believed to be due to water produced from the extraction of anionic oxygen atoms by the acidic protons[45]. Although some decomposition of the anion undoubtedly occurs photoacoustic FTIR spectra show that the peaks characteristic of the Keggin structure remain in the spectra[35]. Unsupported HPMo has been found to be degraded at temperatures higher than 350°C[31], presumably though decomposition to the constituent oxides

$$H_3PMo_{12}O_{40} \rightarrow 1/2\ P_2O_5 + 12\ MoO_3 + 3/2\ H_2O$$

X-ray photoelectron spectroscopic studies for various loadings of HPMo on SiO_2 indicate a dispersion of a small number of heteropoly anions prior to the appearance of larger particles at approximately 23 wt% HPMo[42]. Evidence for the presence of the heteropoly anion on the catalysts pretreated to temperatures as high as 550°C has been obtained from ^{31}P NMR spectra. All the recorded spectra show one relatively broad peak. The NMR peak positions and full widths at half maximum (FWHM) for a 23 wt% HPMo/SiO_2 catalyst remain at -8 ppm and approximately 1.2 Hz, respectively, after pretreatments for 16 h, at various temperatures up to 550°C, while the peak positions and FWHM change markedly when the temperature is raised to 600°C[42].

Experiments in which the protons of HPMo/SiO_2 have been replaced by other cations such as cesium suggest that the protons play a direct role in the partial oxidation process. However in view of the aforementioned temperature-programmed desorption results it appears to be more reasonable to suggest that the protons are employed, at least initially, to generate oxygen vacancies which are apparently a requirement for the catalysis of the oxidation process.

$$PMo_{12}O_{40}{}^{-3} + 2H^+ \rightarrow PMo_{12}O_{39}\square^{-2} + H_2O$$

As shown, HPMo/SiO_2 is capable of functioning catalytically in the conversion of ethane in the absence of an oxidant, although the process, as expected, is shortlived. These observations suggest that, at least under the present conditions, the oxidant is primarily serving as a regenerator of active sites, possibly through the oxidation of molybdenum atoms occupying the peripheral metal positions in the anion.

$$Mo^{5+} + N_2O \rightarrow Mo^{6+} - O^- + N_2$$

Since free radical mechanisms are frequently invoked for oxidation processes, the heteropoly anion may generate the ethyl radical

$$O^- + C_2H_6 \rightarrow C_2H_5\cdot + OH^-$$

which radical may remain on the surface to ethylate the heteropoly anion.

$$C_2H_5\cdot + O^- \rightarrow C_2H_5 - O^-$$

Since acetaldehyde appears to be the analogue in ethane oxidation chemistry of formaldehyde in methane oxidation chemistry, it is tempting to suggest that a conclusion drawn for the formation of the latter species is applicable to the former. Mauti and Mims have provided convincing evidence with MoO_3/SiO_2 as catalyst in the conversion of methane that formaldehyde is formed from methyl radicals which remain on the surface of the catalyst[46].

Although there is no evidence for the formation of ethanol from ethane with the present catalyst, small quantities of methanol have been found from methane. It has been proposed for silica-supported alkali molybdate catalysts that the aforementioned ethoxy group may react with water to generate ethanol and an hydroxide ion[24]. However, in the present work, in view of the high selectivity to acetaldehyde from ethylene and HPMo/SiO_2, the possibility that ethanol may not be a precursor to ethylene cannot be discounted.

ACKNOWLEDGEMENT

The financial support of the Natural Sciences and Engineering Research Council of Canada is gratefully acknowledged.

REFERENCES

1. See, for example, V.D. Sokolovskii and E.A. Mamedov, *Catal. Today.* 14:331 (1992).

2. M.B. Ward, M.J. Lin and J.H. Lunsford, *J. Catal.* 50:306 (1977).

3. E.M. Thorsteinson, T.P. Wilson, F.G. Young and P.H. Kasai, *J. Catal.* 52:116 (1978).

4. T.Y. Yang and J.H. Lunsford, *J. Catal.* 63:505 (1980).

5. M. Iwamoto, T. Taga and S. Kagawa, *Chem. Lett.* 1496 (1982).

6. L. Mendelovici and J.H. Lunsford, *J. Catal.* 94:37 (1985).

7. A. Argent and P.G. Harrison, *J. Chem. Soc., Chem. Commun.* 1058 (1986).

8. E. Morales and J.H. Lunsford, *J. Catal.* 118:255 (1989).

9. S.T. Oyama and G.H. Somorjai, *J. Phys. Chem.* 94:5022 (1990).

10. S.T. Oyama and G.A. Somorjai, *J. Phys. Chem.* 94:5029 (1990).

11. A. Erdöhelyi and F. Solymosi, *J. Catal.* 123:31 (1990).

12. Y. Murakami, K. Otsuka, Y. Wada and A. Morikawa, *Bull. Chem. Soc. Jpn.* 63:340 (1990).

13. R. Burch and S.C. Tsang, *Appl. Catal.* 65:259 (1990).

14. J.C. McCarty, A.B. McEwan and M.A. Quinlan, *in* "New Developments in Selective Oxidation" (G. Centi and F. Trifiro, Ed.), Studies in Surface Science and Catalysis, Vol. 55, p. 405, Elsevier, Amsterdam, 1990.

15. R. Burch and R. Swarnakar, *Appl. Catal.* 75:321 (1991).

16. A. Erdöhelyi and F. Solymosi, *J. Catal.* 129:497 (1991).

17. S.J. Conway and J.H. Lunsford, *J. Catal.* 131:513 (1991).

18. A. Erdöhelyi, F. Mate and F. Solymosi, *Catal. Letters*, 8:229 (1991).

19. E.M. Kennedy and N.W. Cant, *Appl. Catal.* 75:321 (1991).

20. S.J. Conway, D.J. Wang, and J.H. Lunsford, *Appl. Catal.* 79:L1 (1991).

21. R.G. Gaziev, A.D. Berman, and O.V. Krylov, *Kin. and Catal.* 32:587 (1991).

22. J. Lebars, J.C. Vedrine, A. Auroux, B. Pommier, and G.M. Pajonk, *J. Phys. Chem.* 96:2217 (1992).

23. S. Trautmann and M. Baerns, *J. Catal.* 136:613 (1992).

24. A. Erdöhelyi, F. Mate, and F. Solymosi, *J. Catal.* 135:563 (1992).

25. K. Wada, Y. Watanabe, F. Saitoh, and T. Suzuki, *Appl. Catal.* 88:23 (1992).

26. J. LeBars, J.C. Vedrine, A. Anroux, S. Trautmann, and M. Baerns, *Appl. Catal.* A88:179 (1992).

27. E.M. Kennedy and N.W. Cant, *Appl. Catal.* A87:171 (1992).

28. M. Merzouki, B. Taouk, B. Monceaux, E. Bordes, and Courtine, P., *in* "New Developments in Selective Oxidation by Heterogeneous Catalysis" (P. Ruiz and B. Delmon, Eds.), Stud. Surf. Sci. Catal., Vol. 72, p. 165, Elsevier, Amsterdam, 1992.

29. K. Otsuka, Y. Uragami, and M. Hatano, *Catal. Today* 13:667 (1992).

30. R. Burch and E.M. Crabb, *Appl. Catal.* 97:49 (1993).

31. M.T. Pope, "Heteropoly and Isopoly Oxometalates", Springer-Verlag, Berlin, 1983.

32. G.M. Brown, M.R. Noe-Spirlet, W.R. Busing, and H.A. Levy, *Acta Crystallogr.* B33:1038 (1977).

33. G.B. McGarvey and J.B. Moffat, *Catal. Letters* 16:173 (1992).

34. G.B. McGarvey, N.J. Taylor, and J.B. Moffat, *J. Mol. Catal.* 80:59 (1993).

35. J.G. Highfield and J.B. Moffat, *J. Catal.* 88:177 (1984); ibid 89:185 (1984).

36. S. Kasztelan and J.B. Moffat, *J. Catal.* 106:512 (1987); 109:206 (1988); 112:54 (1988); 116:82 (1989).

37. S. Ahmed and J.B. Moffat, *Appl. Catal.* 40:101 (1988).

38. S. Kasztelan and J.B. Moffat, *in* "Proc. 9th Intl. Congr. Catal." (M.J. Phillips and M. Ternan, Eds.), p. 883, Chemical Inst. Canada, Ottawa, 1988.

39. S. Kasztelan and J.B. Moffat, *J. Chem. Soc., Chem. Commun.* 1663 (1987).

40. J.B. McMonagle and J.B. Moffat, *J. Catal.* 91:132 (1985).

41. S. Kasztelan, E. Payen, and J.B. Moffat, *J. Catal.* 112:320 (1988).

42. S. Kasztelan, E. Payen, and J.B. Moffat, *J. Catal.* 125:45 (1990), ibid. 128:479 (1991).

43. E. Payen, S. Kasztelan, and J.B. Moffat, *J. Chem. Soc., Faraday Trans.* 88:2263 (1992).

44. H. Hayashi and J.B. Moffat, *J. Catal.* 77:473 (1982).

45. B.K. Hodnett and J.B. Moffat, *J. Catal.* 88:253 (1984).

46. R. Mauti and C. Mims, to be published.

PARTIAL OXIDATION OF METHANE BY MOLECULAR OXYGEN OVER SUPPORTED V₂O₅ CATALYSTS: A CATALYTIC AND *in situ* RAMAN SPECTROSCOPY STUDY

Qun Sun,[1] Jih-Mirn Jehng,[2]*Hangchun Hu,[1] Richard G. Herman,[1]
Israel E. Wachs,[2] and Kamil Klier[1]

Zettlemoyer Center for Surface Studies
[1]Department of Chemistry
[2]Department of Chemical Engineering
Lehigh University, Bethlehem, PA 18015

Introduction

The direct conversion of methane to methanol and formaldehyde *via* partial oxidation is still a very challenging research area in fundamental heterogeneous catalysis. Many catalyst systems have been investigated for this process and review articles are available in the literature.[1,2] Silica supported V_2O_5 and MoO_3 have been studied extensively ,[3-6] and V_2O_5/SiO_2 was found to be one of the most active and selective catalysts for methane partial oxidation to formaldehyde either by using N_2O[3] or molecular oxygen[4,5] as oxidants. Besides steady-state catalytic tests, there are very few studies that have focused on correlations between the structures of catalysts and their catalytic performances or establishing the nature of the active sites for methane partial oxidation.

In the present study, single component supported V_2O_5 systems (V_2O_5/SiO_2, V_2O_5/TiO_2 and V_2O_5/SnO_2) and mixed oxide systems ($V_2O_5/TiO_2/SiO_2$, $V_2O_5/SnO_2/SiO_2$ and $V_2O_5/MoO_3/SiO_2$) were tested for methane partial oxidation. The influence of vanadia loading and specific oxide support were examined. *In situ* Raman spectra were recorded for the first time under methane partial oxidation reaction conditions with these catalysts, and correlations between the surface vanadia structures and their catalytic performances were made. A new reaction scheme for the partial oxidation of methane to formaldehyde over the supported V_2O_5 catalysts is proposed.

*   Current address: Department of Chemical Engineering, National Chung Hsing University, 250 Kuo Kuang Rd., Taichung 402, Taiwan, R.O.C.

Experimental Section

Catalyst preparations. Amorphous SiO_2 (Cabosil EH-5, surface area=380 m^2/g) and TiO_2 (Degussa P-25, surface area=55 m^2/g) were used as received as the catalyst supports. SnO_2 was also employed as a support and was made from tin(II) acetate (Aldrich) by hydrolysis; after calcination at 450°C its surface area was 20 m^2/g. The incipient-wetness impregnation with solutions of different precursors was the general method used in preparing the supported catalysts for this study. A toluene solution of titanium (IV) isopropoxide $(Ti[OCH(CH_3)_2]_4)$ was used for making TiO_2/SiO_2 samples. A methanol solution of vanadium (VI) triisopropoxide oxide $(VO[i\text{-}OC_3H_7]_3)$ was used for making supported vanadium oxide catalysts. SnO_2/SiO_2 samples were prepared from an aqueous solution of colloidal tin (IV) oxide (SnO_2) in H_2O under ambient condition. An aqueous solution of ammonium heptamolybdate $((NH_4)_6Mo_7O_{24}\cdot4H_2O$, Matheson Coleman & Bell) was used for preparing supported molybdenum oxide catalysts. After impregnation, each catalyst was dried at room temperature, at 120°C overnight, and then calcined at 500°C for four hours under flowing air.

Catalytic testing. Catalytic testing was carried out in a fixed-bed continuous flow quartz reactor.[6] A standard reactant mixture of CH_4/air (1.5/1.0) was used at ambient pressure. The principal products analyzed by on-line sampling using gas chromatography were CO_2, C_2 ($C_2H_6 + C_2H_4$), CO and H_2O. Formaldehyde was condensed from the exit stream with dual water scrubbers and quantitatively determined by iodometric titration.

***In situ* Raman spectroscopy.** The *in situ* Raman spectrometer system consists of a quartz cell and sample holder, a triple-grating spectrometer (Spex, Model 1877), a photodiode array detector (EG&G , Princeton Applied Research, Model 1420), and an argon ion laser (Spectra-Physics, Model 165).[7,8] The quartz cell was capable of operating up to 600 °C, and reactant gas was introduced into the cell at a rate of 100-300 cc/min at atmospheric pressure. The Raman spectra under reaction conditions were obtained using following procedures: Raman spectra of dehydrated samples were collected after heating the samples to 500°C in a flow of pure oxygen gas (Linde Specialty Grade, 99.99% purity) for 30 min. A flowing gas mixture of CH_4/O_2 (10/1) was then introduced into the cell, and the Raman spectra were collected again upon reaching steady state. After the above treatments, the samples were further sequentially treated with pure oxygen gas and then pure methane gas. The Raman spectra were recorded in the 100-1200 cm^{-1}, region and the overall resolution of the spectra was determined to be better than 1 cm^{-1}.

Results and Discussion

Catalytic Testing. The conventional steady-state catalytic partial oxidation of methane by molecular oxygen was determined over the catalysts under ambient pressure, and the methane conversions and product selectivities are summarized in Table 1. Over the V_2O_5/SiO_2 catalysts, very high formaldehyde space time yields (STY > 1 kg CH_2O/kgcat.hr.) were obtained, even though the conventional single pass yields were still quite low (<2%). Figure 1 shows the formaldehyde selectivities as a function of the methane conversions from four V_2O_5/SiO_2 catalysts with different loadings. For the 1.0wt% catalyst, data were collected at three different temperatures.

Table 1. Methane oxidation by air (CH_4/Air = 1.5/1) over supported metal oxide catalysts.

Catalysts	GHSV (L/kgcat.hr)	Temp. (°C)	Conv. (CH_4%)	STY(CH_2O) (g/kgcat.hr)	Selectivities(C-mol%) CH_2O	C_2's	CO	CO_2
$SiO_2(F)^a$	70,000	630	0.05	24.3	100.0	-	-	-
2%MoO_3/(F)	70,000	630	0.08	37.9	100.0	-	-	-
1%V_2O_5/(F)	70,000	630	9.52	684.9	15.7	1.7	76.4	6.3
3%V_2O_5/(F)	140,000	580	6.86	1,022.0	16.6	0.2	76.8	6.3
5%V_2O_5/(F)	280,000	630	5.60	1,440.0	13.5	0.2	81.3	4.3
1%V_2O_5/3%MoO_3/(F)	70,000	630	8.47	675.2	16.6	2.0	73.5	7.9
TiO_2	70,000	630	1.55	17.6	2.3	-	94.0	3.6
3%TiO_2/(F)	70,000	630	0.31	27.6	17.8	-	71.1	11.1
1%V_2O_5/TiO_2^b	70,000	630	0.82	14.0	3.3	1.2	73.0	22.5
1%V_2O_5/3%TiO_2/(F)	70,000	630	1.07	101.3	18.6	-	76.6	4.8
3%V_2O_5/3%TiO_2/(F)	70,000	630	2.30	150.0	12.5	-	82.2	5.3
SnO_2	70,000	530	8.10	2.3	0.1	-	8.9	90.4
3%SnO_2/(F)	70,000	630	1.60	8.8	1.1	13.7	8.7	76.3
1%V_2O_5/SnO_2^c	70,000	530	7.60	-	-	-	13.4	83.6
1%V_2O_5/3%SnO_2/(F)	35,000	630	2.00	17.8	3.9	-	77.2	18.9

^{a} SiO_2(fumed cabosil), ^{b} TiO_2 as support, ^{c} SnO_2 as support.

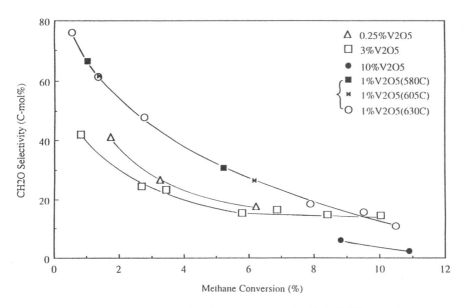

Figure 1. CH_2O selectivity vs. CH_4 conversion for V_2O_5/SiO_2 catalysts.

***In situ* Raman Studies.** The Raman spectra of 1wt% V_2O_5 supported on SiO_2, TiO_2, SnO_2, and 3wt% TiO_2/SiO_2 under methane oxidation reaction at 500°C were recorded by following the procedures described in the experimental section. As an example spectra obtained with the V_2O_5/TiO_2 catalyst are shown in Figure 2. Upon oxygen treatment, dehydrated surface monomeric VO_4 species with a Raman band in the 1027-1034 cm^{-1} region are predominantly present on all the samples (top spectrum in Figure 2).

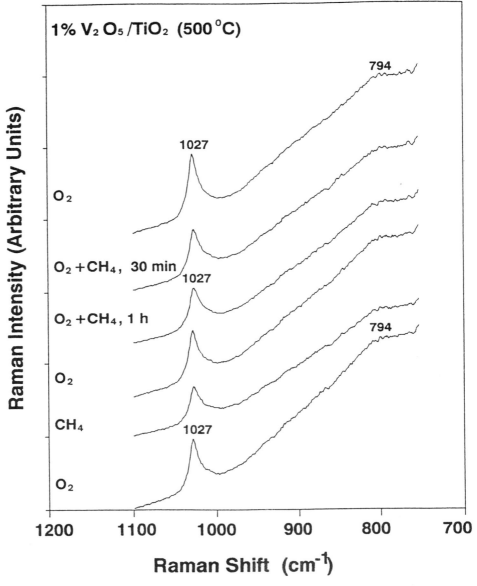

Figure 2. *In situ* Raman spectra of V_2O_5/TiO_2 catalyst.

Under methane oxidation reaction conditions, the Raman band intensities of the surface vanadium oxide species decreased in the V_2O_5/TiO_2 and V_2O_5/SnO_2 systems due to the reduction of the surface vanadium oxide species under the somewhat reducing methane oxidation environment, but no significant changes were observed in the V_2O_5/SiO_2 and $V_2O_5/TiO_2/SiO_2$ systems. The surface vanadium oxide species of the V_2O_5/SnO_2 system could be further reduced in a total methane environment, and forms a reduced surface vanadia phase with a weak and broad Raman band at ~850 cm^{-1}. The original surface vanadium oxide species can be restored by flowing pure oxygen into the cell to reoxidize the reduced surface vanadia phase.

V_2O_5/SiO_2. The data in Table 2 indicate that the TOF of methane conversion do not change significantly as the V_2O_5 loading was increased from 0.25% to 5.0%.

Table 2. The turn over numbers (T.O.N.) of methane conversion over V_2O_5/SiO_2 catalysts.

V_2O_5 loadings (wt.%)	0.25	1.00	3.00	5.00
GHSV (L/kgcat.hr)	16,200	70,000	350,000	280,000
CH$_4$ conversion (%)	1.7	1.0	0.8	1.1
T.O.N. (10^{-2} s^{-1})	6.8	4.5	5.9	3.8

This suggests that the activation of methane only needs one active site, the concentration of which increases linearly with the V_2O_5 loading. Figure 1 shows that formaldehyde selectivity is very sensitive to the methane conversion level, which is well recognized in the literature,[5] and also very sensitive to the V_2O_5 loadings. On the other hand, it is rather insensitive to the reaction temperature at least within the temperature region tested (580-630°C). Figure 1 also indicates the presence of an optimized V_2O_5 loading for methane partial oxidation to formaldehyde over the V_2O_5/SiO_2 catalysts, and over the range investigated 1%V_2O_5 was the optimum loading for formaldehyde selectivity. The *in situ* Raman spectra obtained under the reaction conditions show that the surface vanadium (V) oxides are the predominant species even under a pure flowing methane atmosphere.

V_2O_5/TiO_2 and $V_2O_5/TiO_2/SiO_2$. Analogous to the V_2O_5/SiO_2 catalyst, the V_2O_5/TiO_2 catalyst was found to posses well-dispersed surface vanadia species over the TiO$_2$ support for low loadings. The data in Table 1 reveals that for methane oxidation, the 1wt%V_2O_5/SiO_2 catalyst was much more active than the 1wt%V_2O_5/TiO_2 catalyst. At the same reaction conditions, methane conversion over the V_2O_5/SiO_2 catalyst was about one order of magnitude higher than over the V_2O_5/TiO_2. It is noted that the catalytic performances of the 1%V_2O_5/TiO_2 and the 1%$V_2O_5/3\%TiO_2/SiO_2$ catalysts are very similar, except that the 1%$V_2O_5/3\%TiO_2/SiO_2$ catalyst had higher selectivity to the formaldehyde production. These results are consistent with the structural information that the vanadia overlayer is coordinated to the titania overlayer, which results in behavior similarly to the V_2O_5/TiO_2 catalyst for the methane partial oxidation. The *in situ* Raman spectra (Figure 2) indicate that the surface vanadium (V) oxides have been partially reduced (35-40%) under reaction conditions. The reduced surface vanadium oxide species do not have the 1027 cm^{-1} Raman band attributed to the stretch mode of the terminal V=O bond of vanadium (V) oxide. Under the steady-state reaction conditions, a significant amount of the reduced vanadium oxide sites could provide the sites for the adsorption of dioxygen species and, therefore, accelerate the deep oxidation of methane.

V$_2$O$_5$/SnO$_2$ and V$_2$O$_5$/SnO$_2$/SiO$_2$. The 1%V$_2$O$_5$/SnO$_2$ was extremely active for methane oxidation (Table 1), however, it only produced deep oxidation products (CO$_x$) and predominantly CO$_2$. The *in situ* Raman study demonstrated that V$_2$O$_5$, as well as the SnO$_2$ support, is largely reduced under the methane oxidation reaction conditions. Similar to the V$_2$O$_5$/TiO$_2$ catalyst, these reduced sites could be promoting the deep oxidation of methane to CO$_2$.

Possible Reaction Mechanism. In the field of hydrocarbon selective partial oxidation research, it is of interest to determine the active sites that control the catalytic activity and the product selectivities. Sachtler and De Boer[9] correlated the catalytic selectivity with the catalyst reducibility for the oxidation of propylene. That is, in general, the higher the reducibility of the catalysts, the higher the activity and lower the selectivity. Bielanski and Haber[10] explained selectivity by postulating that lattice oxygen was responsible for partial oxidation, while adsorbed ionic or radical oxygen species caused total oxidation.

For the dehydrated state, the vanadium (V) oxide is bound to the SiO$_2$ surface *via* three V-O-Si bridging bonds and a terminatal V=O double bond. These fully reduced bridging and double bonded oxygen species can be viewed as surface lattice oxygens.[10] The bond order of the V=O was found to vary only by a very small amount over different supports.[7] Therefore, the V=O bond can not be responsible for the activity differences for V$_2$O$_5$ over different catalyst supports. However, the bridging oxygen bond should vary with the support and could be one of the determining factors for the initial activity of the catalysts for the methane activation. Once the methane conversion proceeds, some vanadium (V) oxide species will be reduced to the lower oxidation states and provide the sites for oxygen adsorption. The population of these reduced species should depend on the catalyst supports. The present *in situ* Raman studies clearly show that these reduced species are unstable over the SiO$_2$ surface but are much more stable over the TiO$_2$ and SnO$_2$ supports under the reaction conditions employed. In the case of V$_2$O$_5$/SnO$_2$, even the SnO$_2$ support was largely reduced. The stability trend of the reduced vanadium oxides over these catalysts is V$_2$O$_x$/SnO$_2$ > V$_2$O$_x$/TiO$_2$ > V$_2$O$_x$/SiO$_2$ (where x < 5).

Based on the above results and discussion, a new reaction scheme for par   tial oxidation of methane to formaldehyde is proposed (Reaction Scheme 1). S$_1$ is the O$^-$ site, such as Si-O$^\bullet$ and $^-$O-Si-O$^\bullet$ or V^{IV}-O$^\bullet$, which is capable of H-abstraction from methane to produce methyl radicals. Once a methyl radical has formed, the V=O double bond is the key to converting it to the methoxy complex and to the selective production of formaldehyde. Another pathway to formaldehyde could proceed through the direct adsorption of methane on the M-O-N bridging oxygen site (where M is a transition metal and N is the support element) to form the methoxy complex, which is probably the case for methane conversion to formaldehyde over the pure silica supports [6]. In fact, most selective catalysts for methane conversion to formaldehyde possess M=O double bonds, e.g. MoO$_3$, V$_2$O$_5$, and Re$_2$O$_7$. A systematic study of the methane partial oxidation to formaldehyde over a range of SiO$_2$-supported transition metal oxide catalysts is currently underway in our laboratories.

Pure SiO$_2$ alone produces formaldehyde, as well as   C$_2$ hydrocarbons, at low methane conversions at elevated temperatures, parallel reaction pathways for formation of formaldehyde and C$_2$ products have been proposed.[6] Unlike the V$_2$O$_5$ or MoO$_3$/SiO$_2$ catalysts, there is no double bond M=O species over the SiO$_2$ surface, and once formed the methyl radicals are likely to couple in the gas phase to produce ethane. Because of the capability of the M=O bond to convert methyl radical to formaldehyde, at very low conversions (e.g. <1%) the selectivities to formaldehyde approach 100% over the V$_2$O$_5$/SiO$_2$ and MoO$_3$/SiO$_2$ catalysts.[4,5] Another very unique property of the SiO$_2$ support is its low surface acidity. In comparison with other common supports like TiO$_2$, γ-Al$_2$O$_3$, and ZrO$_2$ that all have quite high densities of surface Lewis acidic sites, pure SiO$_2$ is almost free of Lewis acidic sites.[11] This could be one of the major differences between SiO$_2$ and TiO$_2$ supports, which leads to the relatively higher selectivity to formaldehyde for methane partial oxidation over the pure SiO_2 and SiO_2-supported catalysts, while the Lewis acidic sites on TiO_2 promotes the formation of CO. This is supported by a double bed experiment, in which a V_2O_5/SiO_2 catalyst bed was followed by TiO_2 for the methane partial oxidation, where it was noticed that the formaldehyde produced by the upstream V_2O_5/SiO_2 catalyst was decomposed by the downstream TiO_2.

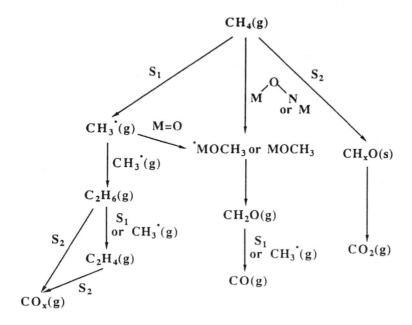

S_1: Hydrogen abstraction sites, $O^-(s)$, or $O_2^{2-}(s)$

S_2: Oxygen insertion sites, $O_2^-(s)$, $O_3^-(s)$, or $O_2^{2-}(s)$

Reaction Scheme 1

Two effects should be considered to explain the optimized V_2O_5 loading observed at $1\%V_2O_5/SiO_2$ for the formaldehyde production in the partial oxidation of methane. Since the surface V=O bond is capable of converting ${}^\bullet CH_3$ radical to formaldehyde, increasing the V_2O_5 loading should favor the formaldehyde production at relatively low loadings. On the other hand, even though the pure SiO_2 surface is free of Lewis acidic sites, once multivalance transition metal oxides, such as WO_3, Nb_2O_5, and V_2O_5, are dispersed on to the SiO_2 surface, Lewis acidic sites can be generated.[11,12] The acidity of the V_2O_5/SiO_2 catalyst was found to increase with the V_2O_5 loading.[12] The Lewis acidic sites on the higher loading catalysts might have promoted the decomposition of the formaldehyde, and, therefore, the 1wt% V_2O_5/SiO_2 catalyst gives maximum formaldehyde selectivity at the same level of methane conversions.

Acknowledgements

This research was supported in part by the Union Carbide Chemical & Plastics, Inc. and The National Science Fundation.

References

1. Pitchai, R., and Klier, K., *Catal. Rev. -Sci. Eng.*, <u>28</u>, 13 (1986).
2. Brown, M.J., and Parkyns, N.D., *Catal. Today*, <u>8</u>, 305 (1991).
3. Liu, H.-F., Liu, R.-S., Liew, K.Y., Johnson, R.E., and Lunsford, J.H., *J. Am. Chem. Soc.*, <u>106</u>, 4117 (1984).
4. Spencer, N.D., *J. Catal.*, <u>109</u>, 187 (1988).
5. Spencer, N.D., and Pereira, C.J., *J. Catal.*, <u>116</u>, 399 (1989).
6. Sun, Q., Herman, R.G., and Klier, K., *Catal. Lett.*, <u>16</u>, 251(1992).
7. Deo, G., and Wachs, I.E., *J. Catal.*, <u>129</u>, 137 (1991).
8. Das, D., Eckert, H., Hu, H., Wachs, I.E., and Feher, F.J., *J. Phys. Chem.*, <u>97</u>, 8240 (1993).
9. Sachtler, W.M.H., and de Boer, N.D., *Proc. 3rd Intern. Cong. Catal.*, Amsterdam, 1964, <u>1</u>, 252 (1965).
10. Bielanski, A., and Haber, J., *Catal. Rev. -Sci. Eng.*, <u>19</u>, 1 (1979).
11. Datka, J., Turek, A.M., Jehng, J.M., and Wachs, I.E., *J. Catal.*, <u>135</u>, 186 (1992).
12. Le Bars, J., Vedrine, J.C., Auroux, A., Trautmann, S., and Baerns, M., *Appl. Catal. A*, <u>88</u>, 179 (1992).

PARTIAL OXIDATION AND CRACKING OF ALKANES OVER NOBLE METAL COATED MONOLITHS

M. Huff[1] and L.D. Schmidt[2]

[1]Department of Chemical Engineering
University of Delaware
Newark, Delaware 19716

[2]Department of Chemical Engineering and Materials Science
University of Minnesota
Minneapolis, Minnesota 55455

ABSTRACT

Partial oxidation of propane and n-butane has been used to selectively produce ethylene and propylene. This autothermal process has been studied at atmospheric pressure over Pt coated ceramic foam monoliths at contact times less that 5 milliseconds. Olefins and synthesis gas (CO and H_2) are the dominant products with no carbon deposition or catalyst deactivation observed over several days of operation.

At contact times of only 5 milliseconds, olefins are produced with selectivities up to 70% at nearly 100% conversion of the alkane. Higher temperatures favor ethylene production over propylene production. Unlike the results in homogeneous pyrolysis, only very small amounts of acetylene, ethane, and other high molecular weight species are observed. This suggests that this process follows a very different mechanism. This mechanism can be explained quite simply. The reactions are initiated by hydrogen abstraction by surface oxygen to form an alkyl group on the surface. This alkyl group then undergoes a ß-scission reaction to form an olefin.

INTRODUCTION

Liquefied petroleum gas (propane and butanes) is abundant and useful as a precursor for short chain olefin production[1]. Ethylene, propylene, and butylenes are currently produced industrially by thermal pyrolysis (steam cracking) or ethane, propane, and butane. This process must be run at elevated temperatures due to equilibrium considerations and steam is added to the feed to reduce carbon deposition.

Methane and Alkane Conversion Chemistry
Edited by M. M. Bhasin and D. W. Slocum, Plenum Press, New York, 1995

Table 1. Reactions of Propane

			ΔT_{ad} (°C)	C_3H_8/O_2	$\Delta H°$ (kJ/mol)	K_{eq} 1200 K
(1)	$C_3H_8 + 5O_2 \rightarrow 3CO_2 + 4H_2O$	complete combustion	2800	0.20	-2043	$>10^{38}$
(2)	$C_3H_8 + \frac{3}{2}O_2 \rightarrow 3CO + 4H_2$	partial oxidation to syngas	900	0.67	-227	$>10^{38}$
(3)	$C_3H_8 + \frac{1}{2}O_2 \rightarrow C_3H_6 + H_2O$	oxidative dehydrogenation	1044	2.00	-118	10^9
(4)	$C_3H_8 \rightarrow C_3H_6 + H_2$	dehydrogenation	-	-	+124	10^1
(5)	$C_3H_8 \rightarrow C_2H_4 + CH_4$	cracking	-	-	+83	10^3

This work investigates a novel catalyst for the production of olefins, namely, oxidation over monoliths. The possible reactions of propane and n-butane are listed in Tables 1 and 2 with their corresponding heats of reaction and equilibrium constants. The reactants and products can also react to form solid carbon, C_s. These reactions are listed in Table 3. The reactions in the tables are numbered and will be referenced by these numbers in the text.

CONVERSION OF PROPANE AND N-BUTANE TO OLEFINS

There are both homogeneous[2,3] and heterogeneous[4,5] routes to the production of ethylene and propylene from propane. The homogeneous process, thermal pyrolysis (eq. 4 and 5), can achieve 40% selectivity to ethylene and 20% selectivity to propylene at 80% propane conversion[2]. However, in order to achieve this, the reactor must be externally heated to 800°C and long residence times (>0.3 seconds) are requires for this equilibrium process.

Many researchers have tried to utilize the exothermicity of oxidative dehydrogenation of propane (eq. 3) to improve on the shortcomings of the thermal process[6,7]. The oxidative dehydrogenation process can be quite selective (~65% propylene) over V-Mg-O catalysts, but only at very low propane conversion (<15%). At higher conversions, the selectivity drops quickly[5]. This trade off between selectivity and conversion severely limits the industrial value of this process.

Like the propane system, olefins can be produced from n-butane by both homogeneous pyrolytic routes[8,9] and heterogeneous oxidative dehydrogenation and cracking over oxide catalysts[10,11]. In the thermal pyrolysis of n-butane (eq. 10-12), propylene and ethylene[8] are produced with ~50% and ~20% selectivity, respectively, at ~40% n-butane conversion[9]. For these results, the reactor must be externally heated to ≥750°C. Other products of this process are listed in Table 4. At even higher temperatures, the n-butane conversion increases, but at the expense of olefin selectivity[12].

Oxidative dehydrogenation or cracking of n-butane to olefins (eq. 8 and 9) over oxide catalysts can be highly selective to butylene. Selectivities of 50-65% butylene over a V_2O_5/SiO_2 catalyst can be achieved at low conversion[10]. However, as in the propane case, the selectivity decreases rapidly at conversion >20%.

EXPERIMENTAL

The reactor and experimental apparatus have been described previously for similar processes involving synthesis gas production by direct oxidation of methane[13] and for oxidative dehydrogenation of ethane[14].

Table 2. Reactions of n-Butane

			ΔT_{ad} (°C)	C_4H_{10}/O_2	$\Delta H°$ (kJ/mol)	K_{eq} 1200 K
(6)	$C_4H_{10} + \frac{13}{2}O_2 \rightarrow 4CO_2 + 5H_2O$	complete combustion	2700	0.15	-2657	$>10^{38}$
(7)	$C_4H_{10} + 2O_2 \rightarrow 4CO + 5H_2$	partial oxidation to syngas	1688	0.50	-568	$>10^{38}$
(8)	$C_4H_{10} + \frac{1}{2}O_2 \rightarrow C_4H_8 + H_2O$	oxidative dehydrogenation	778	2.00	-116	10^9
(9)	$C_4H_{10} + \frac{1}{2}O_2 \rightarrow 2C_2H_4 + H_2O$	oxidative cracking	80	2.00	-12	10^{11}
(10)	$C_4H_{10} \rightarrow C_4H_8 + H_2$	dehydrogenation	-	-	+126	10^1
(11)	$C_4H_{10} \rightarrow C_3H_6 + CH_4$	cracking	-	-	+70	10^4
(12)	$C_4H_{10} \rightarrow C_2H_6 + C_2H_4$	cracking	-	-	+92	10^3

The data shown consists of alkane conversions and product selectivities. In all cases, the oxygen was completely consumed. The selectivities of the carbon containing species are calculated on a carbon atom basis. In other words, the moles of each product have been scaled by the number of carbon atoms in that product. Likewise, the H_2 and H_2O selectivities are calculated on a hydrogen atom basis. This method implicitly accounts for the change in number of moles due to reaction.

RESULTS

The data presented here was all collected using the same catalyst sample. The catalyst consists of 5.1 wt. % Pt loaded on a 45 ppi (pores per inch) foam α-alumina monolith. In the next sections, we will discuss the product distributions dependence on reactant composition and reaction temperature.

Propane

Air Oxidation. Figure 1 shows the variation of carbon atom and hydrogen atom selectivities and propane conversion for the oxidation of propane in air over the Pt catalyst as a function of the C_3H_8/O_2 ratio in the feed. In these experiments, the relative amounts of propane and air were varied while maintaining a fixed total flow of 5 SLPM with room temperature feed.

As indicated in Table 1, as the C_3H_8/O_2 ratio increases from 0.2 to 0.67, production should shift from CO_2 and H_2O (eq. 1) to CO and H_2 (eq. 2). Experiments were not conducted at C_3H_8/O_2 ratios less than 0.5 because these compositions lie within the flammability limits. According to the stoichiometry listed in Table 1, production should then shift to C_3H_6

Table 3. Reactions Producing Solid Carbon

			$\Delta H°$ (kJ/mol)	K_{eq} 1200 K	K_{exp} 1200 K
(13)	$C_3H_6 \rightarrow 3C_s + 3H_2$	cracking	-20	10^9	10^{-3}
(14)	$C_2H_4 \rightarrow 2C_s + 2H_2$	cracking	-53	10^5	10^{-2}
(15)	$2CO \rightarrow C_s + CO_2$	CO disproportionation	-172	10^{-2}	10
(16)	$CO + H_2 \rightarrow C_s + H_2O$	reverse steam reforming of carbon	-131	10^{-2}	10^2

as the C_3H_8/O_2 ratio approached 2.0. Figure 1 reflects these predicted trends somewhat; at C_3H_8/O_2 ratios less than 0.7, CO and H_2 are the dominant products.

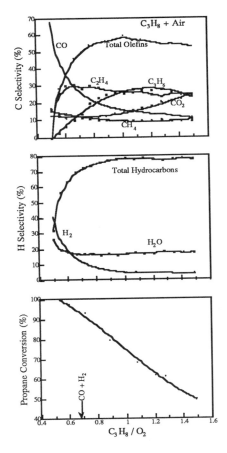

Figure 1. Carbon selectivity, hydrogen selectivity, and propane conversion for a 45 ppi x 1 cm, 5.1 wt % Pt foam monolith as a function of C_3H_8/O_2 ratio in the feed at a total feed flow rate of 5 SLPM in an autothermal reactor at a pressure of 1.4 atm.

However, at the stoichiometric composition for the production of synthesis gas (C_3H_8/O_2 =0.67) we observe *30% selectivity* to ethylene. In fact, at C_3H_8/O_2 ratios between 0.67 and 1.0, ethylene is the dominant product. Ethylene selectivity peaks at ~30% at the synthesis gas stoichiometry with a propane conversion >95%. At even richer compositions, propylene selectivity peaks at ~30% near a C_3H_8/O_2 ratio of 1.2 with a propane conversion of ~65%. At C_3H_8/O_2 ratios >0.8, the total olefin production (C_2H_4, C_3H_6, and C_4H_8) remains fairly constant with a selectivity of 55-60%. This olefin production is surprising since thermodynamics predicts the production of only CO, H_2, and graphite in this composition and temperature region. Figure 1 also shows that the ratio of the ethylene selectivity to the methane selectivity is nearly 2:1 on a carbon atom basis. This corresponds to one mole of ethylene formed for every mole of methane and supports the unimolecular cracking reaction (eq. 5).

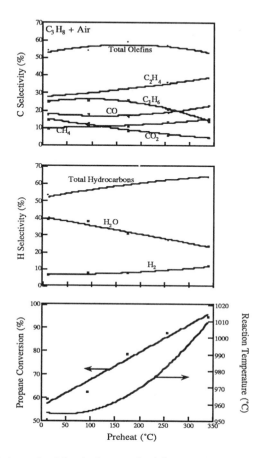

Figure 2. Carbon selectivity, hydrogen selectivity, propane conversion, and reaction temperature for a 45 ppi x 1 cm, 5.1 wt % Pt foam monolith as a function of preheat temperature. The total flow rate was maintained at 5 SLPM. The C_3H_8/O_2 ratio was 1.0 at a pressure of 1.4 atm.

Preheat. In Figure 2, we show the effect of preheating the reactants on selectivities, conversion, and reaction temperature. Propane and air at a C_3H_8/O_2 ratio of 1.0 were fed at a total flow rate of 5 SLPM. By preheating the reactants up to 350°C above ambient, before reaching the reaction zone, the reaction temperature increased to 1010°C. This pushed the propane conversion in excess of 90%, while the ethylene selectivity also improved. Unfortunately, the propylene selectivity decreased at higher temperatures. Because of this trade-off between ethylene and propylene, the total olefin production remained fairly constant as the temperature increased between 55 and 60% selectivity.

n-Butane

Air Oxidation. In Figure 3, we show carbon and hydrogen atom selectivities, n-butane conversion, and the reaction temperature for the oxidation of n-butane in air over a Pt catalyst as a function of the C_4H_{10}/O_2 ratio. The relative amounts of n-butane and air were varied while maintaining a fixed total flow of 5 SLPM with room temperature feed.

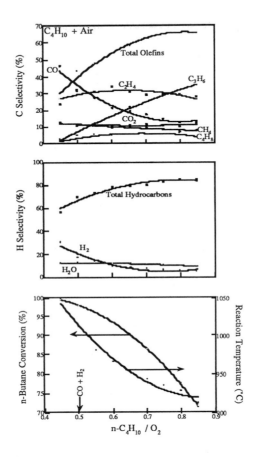

Figure 3. Carbon selectivity, hydrogen selectivity, n-butane conversion, and reaction temperature for a 45 ppi x 1 cm, 5.1 wt % Pt foam monolith as a function of C_4H_{10}/O_2 ratio in the feed at a total feed flow rate of 5 SLPM in an autothermal reactor at a pressure of 1.4 atm.

As listed in Table 2, the products of n-butane oxidation should shift from CO_2 and H_2O to CO and H_2 and finally to olefins as the C_4H_{10}/O_2 ratio increases from 0.15 to 2.0. Figure 3 roughly illustrates this trend. Experiments were not conducted in the CO_2 and H_2O forming region due to the danger of flames and explosions.

Table 4. Product Distribution in n-Butane Conversion on a Carbon Atom Basis

	Conversion	CH_4	C_2H_6	C_2H_4	C_3H_6	C_2H_2, C_2H_6, BTX	CO	CO_2
Pyrolysis	40	18	10	20	50	2	<0.2	<0.2
Oxidative Dehydrogenation	90	10	1	35	20	<0.1	20	10

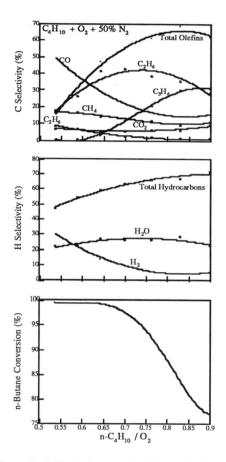

Figure 4. Carbon selectivity, hydrogen selectivity, and n-butane conversion for a 45 ppi x 1 cm, 5.1 wt % Pt foam monolith as a function of C_4H_{10}/O_2 ratio in the feed with 50% N_2 diluent at a total feed flow rate of 5 SLPM in an autothermal reactor at a pressure of 1.4 atm.

For C_4H_{10}/O_2 ratios less than 0.5, CO and H_2 are the dominant products. At C_4H_{10}/O_2 ratios $\geq$ 0.5, olefins are the dominant products. Ethylene selectivity peaks at ~32% at a C_4H_{10}/O_2 ratio of 0.65 with a n-butane conversion of 90% and propylene selectivity increases at the higher C_4H_{10}/O_2 ratios reaching ~40% near a C_4H_{10}/O_2 ratio of 0.85 with a n-butane conversion of ~72%. Contrary to the reactions listed in Table 2, butylene is never a dominant product. Total olefin selectivity increases as the C_4H_{10}/O_2 ratio increases and reaches a plateau at ~65% selectivity.

Oxygen Oxidation. Figure 4 illustrates the variation in selectivities and conversion with the C_4H_{10}/O_2 ratio at a fixed level of 50% N_2 dilution. Similar to Figure 3, ethylene selectivity peaks near a C_4H_{10}/O_2 ratio of 0.7 but now reaches 42% at nearly 100% butane conversion. Also, as observed in the air experiments, the propylene selectivity increases as the C_4H_{10}/O_2 ratio increases. However, the propylene selectivity is slightly lower in the oxygen enriched experiments than it was in the air experiments.

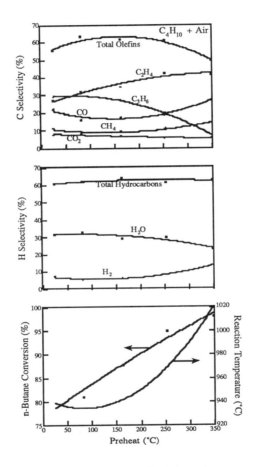

Figure 5. Carbon selectivity, hydrogen selectivity, n-butane conversion, and reaction temperature for a 45 ppi x 1 cm, 5.1 wt % Pt foam monolith as a function of preheat temperature. The total flow rate was maintained at 5 SLPM. The C_4H_{10}/O_2 ratio was 0.7 at a pressure of 1.4 atm.

Preheat. In Figure 5, we show the effect of preheating the reactants on the selectivities, conversion, and reaction temperature for butane oxidation in air at a C_4H_{10}/O_2 ratio of 0.7 at a total flow rate of 5 SLPM. The reaction temperature increased to 1020°C which drove the n-butane conversion to >95%. At this higher conversion and higher temperature, the ethylene selectivity improved, exceeding 40%, and the propylene selectivity rapidly fell.

DISCUSSION

These experiments have produced up to 70% selectivity to olefins at nearly 100% conversion of propane or n-butane and the catalyst has displayed no deactivation over several weeks of operation. A reaction mechanism must address several aspects of these results including (1) the absence of carbon deposition under conditions where it is thermodynamically favorable, (2) the similarity of the product distributions for either propane or n-butane fuel, and (3) the ethylene production increase with temperature while propylene production decreases. Figure 6 shows reaction steps that we believe can explain these results for butane oxidation. Similar steps exist for propane.

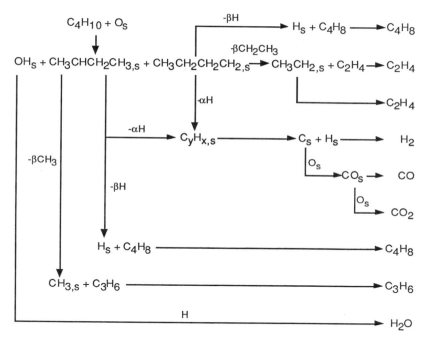

Figure 6. Proposed surface reactions in n-butane oxidation. The gaseous species produced are indicated at the right.

Thermal Pyrolysis versus Catalytic Oxidation

The high selectivities to ethylene and propylene cannot be explained by a pyrolysis mechanism. The homogeneous pyrolysis mechanism predicts significant production of acetylene, butadiene, and aromatics. These products can account for as much as 15% of the product distribution[2]. None of these undesirable products are detected in these experiments and must be present in quantities <0.1%. As shown in Table 4, the product selectivity distribution obtained in thermal pyrolysis of n-butane[8] does not agree with the product selectivity distribution observed here. We see significantly more C_2H_4 and less CH_4, C_2H_6, and C_3H_6 than the pyrolysis data. Also, our conversions are 90% versus 40% in pyrolysis.

Reaction Steps

The reaction steps detailed in Figure 6 for n-butane oxidation can be simplified to five essential steps. These steps also apply to the propane system.

$$alkane \xrightarrow{r_i} \begin{matrix} surface \\ alkyl \end{matrix} \xrightarrow{r_\beta} olefin \xrightarrow{r_{crack}} \begin{matrix} smaller \\ olefin \end{matrix} \qquad (17)$$

$$\xrightarrow{r_\alpha} C_s \xrightarrow{r_{CO}} CO$$

In this scenario, olefins are formed by both ß-hydrogen and ß-alkyl elimination.

Initiation. At the conditions of this reaction, the front face of the monolith should be covered with nearly a monolayer of adsorbed oxygen atoms. The most likely initiation step for this reaction is hydrogen abstraction by surface oxygen, O_s,

$$C_3H_8 + O_s \rightarrow C_3H_{7,s} + OH_s \qquad (18)$$

$$C_4H_{10} + O_s \rightarrow C_4H_{9,s} + OH_s \qquad (19)$$

to form a surface alkyl group and hydroxyl group.

Since these experiments are conducted in the fuel rich regime, they are oxygen limited. In fact, O_2 is always completely consumed in the results discussed here. The adsorbed oxygen, O_s, is completely depleted at some distance into the catalyst. Based on experiments with differing contact times, we estimate this distance to be ~1 mm in these cases. The remainder of the catalyst provides a hot surface for secondary reactions to take place.

ß-Hydroxyl and Alkyl Elimination. Both the propyl and butyl groups may be adsorbed at a primary carbon atom or at a secondary carbon. Assuming that bonds breaking and reforming at 1000°C in less than 5 milliseconds is unlikely, then the only way an olefin can be formed from an adsorbed alkyl is by a ß-elimination reaction, eq. 17. Olefin formation *cannot occur* by either α-hydrogen elimination or α-alkyl elimination. As shown in eq. 17, these mechanisms would lead to the complete decomposition of the alkyl to C_s and H_s which may then form CO and H_2 or remain on the surface as coke. There is evidence that the more noble metals prefer ß-elimination reactions over α-elimination[15-17].

A primary propyl can undergo either ß-hydrogen elimination to form propylene or ß-alkyl elimination to form ethylene and adsorbed methyl.

$$CH_3CH_2CH_{2,s} \rightarrow CH_{3,s} + C_2H_4 \qquad (20)$$

Ethylene is preferentially formed because the C-C bond is somewhat weaker than the C-H bond. The only ß-elimination reaction available to a secondary propyl group is ß-hydrogen elimination resulting in propylene production.

$$CH_3CHCH_{3,s} \rightarrow H_s + C_3H_6 \qquad (21)$$

Likewise, a primary butyl can undergo either ß-hydrogen elimination to form butylene of ß-alkyl elimination to form ethylene.

$$CH_3CH_2CH_2CH_{2,s} \rightarrow C_2H_{5,s} + C_2H_4 \qquad (22)$$

Ethylene formation is preferred in this case since the C-C bond is more vulnerable than the C-H bond. The secondary butyl can undergo either ß-hydrogen elimination to form butylene or ß-alkyl elimination to form propylene.

$$CH_3CHCH_2CH_{3,s} \rightarrow CH_{3,s} + C_3H_6 \qquad (23)$$

Similarly, propylene formation is preferred due to the preference for ß-alkyl elimination over ß-hydrogen elimination reactions. This explains the lack of butylene as a major reaction product.

Equilibrium

These cannot be equilibrium reactions since the equilibrium product distribution is not obtained. At all fuel/O_2 ration greater than the stoichiometric ratio for production of synthesis gas, thermodynamics predicts carbon deposition. In fact at a C_3H_8/O_2 ratio of 1, at equilibrium 30% of the reacting carbon should remain on the surface as coke. Since even a fraction of this rate of coking would deactivate the catalyst within minutes, and this deactivation is not observed over several hours of operation, then equilibrium is not being approached.

There are at least four primary routes to the production of C_s including olefin cracking (eq. 13 and 14), the Boudouard reaction (eq. 15) and reverse steam reforming (eq. 16). These reactions are listed in Table 3 with their equilibrium constants and experimental values, K_{exp}, of the equilibrium constants (ratios of partial pressures) obtained under reaction conditions. These K_{exp}'s can be calculated as

$$K_{C_3H_6} = \frac{P_{H_2}^3 a_C}{P_{C_3H_6}} \qquad (24)$$

$$K_{C_2H_4} = \frac{P_{H_2}^2 a_C}{P_{C_2H_4}} \qquad (25)$$

$$K_{CO} = \frac{P_{CO_2} a_C}{P_{CO}^2} \qquad (26)$$

$$K_{H_2O} = \frac{P_{H_2O} a_C}{P_{CO} P_{H_2}} \qquad (27)$$

where P_i are partial pressures and a_c is the activity of solid carbon. The values in Table 3 were obtained assuming $a_c=1$, for equilibrium in solid graphite formation.

Since K_{eq} is many orders of magnitude greater than K_{exp} for the olefin cracking reactions (eq. 13 and 14), these reactions cannot attain equilibrium if the catalyst does not deactivate. For the Boudouard reaction (eq. 15) and reverse steam reforming (eq. 16), however, $K_{eq}<K_{exp}$ suggesting that the reverse of these reactions takes place. Although carbon may be deposited on the catalyst surface by hydrocarbon cracking (eq. 13 and 14), the CO_2 and H_2O

partial pressures are sufficiently high to completely suppress this coking and continuously remove any coke from the surface by either CO_2 reforming or steam reforming (eq. 15 and 16).

CONCLUSIONS

Ethylene and propylene are produced from propane and n-butane at high conversions of hydrocarbon and total conversion of oxygen at atmospheric pressure over Pt coated monoliths with residence times on the order of milliseconds. The reaction is initiated by hydrogen abstraction by surface oxygen to form a surface alkyl group. On Pt, the alkyl group undergoes a ß-elimination reaction to form an olefin. Due to relative bond strengths, ß-alkyl eliminations are preferred over ß-hydrogen eliminations.

The extremely short contact times and the non-equilibrium production of H_2O and CO_2 suppress the formation of coke and lead to significant olefin formation at high alkane conversions. The formation of simple product distributions and the absence of small alkanes argues that these processes occur primarily on the surface.

ACKNOWLEDGMENTS

This work was partially supported by DOE under Grant DE-FG02-88ER13878-AO2. This is an abbreviated version of a paper published in the Journal of Catalysis, "Production of Olefins by Oxidative Dehydrogenation of Propane and Butane over Monoliths at Short Contact Times," *J. Catal.* **149**, 127-141 (1994).

REFERENCES

1. Seshan, K., Swaan, H. M., Smits, R. H. H., van Omens, J. G., and Ross, J. R. H. In *New Developments in Selective Oxidation*; G. Centi and F. Trifiro, Ed.; Elsevier Science Publishers B.V.: Amsterdam, ; pp. 505-512 (1990).
2. McConnell, C. F., and Head, B. D. In *Pyrolysis: Theory and Industrial Practice*; L. F. Albright, B. L. Crynes and W. H. Corcoran, Ed.; Academic Press: New York, ; pp. 25-45 (1983).
3. Sundaram, K. M., and Froment, G. F., *Ind. Eng. Chem. Fundam.* **17**, 174-182 (1978).
4. Nguyen, K. T., and Kung, H. H., *J. Catal.* **122**, 415-428 (1990).
5. Chaar, M. A., Patel, D., and Kung, H. H., *J. Catal.* **109**, 463-467 (1988).
6. Burch, R., and Crabb, E. M., *Appl. Catal. A* **100**, 111-130 (1993).
7. Smits, R. H. H., Seshan, K., and Ross, J. R. H., *J. Chem. Soc., Chem. Commun.* 558-559 (1991).
8. Corcoran, W. H. In *Pyrolysis: Theory and Industrial Practice*; L. F. Albright, B. L. Crynes and W. H. Corcoran, Ed.; Academic Press: New York, ; pp. 47-68 (1983).
9. Delzer, G. A., and Kolts, J. H. In *Novel Production Methods for Ethylene, Light Hydrocarbons, and Aromatics*; L. F. Albright, B. L. Crynes and S. Nowak, Ed.; Marcel Dekkar, Inc.: New York, ; pp. 41 (1992).
10. Owens, L., and Kung, H. H., *J. Catal.* **144**, 202-213 (1993).
11. Patel, D., Andersen, P. J., and Kung, H. H., *J. Catal.* **125**, 132-142 (1990).
12. Heiman, J. C. In *Pyrolysis: Theory and Industrial Practice*; L. F. Albright, B. L. Crynes and W. H. Corcoran, Ed.; Academic Press: New York, ; pp. 365-375 (1983).

13. Hickman, D. A., and Schmidt, L. D., *J. Catal.* **138**, 267-282 (1992).
14. Huff, M., and Schmidt, L. D., *J. Phys. Chem.* **97**, 11815-11822 (1993).
15. Jenks, C. J., Chiang, C.-M., and Bent, B. E., *J. Am. Chem. Soc.* **113**, 6308-6309 (1991).
16. Zaera, F., *Surf. Sci.* **219**, 453-466 (1989).
17. Zaera, F., *Acc. Chem. Res.* **25**, 260-265 (1992).

STRUCTURAL FEATURES OF SILICA-SUPPORTED VANADIA CATALYSTS AND THEIR RELEVANCE IN THE SELECTIVE OXIDATION OF METHANE TO FORMALDEHYDE

Marisol Faraldos, Miguel A. Bañares, James A. Anderson, and José Luís G. Fierro

Instituto de Catálisis y Petroleoquímica, C.S.I.C., Campus UAM, Cantoblanco, 28049 Madrid, Spain

INTRODUCTION

The direct conversion of methane to formaldehyde and methanol in a single catalytic step in sufficiently high yield is an extremely attractive process for the conversion of natural gas to other useful fuels and chemicals. Both the homogeneous and heterogeneous processes have been studied under varied conditions[1-3] although progress towards obtaining a yield that would make such a process industrially viable has been very slow[4-22]. Nearly all of these studies have involved redox oxides[4,5,7], and more specifically, silica supported heteropolyacids[6] and metal oxide catalysts[8-22]. From this large body of work, there is a considerable range of conditions over which the CH_4/oxidant stoichiometry has been varied (20 and 0.5), although the temperature has been kept in a narrow window (723-923 K) as a result of the extremely low reaction rate at low temperatures while above 900 K CO_x formation becomes much more dominant. These limited yields arise from the unfavourable ratio of the reactivity of formaldehyde and methanol to that of methane. As shown by Spencer et al.[8,9], the rate constant for the HCHO oxidation to CO derived from their data is 50-100 times higher than the rate constant for methane conversion.

It would appear that the selection of appropriate conditions and the precise reactor design, which minimizes the combustion or gas phase reaction, are critical in obtaining reliable data. Following this strategy, we have investigated the CH_4 oxidation on silica-supported MoO_3 catalysts, and found that the activity depends markedly on the MoO_3 content[20,21,23] and that the selective oxidation proceeds by a Mars-van Krevelen mechanism in which the CH_4 molecule is oxidised by lattice oxygen while the oxygen consumed is restored by dioxygen from gas phase[21]. The same applies for silica-supported vanadia catalysts although the binding energy of lattice oxygen is lower in V_2O_5 than in MoO_3[25]. Accordingly, this paper is aimed at evaluating the activity of CH_4 partial oxidation on silica-supported V_2O_5 catalysts and at describing how the catalytic performance can be correlated with fundamental structural and surface parameters of the supported oxide phases.

EXPERIMENTAL

A commercial silica (Aerosil 200), particle size 12 nm, BET area 200 m^2 g^{-1} and composition SiO_2>98.3%, Al_2O_3 = 1.3%, Fe_2O_3<0.01% and Na_2O ca. 0.05% was used as carrier. This was impregnated with an aqueous solution of ammonium metavanadate ((NH_4)VO_3 Merck reagent grade) in appropriate amounts to give surface concentrations in the range 0.3-3.0 V atoms per square nm of the silica surface, assuming a complete dispersion of the supported vanadium oxide. They will be referred to as xV, where x indicates the surface concentration in V atoms/nm^2 of the carrier. The impregnates were dried at 383 K and calcined in two steps: 623 K for 2 h and 923 K for 5 h. Prior to use in the reaction the catalysts were pelleted and sieved in the range 0.42-0.50 mm.

Activity measurements were carried out at atmospheric pressure in a 8 mm OD fixed bed quartz microcatalytic reactor by co-feeding CH_4 (99.95% vol.) and O_2 (99.98% vol.) without diluent. The CH_4:O_2 ratio was adjusted in the range 3-10 molar by means of mass flow controllers and the methane residence time adjusted to 0.4 or 2 g mol^{-1}. The reactor effluents were analysed by on-line GC using a Konik 3000HR gas chromatograph fitted with a thermal conductivity detector. Chromosorb 107 and Molecular Sieve 5A packed columns using a column isolation analysis system were used.

Catalysts reducibility and the initial rate of reduction were obtained from reduction isotherms performed at 823 K on a Cahn 2000 microbalance operating at a sensitivity of 10 µg. Catalyst samples were firstly heated at 4 K min^{-1} to 823 K in a flow (60 cm^3 min^{-1}) of helium (99.997% vol.) before measuring the kinetic reduction curves in a flow (60 cm^3 min^{-1}) of H_2 (99.995% vol.) at the same temperature. Weight changes as a function of time were collected by microprocessor and then differentiated at zero time thus allowing calculation of the initial rate of reduction.

Low temperature oxygen chemisorption (LTOC) measurements were performed on H_2-reduced catalysts. For this purpose the catalysts (about 0.10 g) were flushed in helium (H_2O and O_2 levels lower than 1 ppm) and heated up to 773 K. Then the gas flow was switched to a mixture of 10% vol. H_2 in He and kept at the same temperature for one hour. After flushing with Ar for 15 min, the reduced catalysts were cooled down to 195 K at which temperature the LTOC test is performed. Pulses of 10% O_2 (99.996% vol.) in He were injected in the He carrier gas flow and passed through the catalyst and a TCD. The total oxygen consumed was calculated when the effluents peaks are increased to constant size (less than 1% difference between consecutive peaks).

X-ray photoelectron spectra were obtained using a Fisons ESCALAB Mk II 200R spectrometer fitted with a hemispherical electron analyzer and a Mg anode X-ray exciting source (MgKα = 1253.6 eV). The residual pressure in the ion-pumped analysis chamber was maintained below 3x10^{-9} Torr during data acquisition. Each spectral region was signal averaged for a number of scans to obtain good signal to noise ratios. A binding energy (BE) of 103.4 eV, corresponding to the Si_{2p} peak, was used as an internal standard. The spectra were collected in a PDP computer and the peak areas measured assuming a curved background.

RESULTS AND DISCUSSION

Some of the chemical and physical properties of the supported metal oxides are contained in Table 1. Samples show a decrease in BET surface area with an increase in vanadium oxide loading. The same trend has been observed by Nag et al. [25] who found that up to 3.6V nm^{-2} BET area decreased continuously and then it became independent of vanadia loading.

Reduction by Hydrogen

Kinetic curves of reduction in a H_2 flow were recorded for periods long enough until no appreciable further weight loss was detected. For all the catalyst, the main proportion of this weight loss occurred during the first few minutes in H_2 at 823 K followed by a further slow gradual reduction, reaching stability after approximately 2 h. The initial rate of reduction (R_0) was also measured as an indicator of the vanadia dispersion (Table 1). The catalysts

Table 1. Composition and Surface Characteristics of the Catalysts

Catalyst	$V/nm^{2 a}$ (exp)	S_{BET} (m^2/g)	R_o^b (O/V/min)	$\alpha_{t=2h}^c$ (O/V)
0.3V	0.3	180	0.076	0.94
0.4V	0.4	173	0.101	0.86
0.7V	0.7	165	0.143	0.88
1.0V	1.0	153	0.150	0.75
1.4V	1.4	148	0.156	0.84
1.9V	1.9	132	0.193	0.95
3.0V	3.0	83	0.050	0.56

[a] Vanadia concentration is expressed as vanadium atoms per square nanometer of silica.
[b] R_o is the initial rate of reduction by hydrogen.
[c] Extent of the reduction of vanadia after 2 h in a hydrogen flow.

showed a progressive increase in R_o up to vanadium contents ca. 0.7 V/nm^2, followed by a flat region up to 1.9 V/nm^2 and then falling off. Table 1 also includes the extent of vanadia reduction ($\alpha_{t=2h}$), expressed as the number of O-atoms removed per V_2O_5 entity, after 2 h in hydrogen stream. With the exception of the 3.0V catalysts, $\alpha_{t=2h}$ values are in the range 0.75-0.95, the average reduction of vanadium approaching the stoichiometric reduction V^{5+} to V^{3+}.

Surface Analysis (XPS)

The V_{2p} and Si_{2p} core level spectra were recorded for the xV calcined catalyst series. In the V_{2p} peaks there is overlaping between the less intense $V_{2p1/2}$ peak and the $O_{1s3,4}$ satellite interfering peak, whose origin lies in the $MgK\alpha_{3,4}$ line because non-monochromatic source was used. Accordingly, only the $V_{2p3/2}$ peak was considered in calculations, even after peak deconvolution. Some surface reduction of the catalysts seems to occur as indicated not only by the change in colour of the sample from yellow to green after analysis but also from the average BE value of the $V_{2p3/2}$ peak somewhat below 516.8 eV. The relative dispersion of V_2O_5 on the silica surface was determined by comparing the intensity of the signal corresponding to the V_{2p} peaks with the intensity of Si_{2p} peak of the silica carrier. The V/Si XPS intensity ratios have been calculated and represented in Figure 1 as a function of vanadia concentration. The intensity ratio increases steeply with vanadium content up to ca. 1 V/nm^2 and more slowly up to 1.9 V/nm^2. Slower increase in V/Si XPS intensities ratio with vanadium content must originate from the presence of very small agregates of vanadium oxide. Further increase in vanadium content gave rise to an abrupt increase in the V/Si intensity ratio. Note again that the xV samples giving the largest V/Si XPS intensity ratios are those in which the BET area was strongly depleted.

Oxygen Chemisorption

The BET areas, initial reduction rates (R_o) and V/Si XPS intensity ratios revealed that the dispersion of the vanadia phase in the various catalysts studied might differ substantially. Since dispersion is an important parameter influencing the catalyst activity, LTOC is used to determine the relative site density of the vanadia phase present in the catalysts. The extent of oxygen chemisorption at 195 K as a function of the vanadia loading is given in Figure 1. As the SiO_2 support itself showed no adsorption, no correction before reporting the results was necessary. As shown in Figure 1, the oxygen uptake increases almost linearly with increasing vanadia loading up to about 1V/nm^2 and then levels off with higher vanadium oxide loading. This saturation level is indicative of larger vanadium oxide aggregates. These results are consistent with those reported by Reddy et al.[26] who found saturation level of chemisorbed

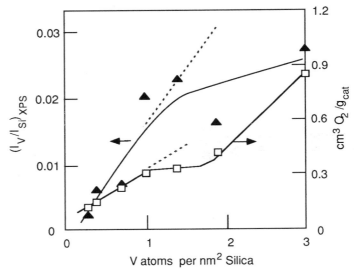

Figure 1. Influence of the vanadium content on the extent of oxygen chemisorption at 195 K on prereduced catalysts at 773 K; and V/Si XPS intensity ratios.

oxygen at 195 K on prereduced V_2O_5/TiO_2-ZrO_2 catalysts at vanadium contents close to 2 V/nm^2. If an stoichiometry between chemisorbed oxygen and reduced vanadium sites: $O_{2\ chem}/V_s = 1$ is defined[24], the number of exposed vanadium sites can be determined. Following this reasoning, the turnover frequency (TOF) for the methane partial oxidation can been calculated.

Catalytic Activity

First of all, the contribution of the gas-phase reaction has been evaluated by performing a series of experiments with the empty reactor. The data obtained with the 6 mm OD reactor and the results obtained in other series of experiments performed using empty and SiC filled 12 mm OD reactor confirmed that the contribution of the gas-phase reaction in the explored temperature range is completely negligible, only trace amounts of dimerization products, mainly ethane, were detected at temperatures above 873 K. Next blank experiments were performed with bare SiO_2 (Aerosil) since there are claims in the literature[15,20] that HCHO can readily be formed on several commercial silica substrates. The main relevant results obtained in the temperature range 823-953 K, contact time W/F = 2.0 g.h/mol and $CH_4/O_2 = 10$ (molar ratio) in the feed, is that no HCHO was observed; carbon oxides and water were the major reaction products, with only trace amounts of ethane (dimerization) at the highest temperatures.

One fundamental issue generally accepted is the direct participation of lattice oxygen of the redox oxide in the formation of HCHO. In order to understand the role played by lattice oxygen of V_2O_5 oxides in the selective oxidation of methane, the catalytic behaviours in terms of: (i) methane conversion-temperature trend; (ii) methane conversion-contact time trend; and (iii) selectivity and conversion-V content plot, have been examined. Activity and selectivity data obtained for a given set of experimental conditions were practically the same when experiments were duplicated. In respect to point (i), the increase in temperature resulted, as expected, in an increase in the methane and oxygen conversions but in a decrease in the selectivity towards HCHO in favour of CO_x oxidation products. The dependence of the contact time (W/F) on the CH_4 conversion (ii) illustrates the same trend. Selectivities *vs.* CH_4 conversion are plotted in Figure 2A. In agreement with literature findings[8,20,22,24], HCHO

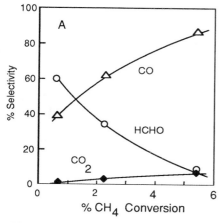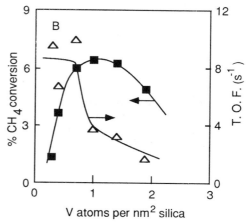

Figure 2. (A) Activity and product distributions for methane partial oxidation, CH_4/O_2 = 10:1, W/F = 2.0 g.h/mol as a function of the vanadium content of the catalysts: ●, methane conversion; ○, CO selectivity; △, HCHO selectivity; □, CO_2 selectivity. (B), Methane conversion and apparent TOF number calculated from LTOC data. Reaction condions: 863 K, CH_4/O_2 = 10:1, W/F = 2.0 g.h/mol.

selectivity decreased rapidly with methane conversion indicating that formaldehyde is a primary product. Simultaneously, selectivity to carbon oxides tends to zero at the limit of zero methane conversion and increased with methane conversion implying that carbon oxides, and more specifically CO, are secondary products.

In order to rationalize the activity pattern in terms of surface parameters, the apparent TOF numbers have been calculated and plotted in Figure 2B. In this calculation the reaction conditions given in Figure 2A and the O_2 uptakes compiled in Figure 1 have been considered. From this plot two activity regions are clearly distinguished.The remarkable change in apparent TOF number with surface vanadium oxide loading suggests that the active sites for methane conversion are not related to exposed vanadium sites. Characterization results clearly suggest that there are structural changes of vanadia phase with surface vanadium oxide loading. The initial rates of reduction decrease with the increase of the thickness of the V_2O_5 crystal[27]. The plateau for R_0 values between ca. 1 and 2 V/nm² suggests the presence of microcrystalline V_2O_5. These V_2O_5 microcrystals must have a thickness below ca. 2 nm as the constant values of the XPS vanadium-to-silicon intensity ratio and R_0 in this loading range suggest. The abrupt decrease of R_0 and reducibility along with the significant increase of the XPS V/Si intensity ratio suggests the onset of larger V_2O_5 crystals covering the silica support. It appears that the change in catalytic activity is related to the appearance of vanadia species other than dispersed surface vanadium oxide. The aggregation of vanadium into V_2O_5 microcrystals is the main feature affecting catalytic activity of V_2O_5/SiO_2 materials. Therefore, it appears that activity is related to dispersed surface vanadium oxide species rather than to exposed vanadium sites. The relevance of surface oxide species in this reaction has also been observed on other catalytic systems[28].Further characterization of this system by Raman spectroscopy provides additional evidence for this model[29]. All these results nicely illustrate that surface vanadium oxide species in which V=O terminal groups exist are more active than the tridimensional V_2O_5 phase.

CONCLUSIONS

Characterization of silica-supported vanadia catalysts shows that the nature of surface vanadium oxide is determined by its coverage on the silica surface. Dispersed surface vanadium oxide species are dominant below ca. 1 V/nm² while small aggregates of vanadium oxide are formed in the 1-2 V/nm² range. Bulk crysttalline vanadia dominates at vanadium oxide surface coverage higher than 2 V/nm². Catalytic results determine that dispersed surface vanadium oxide species are more active for the conversion of methane than exposed ones, as determined by low temperature oxygen chemisorption technique.

ACKNOWLEDGMENTS

The authors are indebted to Mr. E. Pardo for recording the XP spectra. This work was partly supported by the Commission of the European Communities, EEC Contract No. JOU92-0234.

REFERENCES

1. R. Pitchai, and K. Klier, "Partial oxidation of methane" *Catal. Rev.-Sci.Eng.* 28: 13 (1986).
2. M. Yu Sinev, V. N. Korshak, and O. V. Krylov,"The mechanism of the partial oxidation of methane" *Russ. Chem. Rev.* 58, 22 (1989).
3. J. J. Brown, and N. D. Parkyns, "Progress in the partial oxidation of methane to methanol and formaldehyde" *Catal. Today* 8: 305 (1991).
4. M. M. Khan, and G. A. Somorjai, "A kinetic study of partial oxidation of methane with nitrous oxide on a molybdena-silica catalyst", *J. Catal.* 91: 263 (1985).

5. K. J. Zhen, M. M. Khan, C. H. Mak, K. B. Lewis, and G. A. Somorjai, "Partial oxidation of methane with nitrous oxide over V_2O_5-SiO_2 catalyst", *J. Catal.* 94: 501 (1985).
6. S. Kasztelan, and J. B. Moffat, "The formation of molybdosilicic acid on Mo/SiO_2 catalysts and its relevance to methane partial oxidation", *J. Catal.* 112: 320 (1988).
7. K. Otsuka, and M. Hatano, "The catalysts for the synthesis of formaldehyde by partial oxidation of methane", *J. Catal.* 108: 252 (1987).
8. N. D. Spencer, "Partial oxidation of methane to formaldehyde by means of molecular oxygen", *J. Catal.* 109: 187 (1988).
9. N. D. Spencer, and C. J. Pereira, "V_2O_5-SiO_2-Catalyzed methane partial oxidation with molecular oxygen", *J. Catal.* 116: 399 (1989).
10. Y. Barbaux, A. R. Elamrani, E. Payen, L. Gengembre, B. Grzybowska, and J. P. Bonnelle, "Silica supported molybdena catalysts. Characterization and methane oxidation", *Appl. Catal.* 44: 117 (1988).
11. G. N. Kastanas, G. A. Tsigdinos, and J. Schwank, "Effect of small amounts of ethane on the selective oxidation of methane over silicic acid and quartz surfaces", *Appl. Catal.* 44: 33 (1988).
12. V. Amir-Ebrahimi, and J. J. Rooney, "Selective air oxidation of methane to formaldehyde using silica-supported $MoCl_5/R_4Sn$ olephin methatesis catalysts", *J. Mol. Catal.* 50: L17 (1989).
13. E. Mac Giolla Coda, M. Kennedy, J. B. McMonagle, and K. B. Hodnett, "Oxidation of methane to formaldehyde over molybdena catalysts at ambient pressure: isolation of the selective oxidation product", *Catal. Today* 6: 559 (1990).
14. A. Parmaliana, F. Frusteri, D. Miceli, A. Mezzapica, M. S. Scurrell, and N. Giordano, "A basic approach to evaluate methane partial oxidation catalysts", *J. Catal.* 143: 262 (1993)
15. T. R. Baldwin, R. Burch, G. D. Squire, and S. C. Tsang, "Influence of homogenous gas phase reactions in the partial oxidation of methane to methanol and formaldehyde in the presence of oxide catalysts" *Appl. Catal.* **74**: 137 (1991).
16. Z. Sojka, R. G. Herman, and K. Klier, "Selective oxidation of methane to formaldehyde over doubly copper-iron doped zinc oxide catalysts via a selective shift mechanism", *J. Chem. Soc. Chem. Commun.*, 185 (1991).
17. D. Miceli, F. Arena, A. Parmaliana, M. S. Scurrell, and V. Sokolovskii, "Effect of the metal loading on the activity of silica supported MoO_3 and V_2O_5 catalysts in the selective partial oxidation of methane", *Catal. Lett.* 18: 283 (1993).
18. T. Weng, and E. E. Wolf, "Partial oxidation of methane on Mo/Sn/P silica supported catalysts", *Appl. Catal. A* 96:383 (1993)
19. Q. Sun, J. I. Di Cosimo, R. G. Herman, K. Klier, M. M. Bashin, "Selective oxidation to formaldehyde and C2 hydrocarbons over double layered Sr/La_2O_3 and MoO_3/SiO_2 catalyst bed", *Catal. Lett.* **15**: 371 (1992).
20. M. A. Bañares, and J. L. G. Fierro, "Methane-selective oxidation of silica-supported molybdenum(VI) catalysts" in "Catalytic Selective Oxidation", S. T. Oyama, and J. W. Hightower, editors. ACS, Washington, D. C., 1993
21. M. A. Bañares, A. Guerrero-Ruiz, I. Rodriguez-Ramos, and J. L. G. Fierro,"Mechanistic aspects of the selective oxidation of methane to C_1-oxygenates over MoO_3/SiO_2 catalysts in a single catalytic step", in "New Frontiers in Catalysis". L. Guczi, F. Solymosi and P. Tetenyi, editors. Elsevier, Amsterdam, 1993
22. M. M. Koranne, J. G. Goodwin Jr., and G. Marcellin, "Carbon pathways for the partial oxidation of methane", *J. Phys. Chem.* 97: 673 (1993)
23. M. A. Bañares, J. L. G. Fierro, and J. B. Moffat, "The partial oxidation of methane on MoO_3/SiO_2 catalysts: Influence of the molybdenum content and type of oxidant", *J. Catal.* 142: 406 (1993).

24. S. T. Oyama, G. T. Went, K. B. Lewis, A. T. Bell, and G. A. Somorjai, "Oxygen chemisorption and Laser Raman spectroscopy of unsupported and silica-supported vanadium oxide catalysts", *J. Phys. Chem.* 93: 6786 (1989).

25. N. K. Nag, K. V. R. Chary, and V. S. Subrahmanyam, "Characterization of supported vanadium oxide catalysts by low temperature oxygen chemisorption technique: II. The V_2O_5/SiO_2 system", *Appl. Catal.* 31: 73 (1987).

26. B. M. Reddy, B. Manohar, and E. P. Reddy, "Oxygen chemisorption on titania-zirconia mixed oxide supported vanadium oxide catalysts", *Langmuir* 9: 1781 (1993).

27. R. J. D. Tilley, and B. G. Hyde, "An electron microscopic investigation of the decomposition of V_2O_5", *J. Phys. Chem. Solids* 31: 1613 (1970).

28. M. A. Bañares, N. D. Spencer, M. D. Jones, and I. E. Wachs, "Effect of alkali metal cations on the structure of $Mo(VI)/SiO_2$ catalysts and its relevance to the selective oxidation of methane and methanol", J. Catal. 146: 204 (1994).

29. M. Faraldos, M.A. Bañares, J.A. Anderson, H. Hu, I.E. Wachs and J.L.G. Fierro, "Comparison of silica-supported MoO_3 and V_2O_5 catalysts in the selective partial oxidation of methane", submitted

METHANE CONVERSION IN AC ELECTRIC
DISCHARGES AT AMBIENT CONDITIONS

Rajat Bhatnagar and Richard G. Mallinson

Institute for Gas Utilization Technologies
School of Chemical Engineering and Materials Science
University of Oklahoma
100 East Boyd Street, Room T335
Norman, Oklahoma 73019

INTRODUCTION

Natural gas reserves, with methane as the primary constituent, are found in abundant quantities in different parts of the world. Industrial usage of natural gas is still limited to primarily a combustion fuel. The present industrial processes for the manufacture of methanol and other hydrocarbons from methane are economically feasible in areas where large reserves of natural gas are available. It is also not generally feasible to transport natural gas from remote places to areas of utilization. Therefore, to utilize such reserves economically, the conversion of methane to liquid products has been of great interest to many researchers.

During the course of the 20th century extensive research has been done in the area of alkane oxidation. That research was motivated by a need to understand the chemistry and free radical mechanisms of combustion. Partial oxidation, at elevated pressures and temperatures, has been investigated by several researchers[1-4]. It has been found that the yield of methanol and the other oxygenated products increased with the pressure. Efforts have also been made to find an efficient catalyst for this reaction [5-9]. However, noncatalytic yields are generally comparable to or better than those from the catalytic reactions.

Recent work in this field has examined the effect of surface reactions on the overall yield of methanol. It is generally accepted that "homogeneous systems" may be significantly affected by surface reactions. Gesser et al[3] and Feng et al[4], have achieved methanol selectivities of >70 % at methane conversions of 8 - 10 %. They attributed the high selectivities to the inert reactor lining, which may have minimized the surface reactions. Elsewhere [10], it has also been concluded that the effect of radical destruction at inert surfaces are significant and cannot be neglected.

The formation of ozone and oxides of nitrogen during a lightning discharge was the earliest observation of a chemical reaction initiated by an electrical phenomena. Numerous papers have been published on the subject of the silent discharge[11-22] most of which deal with ozonizers. Ozone formation in the silent electric discharge has been an extensively researched subject both from the chemical and electrical viewpoints. Presently the only gas phase electrical discharge chemical process that has become commercially viable is the production of ozone on a relatively small scale.

Work done in the early part of the twentieth century on methane in an electric discharge reactor led to some patents on several processes [23-24]. These dealt with the formation of formaldehyde from methane and oxygen [23] and various aldehydes from methane and carbon dioxide [24]. More recently Mach and Frost[25] studied the kinetics of methane conversion in a glow discharge. Methane, at pressures of 1 - 2 mm Hg, was fed to the reactor and a high DC voltage was applied to the system. Ethane, ethylene, acetylene and hydrogen along with layers of hydrocarbon on the tube walls were the main products. The extent of decomposition of methane was 98 %. A model was also proposed assuming several reactions involving radicals taking place simultaneously. Fraser et al [26], studied the effect of an electric discharge on nitrogen in the presence of small amounts of methane. That study was undertaken to gauge the potential of high voltage AC discharges as purification systems.

Mallinson et al[26], did some preliminary work on the partial oxidation of methane in alternating electric fields. The effect of several variables was investigated in a number of experimental runs. It was concluded that temperature along with type of waveform had little effect on the process over the range of conditions studied (20-55 °C and 50-300 Hz). Conversion was found to vary linearly with RMS voltage.

In the present study, steady state experiments were conducted to obtain qualitative and quantitative information about the effects of methane to oxygen ratio, residence time and voltage. The effect of these factors were gauged by analyzing conversions, selectivities and reaction rates. Another objective of this study was to determine the reaction pathways to the products formed in this reaction system. We have also quantified the effects using an empirical model for rate expressions which represent the rates of disappearance of methane and oxygen.

EXPERIMENTAL DETAILS

The reactant gases, methane (99.97) and oxygen (99.99), were supplied by Linde. All gases were used without additional purification. The flow rates were controlled by using fast response electronic mass flow controllers (Porter Instruments). The experimental apparatus used for the steady state runs is shown in Figure 1.

The reactant gases were mixed before entering the reactor. The discharge reactor used was made of two concentric cylinders with an annular volume of 225 milliliters. A cross sectional view of the reactor is shown in Figure 2. The inner cylinder was made of glass and a steel foil was pressed on its inner surface. This acted as one of the electrodes. The inner surface of the outer cylinder, which is made of steel acted as the other electrode. The area covered by the foil electrode and the gap distance were fixed at 920 cm^2 and 0.5 mm, respectively, for all the experiments conducted.

A condenser was used to remove water, methanol and other condensible products from the product stream. A cooling mixture of dry ice with acetone was found suitable for this reaction. An additional condenser was provided to insure quantitative recovery of the liquid products. The gaseous product stream was analyzed on-line using a Carle model 400AGC gas chromatograph. The liquid product collected during each run was analyzed by using a Varian 3300 Gas Chromatograph. using Porapak or Haysep Q and R packed columns.

The alternating current voltage was supplied through an Elgar AC Power Supply (Model 251B) with a Wavetek function generator (Model 182A). This feature allowed us to alter the frequency as well as the type of waveform. It was found that the type of waveform had no significant affect on the reaction system and therefore a sinusoidal waveform was used for all experiments. A FLUKE 8050A Digital Multimeter was used to measure the applied voltage. The output from the power supply was then sent through a transformer where the voltage was stepped up to the desired value.

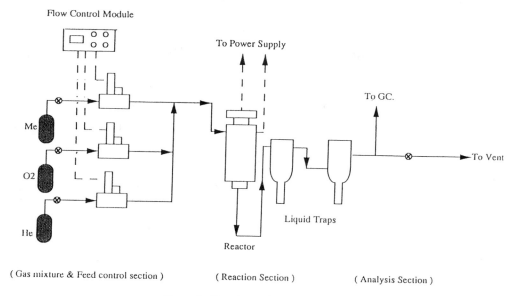

To Power Supply

To GC.

Me

O2

He

To Vent

Liquid Traps

Reactor

(Gas mixture & Feed control section)   (Reaction Section)   (Analysis Section)

Figure 1. Experimental Reaction System.

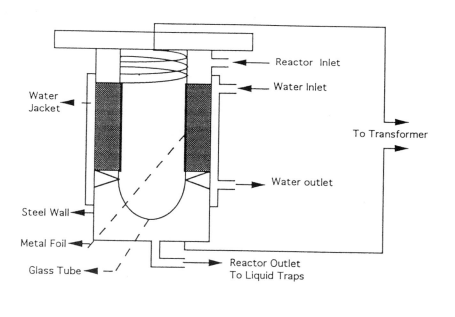

Reactor Inlet

Water Inlet

Water
Jacket

To Transformer

Water outlet

Steel Wall

Metal Foil

Glass Tube

Reactor Outlet
To Liquid Traps

Effective Reactor Volume

Figure 2. Reactor Vessel Cross-section.

The steady state experimental data were obtained over varying experimental conditions. The pressure in the system was maintained at 1 atm for all experiments. The range for the methane to oxygen feed ratio was between 0.5:1 and 25:1. For experiments with lower feed ratios (< 2.5) helium was introduced as a balance gas to avoid the formation of explosive mixtures. There appeared to be no discontinuity in the results with and without helium present. A threshold value for the voltage was observed below which no reaction occurred. This was just below 5 kV, RMS. At this voltage level, the peak voltage is about 7 kV and may be referred to as the "breakdown potential" of the system. The experimental range for the RMS voltage was between 5 kV and 18 kV, RMS. The constraint on the upper value for this study was the limitation of the power supply. The effect of residence time was also examined in this study. The residence time was varied by changing the total flow rate of the feed stream. The range for the total flow rate was between 150 and 1000 cc/min. With an internal volume of the annular space between the electrodes of 225 cm^3, the range of residence times is from 13.5 to 90 seconds.

A Few experiments was carried out with the feed consisting of various observed products combined with oxygen with and without helium and/or with methane. The products which were fed were ethane, methanol and carbon dioxide. These experiments were done to assist in understanding the reaction pathways. Carbon dioxide was introduced in the feed along with methane and oxygen.

Ethane and carbon dioxide were fed using mass flow controllers. Methanol was fed using a constant temperature bubbler. Methanol was kept at 0 °C in the bubbler and a stream of helium was bubbled through it. The temperature was maintained at 0 °C by using an ice water bath. The amount of methanol introduced was calculated by measuring the weight difference of the bubble during the period of time the helium was flowing.

For the purpose of the development of rate models, several experiments were done in which the oxygen to methane ratio was 20. The range of flow rates chosen was 200 cc/min - 800 cc/min. Helium was used as the balance gas in this set of experiments.

The overall conversions are defined as:

$$CH_4 \ Conversion \ = \ \frac{CH_4 \ of \ reactant \ - \ CH_4 \ of \ product}{CH_4 \ of \ reactant} \times \ 100 \ \%$$

$$O_2 \ Conversion \ = \ \frac{O_2 \ of \ reactant \ - \ O_2 \ of \ product}{O_2 \ of \ reactant} \times \ 100 \ \%$$

In general, the selectivity of any product is defined as:

$$Product \ Selectivity \ = \ \frac{Moles \ of \ product \ produced}{Moles \ of \ CH_4 \ converted} \times \ 100 \ \%$$

In this case, the selectivity was defined in terms of carbon i.e. amount of carbon converted to a particular product divided by the amount of carbon converted. In the case of ethane and ethanol there are two carbon atoms present and for these the equation for the selectivity will be:

$$C_2 \ Product \ Selectivity \ = \ \frac{2 \ \times \ Moles \ of \ product \ produced}{Moles \ of \ CH_4 \ converted} \times \ 100 \ \%$$

252

RESULTS AND DISCUSSION

As mentioned earlier, the primary variables examined in this study are Methane to Oxygen ratio, residence time and the voltage. The effects of these variables are discussed here.

Effect of Methane to Oxygen Feed Ratio

The variation of methane and oxygen conversions with the methane to oxygen ratio in the feed is illustrated in Figure 3. As can be seen from the figure, with an increasing fraction of methane in the feed mixture the conversion of methane decreases, whereas the oxygen conversion increases. In the presence of a high voltage electric discharge, pure methane remains essentially inactive while pure oxygen forms ozone. It has also been shown that ozone does not react with methane at room temperature[28]. Evidently, some in situ active form of oxygen must be required to initiate the reaction of methane. As the oxygen content is reduced in the feed, a lower density of oxygen based active species are present to attack the methane molecule. Consequently the methane conversion goes down rapidly with the decrease in oxygen partial pressure.

From the same set of experiments it was found that the rate of methane conversion decreased with an increase in the feed ratio. In this set of experiments, the percentage of methane in the feed was quite high and the percentage decrease in the oxygen concentration controlled the value of the conversion rate of methane. To explore the effect at lower partial pressures of methane, experiments were done where the partial pressure of oxygen was kept constant and the partial pressure of methane was varied. The total flow rate of the feed stream was kept constant by using helium as a balance gas. In this case, the rate of methane conversion was found to increase only slightly with increasing partial pressure of methane. Similar experiments were conducted at constant partial pressure of methane with varying oxygen partial pressure. The methane conversion rate was also relatively insensitive to the change in oxygen partial pressure. These observations are quantified in the Reaction Modelling section.

Figure 4 illustrates the variation of the alcohol and ethane selectivities with increasing feed ratio. There is a slight increase in the selectivity of the alcohols as can be seen from the figure. There is a significant, linear, increase in the selectivity of ethane as the methane percentage is increased in the feed. Ethane selectivity achieves a level of about 20 percent at a methane to oxygen feed ratio of 20. As the concentration of methane is increased, the probability of an activated methane specie (such as methyl) finding another one is relatively more probable than finding oxygen species and further reacting, thereby increasing the likelihood of ethane formation. Essentially no ethylene or hydrogen has been observed in the product streams.

The COx selectivities are shown in Figure 5. At lower values of the methane to oxygen feed ratios, both carbon dioxide and carbon monoxide selectivities decrease with increasing ratio. The carbon oxide selectivities are relatively constant at higher values of the feed ratios, where methane conversion is on the order of a few percent.

Effect of Residence Time

Figure 6 shows the variation of conversion of methane and oxygen with residence time. Both conversions increase with increasing residence time (decreasing total flow rate). The variation of the ratio of the conversion rates of methane and oxygen with the residence time was also calculated and was found to increase slightly with decreasing residence time, indicating that at longer residence times a higher degree of oxidation occurs.

Figure 7 shows the dependence of alcohol and ethane selectivities with residence time. It can be seen that both methanol and ethanol selectivities show a decreasing trend with increasing residence time (decreasing total flow rate) in the range studied. The ethane

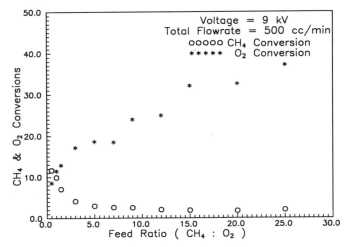

Figure 3. Methane and oxygen conversions as a function of methane to oxygen ratio.

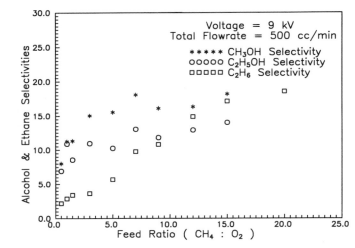

Figure 4. Alcohol and ethane selectivities as a function of methane to oxygen feed ratio.

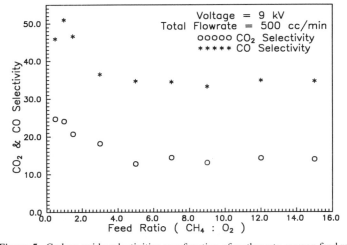

Figure 5. Carbon oxide selectivities as a function of methane to oxygen feed ratio.

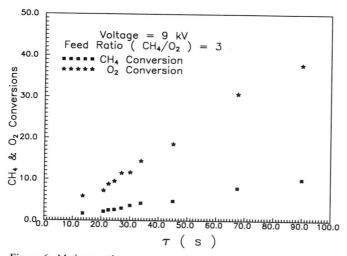

Figure 6. Methane and oxygen conversions as a function of residence time.

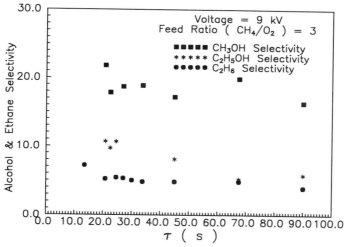

Figure 7. Alcohol and ethane selectivities as a function of residence time.

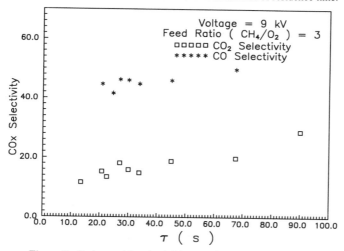

Figure 8. Carbon oxide selectivities as a function of residence time.

selectivity shows a decreasing trend at low residence times, but becomes almost constant at about 4.5 percent for residence times greater than about 20 seconds.

The variation of COx selectivity with residence time is shown in Figure 8. The carbon monoxide selectivity does not change significantly under the conditions studied, but the selectivity for carbon dioxide increases at long residence times. Since the formation of CO_2 requires more moles of oxygen per mole of methane, the ratio of rates between the reactants also may be expected to change with the residence time.

These results suggest that at higher values of residence times the methanol decomposes to carbon oxides and water. The same is the case with ethane. This suggests that a series pathway for the formation of the carbon oxides exists. Since the selectivity of carbon monoxide remains relatively constant in the presence of these other selectivity shifts, a pathway leading to its formation which is parallel to the other products may be suggested. These results also suggest that the pathways for formation of methanol and ethane are parallel. These tentative conclusions do not imply exclusivity for these pathways, but rather that these represent significant contributions to the rates of production and destruction of the various products.

To gain further information about the pathways, a few experiments were done in which methanol was introduced into the feed. The amount of methanol introduced was designed to be similar to that produced when methane and oxygen were reacted at the same conditions. In the experiment done with methanol and oxygen only, the methanol converted went entirely to carbon oxides and water, no ethanol was detected. The conversion was in excess of 75 %. This suggests that the rate of destruction of methanol is quite high under these conditions. Since the experiments with methane and oxygen only under the same conditions lead to some net positive formation rate of methanol, the rate of formation of methanol must also be very high. In the experiment done with methane, oxygen and methanol it was found that the selectivity of methanol was reduced to a great extent. When the carbon oxide selectivities, from this experiment, were compared with the experiment in which only methane and oxygen were fed under the same conditions, it was found that they increased slightly. Of more interest is the fact that the ratio of CO_2/CO is nearly two to one in contrast to the value of this ratio for the standard methane/oxygen feed experiments, where it significantly below one. This indicates that the destruction of methanol via over-oxidation increased. There was essentially no effect on the ethanol selectivity.

Similar experiments were conducted with ethane. In the experiment done with ethane and oxygen only, the products formed consisted mainly of carbon oxides (CO_2/CO about 2/1) and water with some traces of methanol and ethanol. In the experiment done with ethane, methane and oxygen, there was a decrease in the selectivity for ethane and also an increase in the carbon oxide selectivities when compared with the experiment in which methane and oxygen only were fed. This suggests that the destruction of ethane via over-oxidation is also increased.

It can be concluded from these experiments that the primary pathway leading to ethanol is independent of either ethane or methanol. It can be seen from Figure 11 that with increasing residence time the ethanol selectivity drops. This indicates that at higher residence times ethanol decomposes to carbon oxides and water.

Despite the mild decrease in methanol, ethanol and ethane selectivities at longer residence time (and the concomitant higher conversion) the low CO_2/CO ratio persists despite the fact that, at least when no methane is present, the oxidation products of these species greatly favors CO_2.

In experiments done with a feed of carbon dioxide, methane and oxygen, the flow rate of carbon dioxide ranged from 50 cc/min to 250 cc/min in a total flow rate of 500 cc/min. The methane to oxygen ratio was kept at 3:1. The results show that with increasing fraction of carbon dioxide in the feed, the overall selectivity for carbon dioxide decreased. These runs were compared with the experiments in which methane, oxygen and helium were fed at the same conditions, with helium replacing the carbon dioxide. It was found that the selectivity for carbon monoxide increased considerably. The conversions of methane were slightly lower and the selectivity for alcohols also decreased. These results suggest that the presence of excess carbon dioxide favors the formation of carbon monoxide, but it remains unclear if carbon monoxide might be formed from carbon dioxide directly.

Using the information attained from these experiments, in which products were introduced in the feed, and the experiments done earlier, a schematic representation of the pathways to various products is postulated, shown in Figure 9.

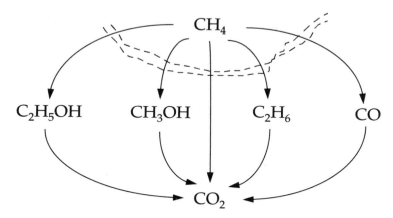

Figure 9. Tentative major reaction pathway schematic.

Effect of Applied Voltage

Figure 10 shows the variation of methane and oxygen conversions with voltage. As can be seen from the figure, both methane and oxygen conversions increase linearly with increasing voltage. Although the maximum conversion is lower than with the residence time experiments, the behavior is otherwise quite similar.

The exact interpretation of the changes to the discharge when the applied voltage is increased is not completely clear. It has been established[17,18] that the effect of increasing the voltage, with a fixed geometry, leads to an increase in the quantity of electricity transferred between the electrodes i.e. an increase in the current. The electrodes have imperfections on the surface and when the voltage is increased to a sufficiently high level, the discharge of electrons takes place from spots which can least support the potential difference. Evidently, an increase in the applied voltage cannot lead to a higher potential difference between the electrodes, but instead, leads to more discharge sites on the surface of the electrode. There is some evidence that the number of electrons per discharge event isn't significantly altered under these conditions. These additional electrons may be supposed to increase the rate of initiation and other electron driven reactions. Consequently, the conversions of the reactants may be expected to increase, as is observed in Figure 10.

Figure 11 shows the dependence of the alcohol and ethane selectivities on the voltage. Unexpectedly, the methanol selectivity shows an increase with increasing voltage while ethanol selectivity goes down. At higher voltages, from 12 to 18 kV RMS, the change in the selectivities is very small, despite the continued linear increase in reactant conversions in this range. This behavior is clearly not analogous to longer residence times. In the discussion of the effect of voltage made above, one might expect an analogous behavior, in which an element of the gas traveling through the reactor is exposed to a higher density of similar discharges when its velocity is slowed (longer residence time) or applied voltage is increased. The variation of ethane selectivity with the voltage is almost negligible, decreasing from about five percent to four percent from the lowest to highest voltage. This result, along with that of ethanol, may

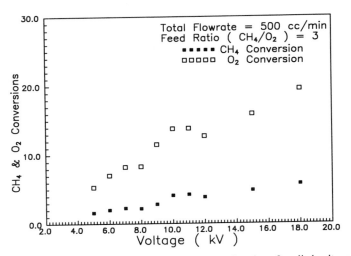

Figure 10. Methane and oxygen conversions as a function of applied voltage.

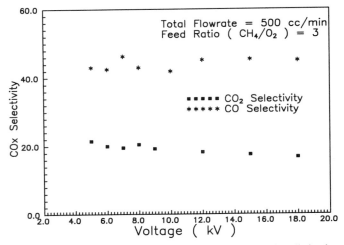

Figure 11. Alcohol and ethane selectivities as a function of applied voltage.

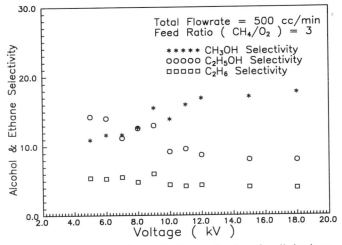

Figure 12. Carbon oxide selectivities as a function of applied voltage.

suggests that larger compounds (or the C-C bond) tend to be more reactive under these conditions. This "selective" destruction of C-C bonds may result in fragments which lead to methanol. The change in the discharge physics which lead to this chemistry is not presently known. Figure 12 shows the dependence of carbon oxides selectivities on the voltage. The carbon monoxide selectivity remains constant while the carbon dioxide selectivity change only slightly, with a decreasing trend at higher voltages.

At higher voltages than achieved here the concentration of active oxygen based species may be expected to increase leading to an increase in the probability of over-oxidation. If true, this may be expected to lead to an increase in carbon oxide selectivities as observed in the case of longer residence time experiments with higher conversion levels.

Reaction Modelling

The conversion rate of methane and oxygen in a plug flow reactor are assumed to be represented by power law model. These equations, where τ is the residence time, are shown below:

$$(-\frac{dP_{CH_4}}{d\tau}) = k_1 \, P_{CH_4}^{a_1} \, P_{O_2}^{b_1} \quad \text{------} \quad (1)$$

$$(-\frac{dP_{O_2}}{d\tau}) = k_2 \, P_{CH_4}^{a_2} \, P_{O_2}^{b_2} \quad \text{------} \quad (2)$$

Since the fractional conversion of methane was $< 10\%$ for most of the experiments, the partial pressure of methane remained essentially constant during the course of the reaction. Thus the term corresponding to the partial pressure of methane can be lumped with the rate constant to obtain an apparent rate constant. Equation (2) becomes:

$$(-\frac{dP_{O_2}}{d\tau}) = k_2' \, P_{O_2}^{b_2} \quad \text{------} \quad (3)$$

Integrating the above equation we have:

$$P_{O_2} = [-k_2'(1-b_2)\tau + P_{O_2 i}^{(1-b_2)}]^{\frac{1}{(1-b_2)}} \quad \text{------} \quad (4)$$

The software package SAS [29] was used to carry out the non linear regression of the above equation. The data points used for the regression included the varying residence time experiments, varying ratio experiments and the varying individual partial pressure experiments. The values of the parameters estimated are given in Table 1.

Table 1. Initial model parameter estimates.

	1	2
k'	0.00197	0.00324
k''	0.000778	0.000013
a	0.377	-0.875
b	0.24	0.38

Equation (1) can be written as:

$$(-\frac{dP_{CH_4}}{d\tau}) = k_1' \, P_{O_2}^{b_1}$$

$$=k_1'[-k_2'(1-b_2)\tau + P_{O_2 i}^{(1-b_2)}]^{\frac{b_1}{(1-b_2)}} \quad \text{-- (5)}$$

Integrating the above equation we get:

$$P_{CH_4} = \frac{k_1'}{k_2'(b_1+1-b_2)}(\, [-k_2'(1-b_2)\tau + P_{O_2 i}^{(1-b_2)}]^{1-\frac{b_1}{(1-b_2)}} - P_{O_2 i}^{(b_1 \cdot 1-b_2)}) + P_{CH_4 i}$$

$$\text{------ (6)}$$

The parameters for Equation (6), obtained by regression are also given in Table 1.

To estimate the exponents for the partial pressure of methane, experiments were conducted in which oxygen was present as a high percentage in the feed in order to maintain a nearly constant oxygen partial pressure. The ratio chosen was 20, i.e. $O_2/CH_4 = 20$. The voltage was kept constant at 9 kV. The range of total flow rates used was from 200 to 800 cc/min. The pseudo-first order equations derived from (1) and (2), in terms of apparent rate constants, are shown below:

$$(-\frac{dP_{CH_4}}{d\tau}) = k_1'' \, P_{CH_4}^{a_1} \quad \text{------ (7)}$$

$$(-\frac{dP_{O_2}}{d\tau}) = k_2'' \, P_{CH_4}^{a_2} \quad \text{------ (8)}$$

The procedure to determine the parameters was similar to the one followed in the previous case. The estimated values of these parameters are also shown in Table 1. The equations (1) & (2) now become:

$$\left(-\frac{dP_{CH_4}}{d\tau}\right) = k_1 \ P_{CH_4}^{0.377} \ P_{O_2}^{0.24} \quad \text{------} \quad (9)$$

$$\left(-\frac{dP_{O_2}}{d\tau}\right) = k_2 \ P_{CH_4}^{-0.875} \ P_{O_2}^{0.38} \quad \text{------} \quad (10)$$

Regressions on these equations are now made to determine the final values of the rate constants:

$$k_1 = 0.001495 \qquad r^2 = 0.993$$
$$k_2 = 0.001446 \qquad r^2 = 0.999$$

These parameters have been used to compare the model estimates with the data for the partial pressure "profiles" of the reactants as a function of residence time. As can be seen from Figures 13 and 14, the estimates are very close to the experimental values. It was also found that the estimated and experimental values of the methane conversion rate for the other experiments were also quite accurate.

The reaction orders, both for the partial pressure of methane and the partial pressure of oxygen, respectively, are small for the conversion rate of methane. The methane conversion rate increases only slightly with an increase in the partial pressure of either reactant. For the oxygen conversion, the dependence on oxygen is again relatively small, but the dependence on methane is significant and negative.

CONCLUSIONS

The investigation of the effects of methane to oxygen ratio, residence time, and applied voltage on the partial oxidation of methane using an electric discharge as the initiator has begun an understanding of the chemical pathways present in this process. A high fraction of oxygen is required in the feed to obtain a high conversion of methane, implying short active species chain lengths. The degree of over-oxidation decreases with increasing feed ratio. It increases slightly at longer residence times and is relatively insensitive to voltage. Parallel pathways appear to exist for the formation of methanol and ethane. Since the selectivity of carbon monoxide remains relatively insensitive to reaction conditions, it may also be formed in a primarily parallel path. A series pathway exists for the formation of carbon dioxide with alcohols and ethane as the intermediate products. The pathway leading to the formation of ethanol appears to be independent of either ethane or methanol. It was also found that carbon dioxide in the feed inhibits further carbon dioxide production in the reactor.

A power law dependence on the reactant partial pressures was determined to adequately fit the experimental conversion data using a plug flow reactor model. The results showed that the rates have a relatively low, positive, dependence on reactant partial pressures except that the rate of oxygen conversion has a significant negative dependence on the methane partial pressure.

Further work is in progress to better understand the reaction pathways of the products through more extensive product reaction experiments which will include isotopic labeling experiments. Further work is also being conducted on the role of the electron physics and reactor geometry on these reactions.

ACKNOWLEDGEMENTS

The assistance of the University of Oklahoma Sarkeys Energy Center and partial support from the US Department of Energy under grant number: DE-FG21-94MC31170 are appreciated.

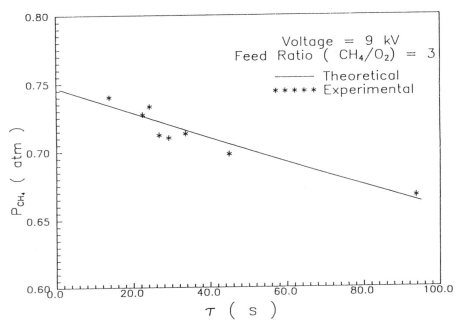

Figure 13. Fit of methane conversion model for effluent partial pressure of methane with data as a function of residence time.

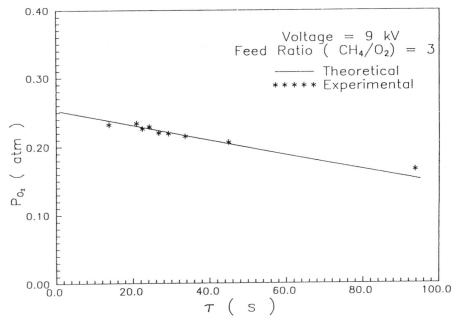

Figure 14. Fit of oxygen conversion model for effluent oxygen partial pressure with data as a function of residence time.

REFERENCES

1. Lott, J.L; Sliepcevich, C.M. Partial Oxidation of Methane at High Pressures. *Ind. Eng. Chem. Process Des. Develop.* **1967**, 6, 67.

2. Hardwicke, N.L; Lott, J.L; Sliepcevich, C.M. Oxidation of Methane at High Pressures. *Ind. Eng. Chem. Process Des. Develop.* **1969**, 8, 133.

3. Gesser, H.D.; Hunter, N.R.; Prakash, C.B. The Direct Conversion of Methane to Methanol by Controlled Oxidation. *Chemical Reviews.* **1985,** 85, 4, 236.

4. Feng, W., F. C. Knopf and K. M. Dooley; Effects of Pressure, Third Bodies, and Temperature Profiling on the Noncatalytic Partial Oxidation of Methane. *Energy & Fuels,* **1994**, 8, 815.

5. Boomer, E.H; Thomas, V. The Oxidation of Methane at High Pressures. *Can J. Research.* **1937**, 15B, 401.

6. Boomer, E.H; Thomas, V. The Oxidation of Methane at High Pressures. *Can J. Research.* **1937**, 15B, 414.

7. Boomer, E.H; Naldrett, S.N. The Oxidation of Methane at High Pressures. *Can J. Research.* **1947**, 15B, 494.

8. Kastanas, G.N.; Tsigdinos, G.A.; Schwank, J. Selective Oxidation of Methane Over Vicor Glass, and Various Silica, Magnesia and Alumina Surfaces. *Appl. Catal.* **1988**, 44, 33.

9. Liu, H.F.; Liu, R.S.; Liew,K.Y.; Johnson,R.E.; Lunsford,J.H. Partial Oxidation of Methane by Nitrous Oxide Over Molybdenum on Silica. *J. Am. Chem. Soc.* **1984,** 106, 4117.

10. Thomas, D.J.; Willi, R.; Baiker, A.; Partial Oxidation of Methane: The Role of Surface Reactions. *Ind. Eng. Chem. Res.* **1992**, 31, 2272.

11. McTaggart, F.K. Plasma Chemistry in Electrical Discharges. Elsevier Publishing Company, Ist Edition. **1967,** 9.

12. Glockler, G.; Lind, S.C. The Electrochemistry of Gases and Other Dielectrics. John Wiley & Sons, Inc. **1939,** 29.

13. Honda, K.; Naito, Y. On the Nature of Silent Electric Discharge. *Journal of the Physical Society Of Japan.* **1955**, 10, 11, 1007.

14. Sugimitsu, H.; Asakura, R.; Okazaki, S.; Suzuki, M. Chemical Reaction in Silent Electric Discharge I. *Bulletin of the Chemical Society Of Japan.* **1970** 43, 1927.

15. Suzuki, M. On the Nature of Chemical Reaction in Silent Discharge. *Proc. Japan Acad.* **1950**, 26, 20.

16. Suzuki, M; Naito, Y. On the Nature of the Chemical Reaction in Silent Electrical Discharge II. *Proc. Japan Acad.* **1952**, 28, 469.

17. Morinaga, K.; Suzuki, M.; The Chemical Reaction in Silent Electric Discharge I. *Bull. Chem. Soc. of Japan.* **1961**, 34, 157.

18. Morinaga, K.; Suzuki, M.; The Chemical Reaction in Silent Electric Discharge II. The Frequency Effect on Ozone Formation. *Bull. Chem. Soc. of Japan.* **1962**, 35, 204.

19. Morinaga, K. The Reaction of Hydrogen and Oxygen Through a Silent Electric Discharge I. The Formation of Hydrogen Peroxide. *Bull. Chem. Soc. of Japan.* **1962,** 35, 345.

20. Khan, A.A. Studies on the Inhomogeneous Nature of the Silent Discharge Reactor. Part 1. Experimental Investigations. *The Canadian Journal of Chemical Engineering.* **1989**, 67, 102.

21. Khan, A.A. Studies on the Inhomogeneous Nature of the Silent Discharge Reactor. Part 2. Kinetic Simulation of a Differential Batch Reactor. *The Canadian Journal of Chemical Engineering.* **1989**, 67, 107.

22. Khan, A.A. Studies on the Inhomogeneous Nature of the Silent Discharge Reactor. Part 3. A Multi-Zoned Flow Model. *The Canadian Journal of Chemical Engineering.* **1989**, 67, 113.

23. Nashan, P. *U.S. Patent*, Process for the Manufacture of Formaldehyde from Methane. **1933.**

24. Finlayson, D.; Plant, J.H.G. *U.S. Patent*, Manufacture of Aliphatic Aldehydes. **1935.**

25. Mach, R.; Drost, H. Kinetics of the Methane Conversion in a Glow Discharge up to Chemical Equilibrium. *Sixth International Sym. on Plasma Chemistry, Montreal, Canada.* **1983**, 251.

26. Fraser, M.E.; Fee, D.A.; Sheinson, R.S. Decomposition of Methane in an AC Discharge. *Plasma Chemistry and Plasma Processing.* **1985**, 5, 163.

27. Mallinson, R. G., C. M. Sliepcevich and S. Rusek. Methane Partial Oxidation in Alternating Electric Fields. *Preprints Am. Chem. Soc. Fuel Chem. Div.*,**1987**, 32(3), 266.

28. Mallinson, R.G.; Sliepcevich, C.M. Field Effect Catalysis Program. *Unpublished results*, **1986**.

29. SAS Statistical Software, SAS Institute, Inc., various versions.

CATALYTIC STUDIES OF METHANE PARTIAL OXIDATION

G. Stewart Walker and Jacek A. Lapszewicz

CSIRO, Division of Coal and Energy Technology
Lucas Heights Research Laboratory
Private Mail Bag 7, Bangor
Australia, NSW 2234

INTRODUCTION

A number of catalysts have been reported in patents and literature that are claimed to produce methanol at selectivities in excess of 75% by direct partial oxidation of methane. Investigations of potential catalysts for the heterogeneous direct conversion of methane to methanol have been undertaken.

This work is a continuation of the systematic investigations of the homogeneous gas phase reaction that have been undertaken in CSIRO[1-5] laboratories. Catalysts investigated include MoO_3:UO_2 - patented by ICI[6], Fe/Sodalite in a novel Bypass Reactor - patented by Sun Chemical Company[7-9] and Stannic Oxide - reported by Anthony and Helton[10]. A comparison of the heterogeneous catalytic, inert packings and gas phase selectivities have been made to illustrate the effect of each catalyst.

EXPERIMENTAL

To accommodate the high space velocities described in the ICI patent[6], a new reactor was designed. The reactor was constructed of stainless steel 316 and lined with a specially-shaped, closefitting quartz glass insert. The internal diameter of the insert was 4mm for the first 200mm length, and then tapered to 2mm i.d. to minimise the residence time of the reaction products.

The catalyst (50 mm bed height) was placed directly above the taper. This ensured that product gases leaving the catalyst bed were accelerated out of the reactor. For the ICI catalyst experiments were carried out using gas flow rates of 250, 2055 and 5138 mlmin[-1], corresponding to GHSVs of 2 432, 20 000 and 50 000 h[-1], respectively.

The tubular reactor was modified in such a way as to repeat the work reported by the Sun Chemical Company[7-9]. This allowed a portion of the gas flow to be drawn off from a variable distance above the catalyst bed, without passing through the catalyst bed. Needle valves at the base of the reactor and on the side arm, were calibrated so that the relative amount of gas drawn off through the bypass and the amount of gas going through the catalyst bed can be controlled.

Lines leading to the G.C. were all insulated and heated to 160°C. By opening selected valves, a sample of gas that had come through the bypass or a sample of gas that had come through the catalyst or a sample of the combined gas could be analysed on the GC.

Preparation of $MoO_3.UO_2$ on SiO_2/Al_2O_3

Ammonium diuranate, $(NH_4)_4(UO_2)_2$, was prepared by precipitation from a solution of uranyl nitrate by addition of excess aqueous ammonia solution.

For a Mo:U ratio of 2:1, $(NH_4)_4(UO_2)_2$ (0.4872g) was dissolved in water and poured over 10g of silica/alumina support in a rotary evaporator flask, and evaporated to dryness. The resultant mixture was calcined at 450°C for 3 hours under hydrogen, and then cooled.

When cold, the hydrogen was switched off, and carbon dioxide was passed over the mixture to purge hydrogen out of the system. 0.4872g of $(NH_4)_6Mo_7O_{24}.4H_2O$ was weighed, dissolved in water, and added to the calcined sample of uranium oxide impregnated silica/alumina, which had been slurried with water in a small rotary evaporator flask. The resultant slurry was evaporated to dryness, and the final catalyst mixture, which had a Mo:U ratio of 1.9:1, collected.

To prepare the catalyst with a Mo:U ratio of 1:2, the above procedure was repeated using 1.48g of silica/alumina support, 0.144g of $(NH_4)_4(UO_2)_2$ and 0.0272g of $(NH_4)_6Mo_7O_{24}.4H_2O$. In this case the final catalyst composition had a Mo:U ratio of 1:2.2.

Fe/Sodalite preparation

Fe/Sodalite (11.9% Fe) was prepared in a similar fashion to that reported in the literature[7-9]. Fe/Sodalite was characterised by standard methods.

Coating of reactor lining with Fe/Sodalite

Fe/sodalite was mixed with sodium silicate solution (RECTAPUR) to form a slurry. The slurry was poured down the inside of the quartz reactor lining until it was coated with a uniform layer. A hot air blower was used to dry the slurry onto the side of the tube.

Preparation of 0.5% Pd on MgO

Catalyst preparation and characterisation of a number of methane activating catalysts have been undertaken. Palladium has been reported to absorb methane at room temperature and produce only methyl radicals above 130°C.[11]

Catalysts containing 0.5% loading of Rh, Ru, Pt and Pd were investigated to determine the extent of hydrogen:deuterium exchange at 500°C. Palladium was found to be the most suitable because it gave 100% exchange of one hydrogen (Table 1), whereas the other metals were found to favour complete H:D exchange.

Catalysts with 0.5% loading of Palladium on magnesium oxide support have been tested, before and after reduction in hydrogen. The palladium catalyst was prepared by an incipient wetness technique using an aqueous solution of $PdCl_2$ (Merck >99%) to produce a 0.5% w/w loading on commercial magnesium oxide (Univar, 97% purity, BET surface area $117m^2g^{-1}$).

Reduction was accomplished by heating the catalyst to 500°C with $100mlmin^{-1}$ of H_2 passing over the catalysts in situ.

Table 1. Deuterium exchange at 500°C over MgO supported catalysts.

Metal (0.5%)	CH_4 Conv. (%)	CD_4 (%)	CD_3H (%)	CD_2H_2 (%)	CH_3D (%)
Rh	92.3	71.5	22.1	–	6.4
Ru	29.3	81.9	11.9	6.1	–
Pt	8.2	56.1	–	–	43.9
Pd	<2.0	–	–	–	100

Preparation of γ-Alumina

γ-alumina (Al_2O_3, Merck, Anhydrous extra pure, >99%, characterised using XRD: surface area = 114.1 m^2g^{-1}; average pore radius = 3.62nm) was compressed into pellets 3.3mm in diameter and 2.3mm in height. 15.64g of these pellets was loaded into the reactor to make a catalysts bed of 26cm depth. The catalyst was calcined in situ for 2 hours at 400°C in air (CIG Instrument Grade) flowing at 3cm^3min^{-1}.

Preparation of Stannic Oxide

Stannic oxide (Merck extra pure, >99%) was compressed into pellets 3.3mm in diameter and 1mm in height. 32.2g of these pellets was loaded into the reactor to make a catalysts bed of 28.5cm depth. The catalyst was calcined in situ for 2 hours at 400°C in air (CIG Instrument Grade) flowing at 30cm^3min^{-1}. The surface area of the Stannic Oxide was determined to be 5.7m^2g^{-1} (BET) which is similar to surface area of 5.8m^2g^{-1} reported for the catalyst used in previous work.

RESULTS

MoO_3:UO_2

The maximum selectivity to methanol was 44%. The patent specified maximum selectivity to methanol of 75% at highest GSHV of 50 000h^{-1}. The contact time of the gas at this space velocity was less than 0.2 seconds. There was no significant difference (Table 2) between the runs with the support or the two formulations of the Mo:U catalyst at GHSV of between 50,000 and 2432 h^{-1}. However, when the reaction is conducted at lower space velocity corresponding to a flow of 200mlmin⁻¹, the catalyst can be seen to have a detrimental effect on the selectivity to methanol.

Table 2. Selectivity at 5.4MPa for various GHSV's for MoO_3:UO_2.

GHSV	T_{wall} °C	Conversion (%)		Selectivity (%)		
		O_2	CH_4	CH_3OH	CO	CO_2
11 h^{-1}	438	92	5	23	59	18
2432 h–¹	466	60	1.3	49	42	8
20 000 h^{-1}	525	93	5	38	53	9
50 000 h^{-1} Mo:U 1:2.3	543	76	2.2	43	52	6
50 000 h^{-1} Mo:U 2:1	537	42	1	39	55	7
Gas Phase	388	84	2.6	47	45	8

Alumina and Stannic Oxide

Filling the reactor with alumina pellets reduced the methanol selectivity from around 30% to 13% and increased the selectivity to CO_2 from around 26% to 55%. Similarly stannic oxide reduced the amount of methanol to 6% with an increase in CO_2 to 56%. This is consistent with reports in the literature of oxidation of methanol and CO to CO_2 by alumina and stannic oxide[12-14]. Table 3 indicates the reduction in methanol selectivity for alumina or stannic oxide compared with the reactor filled with glass spheres.

Table 3. Conversion and Selectivities for Alumina and Stannic Oxide.
5% O_2, F=200 ml min^{-1}, P=3MPa, XO_2 = 60%

Reactor Filling	Conversion %	Selectivity %		
	XO_2	CH_3OH	CO	CO_2
Gas	56	29	47	24
Spheres	64	31	43	26
Alumina	57	13	32	55
SnO_2	58	6	39	56

Fe/Sodalite Catalyst System

A comparison of selectivities at around 60% oxygen conversion indicates (Table 4) that there is no significant difference in the selectivities with the homogeneous gas phase reaction in an empty reactor, in the bypass reactor with the bypass off, and with the bypass on.

Passing the reactants through a reactor with Fe/sodalite coating the reactor lining decreases the selectivity to methanol (35% to 23%) with a resultant increase in CO and a slight increase in CO_2 production (Table 5).

Tables 4 and 5. Oxygen Conversions and Product Selectivities for Fe/Sodalite with and without bypass. 5% O_2 F = 55mlmin^{-1}, P = 5.4MPa, XO_2= 60%

TABLE 4		Conversion %		Selectivity %	
CATALYST	BYPASS	XO_2	CH_3OH	CO	CO_2
NO	NONE	64	35	51	15
NO	FULLY OFF	70	33	51	17
NO	1:1	59	33	51	16

TABLE 5		Conversion %		Selectivity %	
CATALYST	REACTOR LINING	XO_2	CH_3OH	CO	CO_2
NO	NONE	64	35	51	15
YES	Fe/SOD	67	23	59	18

In the reactor with a fixed bed of catalyst and the bypass off, or half on, the selectivity to methanol is greatly reduced to 18% with an increase in CO and CO_2. When the bypass is turned on, the selectivities are similar to the gas phase reaction (Table 6).

Table 6. Oxygen Conversions and Product Selectivities for Fe/Sodalite with and without bypass. 5% O_2 F = 55mlmin^{-1}, P = 5.4MPa, XO_2= 60%

| TABLE 6 | | Conversion % | | Selectivity % | |
CATALYST	BYPASS	XO_2	CH_3OH	CO	CO_2
NO	BLANK	64	35	51	15
YES	FULLY ON	60	27	51	22
YES	1:1	60	17	53	30
YES	FULLY OFF	62	19	53	28

It can be concluded that if the gases pass through the catalyst, the selectivity to methanol is reduced and the selectivity to carbon dioxide is increased. The effects of the bypass seems to be to give an alternative path for the gases that avoids contact with the catalysts.

CONCLUSIONS

The catalysts tested have a detrimental effect on methanol selectivity for a combination of two reasons:

1) The presence of catalysts in the reactor reduces the volume of the reactor and so reduces the residence time of the gases in the heated zone. A higher temperature is required to initiate the reaction and this does not favour methanol production.
   And/Or
2) The presence of metal oxides can actively encourage deep oxidation to CO_2.

Of the two patented catalyst systems investigated the ICI catalyst of MoO_3:UO_2 used high space velocities to minimise the contact of gases with the catalyst and the Sun Chemical Iron Sodalite used a bypass in the reactor that allows some or all of the gases to avoid contact with the catalysts.

None of the catalysts studied have shown the high selectivity of 70% that was claimed in the patents and none have shown an improvement on the homogeneous gas phase selectivities.

REFERENCES

1. Foulds, G.A., Miller, S.A., Walker, G.S. and Gray, B.F. , CSIRO Investigation Report CET/IR016. 1991.
2. Foulds, G.A., Miller, S.A. and Walker, G.S., 'Direct partial oxidation of methane to methanol', CHEMECA 91, 1991, 566-573.
3. Foulds, G.A., Charton, B.G., Walker, G.S. and Gray, B.F., CSIRO Investigation Report CET/IR104., 1992.
4. Foulds G.A., Miller S.A. and Walker G.S., Symposium on Natural Gas Upgrading ll, Presented before the Division of Petroleum Chemistry, Inc., American Chemical Society, San FranciscoMeeting, 1992, April 5-10, 26-33.
5. Foulds G.A., Gray B.F., Griffiths .F. and Walker G.S., Symposium on Natural Gas Upgrading ll, Presented before the Division of Petroleum Chemistry, Inc., American Chemical Society, SanFrancisco Meeting, 1992, April 5-10, 51-60.

6. Dowden, A.D. and Walker, G.T., British Patent 1244001, Aug. 25, 1968.
7. Durante V.A., Walker D.W., Seitzer W.H. and Lyons J.E., Pacifichem '89, Abstract, Conference Proceedings, 1989, 48, 23-26.
8. Durante V.A., Walker D.W., Gussow S.M. and Lyons J.E., United States Patent 4,918,249, April 17, 1990
9. Lyons, J.E., Ellis P.E. and Durante, V.A., Structure-Activity and Selectivity Relationships in Heterogeneous Catalysis, Elsevier Publishers B.V. Amsterdam, 1991, 99-115
10. Anthony, R.G. and Helton T.E., Pacifichem '89, Abstract, Conference Proceedings, 1989, 143
11. Foger, K., Private Communication, 1993.
12. Kurina, L.M., Mekhanizm Katalit. Reaktsii. Materialy 3 Vses. Konf., 1982, 47-50.
13. Bobyshev, A.A. and Radstig, V.A., Geotog. Katal. Mater. Vses. Konf. Katal. Reakts. 3rd 1981,1982, 196.
14. Ai. J., J. Catal., 1978, 54, 426.

SOME CHARACTERISTICS OF THE PARTIAL OXIDATION OF CH$_4$ TO CH$_3$OH AT HIGH PRESSURES

Hyman D. Gesser,[1] Norman R. Hunter,[1] and Albert N. Shigapov [2]

[1] Department of Chemistry, University of Manitoba,
   Winnipeg, MB. Canada R3T 2N2

[2] Institute of Chemistry of Natural Organic Materials, Krasnoyarsk, Russia

INTRODUCTION

The direct conversion of methane to methanol by the partial oxidation process at high pressure has been studied for about 60 years [1,2,3,4] but in the last 5 years more has been published [5-27] than in all previous years. The discrepancies in the results obtained by different workers seems to have been explained by the cool flame characteristics of the system, first described by Yarlaggada *et al.* [23] and later shown by Foulds *et al.* [8] and Charlton *et al.* [28] to exhibit to a hysteresis effect with change in reaction temperature. More recently, high selectively and conversion were obtained by Dooley [29] by using a pre-heater to accommodate the induction period, and a post reactor temperature quencher to prevent further oxidation of the methanol.

It was also shown that a third body effect (the addition of CO$_2$ to the reacting mixture) enhances methanol formation. We now report on the possible cage effect in the reacting system at high pressures P$\geq$100 atm.

EXPERIMENTAL

The basic apparatus has been previously described [21, 22, 25] except for the addition of an FID detector after the TCD detector. A nickel catalyst of 20% Ni on porous glass (pore size 195Å) was placed between the two detectors to reduce CO, HCHO, CO$_2$ CH$_3$OH and other oxygenates to CH$_4$. This simplified the calibration of the gas chromatograph for the on-line analysis of the product stream. The presence of a back pressure in the sample loop meant that the calibration for methanol, water or other liquids by injection was subject to an error which depended on the flow rate of the gas. The reactant mixture was sampled prior to entering the reactor by a needle valve and flow line which by-passed the reactor.

The products detected and analyzed were H_2, CO, CO_2, C_2H_6, C_2H_4, HCHO, C_3H_8, CH_3CHO, methyl formate, formic acid, dimethoxymethane, acetone, dimethyl ether and ethanol.

All experiments were performed in a stainless steel reactor which had previously been tested at high pressures. This reactor was divided into two zones independently heated (pre-heater zone and oven zone). The length of the heating zone was 10 cm for the oven and 11 cm for pre-heater. A Pyrex or quartz liner (0.8 cm o.d.) was inserted to form a close fit. The reaction volume was 5 cm^3. A thermocouple covered by a quartz tube was inserted into reactor. The temperature was controlled by a digital temperature controller (Omega series). The reactor was mounted vertically in a thermobox with upward flow of reactants. Methane and oxygen, both supplied by Matheson, were used in the experiments. The flow rates were maintained by the mass flow controllers (Brooks 5800 series). Gases were premixed before entering the reactor by passing them through a mixing cross filled with Teflon turnings and quartz wool. Check valves (Nupro series) were installed before the mixing cross for uni-directional flow. The reactor pressure was controlled by a back-pressure regulator installed at the exit of the reactor. The temperature of the thermobox was maintained at 120°C. The heated effluent gas was depressurized before entering the analysis loop in the box. The connecting lines after the reactor were heated to 120°C and insulated well to avoid condensation of products. The condensible products were collected in a cold trap filled with chilled ethanol at near -70°C and also analyzed. A quantitative analysis of the gas phase and liquid obtained was carried out by gas chromatography. The gas chromatograph was equipped with TC and FID detectors. Molecular sieve 13X and Porapak Q columns were used (H_2 carrier gas). For H_2 analysis nitrogen was used as carrier gas on a molecular sieve 13X column. The identity of products was confirmed by GC-MS analysis (Finnigan Mat) equipped with a DB-5 capillary column. Data were calculated on the basis of carbon numbers of products and methane reacted. Errors in carbon and oxygen material balances were usually less than 5-7%.

RESULTS AND DISCUSSION

The results of 50 experiments are given in Table 1 and will be discussed under several headings.

Effect of Oxygen Conversion

Dooley [29] found that full oxygen conversion is one of the key factors to high methanol selectivity. It has been observed that, when O_2 conversion was lower than 100%, CH_3OH selectivity is significantly reduced due to further methanol oxidation in the absence of post-reactor cooling. They have reported methanol selectivity as high as 73% (CH_4 75.5%, O_2 5.2%, 50 bar, T=350°C) and 57% (CH_4 54.5%, O_2 9.6%, 50 bar, T=360°C). Our reaction volume (5 cm^3) and furnace heating length (10 cm) were close to the corresponding values (8.1

cm^3 and 12.7 cm) used by Dooley. Therefore, it was of interest to try to repeat these results, although we used pure oxygen instead of air. The first experiments (exp. 32-47) focused on reaching complete oxygen conversion at different pressures and elucidating the effect of oxygen conversion. As one can see from Figure 1 (exp. 32-47) the minimum in methanol selectivity was observed with increase in O_2 conversion.

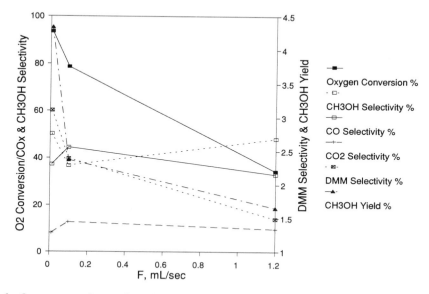

Figure 1. Oxygen conversion, product selectivity and methanol yield versus flow rate (**F**) at P = 740 psi, T = 405 °C, O_2 concentration = 8.12%. (Runs 39-41)

But, when oxygen was practically 100% converted, an increase of methanol selectivity was observed. This effect may be explained by the lowering of further methanol oxidation. It is known that methanol is unstable in a methanol-oxygen mixture [13]. We found an increase in methanol selectivity at different pressures when full oxygen consumption was achieved (exp. 32-34, 36-38, 39-41). The value of 46-50% methanol selectivity was obtained at 8-10% oxygen concentration and near 100% oxygen conversion. For optimal conditions we achieved as high a methanol selectivity as 64% at O_2 concentration of 9.6% (exp. 70), 75% at 4.2% O_2 concentration (exp. 74), and 69% at 5.7% oxygen concentration (exp. 75). These results are similar to those reported by Dooley [29].

Thus, complete oxygen conversion seems to be essential for high selectivity of methanol formation.

Effect of Pressure

Pressure effects were not examined in detail, particularly at low pressures. At full oxygen conversion the increase of pressure from 80 to 102 atm. did not lead to noticeable increase in methanol selectivity (see experiments 32-33 and 36-38). Although CO_X selectivity dropped, the considerable selectivity to oxygen-containing products, such as dimethoxymethane (DMM,

Table 1. Results Obtained In The Direct Oxidation of Methane to Methanol

	Empty Reactor Vol = 5 mL, Pre-heater Vol = 5 mL						Quartz Lined Reactor (*Pyrex Liner) DMM = CH₂(OCH₃)₂				
Reaction Number	Pressure Reactor psi	T Reactor °C	T Pre-heater °C	Flow Rate mL/s	$[O_2]$ Conc. %	CH_4 Conv. %	Select. CH_3OH %	Select. CO %	Select. CO_2 %	Select. DMM %	Yield CH_3OH %
32	1160	405	0	2.12	10.03	11.32	46.4	34.1	11.7	4.8	5.25
33	1160	405	0	0.78	10.03	11.96	48.9	29.7	11.1	7.1	5.84
34	1160	405	0	4.00	10.03	9.91	38.0	40.8	12.6	3.5	3.77
35	1480	405	0	5.00	11.87	13.31	34.6	38.6	14.6	4.4	4.59
36	1480	405	0	1.75	10.09	12.61	49.5	27.3	10.4	7.5	6.24
37	1480	405	0	0.46	10.09	12.93	47.6	25.5	9.2	12.0	6.15
38	1480	405	0	0.08	10.09	13.39	47.3	23.4	8.0	15.3	6.34
39	740	405	0	1.20	8.12	3.44	48.2	33.0	9.6	1.5	1.66
40	740	405	0	0.10	8.12	6.46	36.7	44.5	12.7	2.4	2.37
41	740	405	0	0.01	8.12	8.62	50.3	37.3	8.0	3.1	4.34
42	740	448	0	0.47	8.12	9.07	50.2	34.5	9.5	3.3	4.55
43	740	448	0	4.00	8.12	7.67	40.9	38.3	9.1	1.3	3.14
44	740	448	0	0.19	8.12	9.82	55.0	32.2	7.3	3.8	5.40
45	950	408	0	0.41	7.67	9.93	60.5	27.1	5.8	4.2	6.01
46	950	428	0	0.41	7.67	10.22	60.0	28.3	4.7	4.0	6.14
47	950	445	0	0.41	7.67	10.40	59.2	28.2	4.5	3.8	6.16
48	950	462	0	0.41	7.67	10.51	54.9	30.7	4.3	3.8	5.77
49	1160	405	0	0.54	6.79	9.18	61.6	24.5	8.1	3.4	5.66
50	1160	405	0	0.54	4.71	6.76	64.1	23.7	3.5	5.8	4.34
51	1160	405	0	0.54	3.35	4.94	66.5	22.9	1.7	6.0	3.28
52	1160	448	0	5.10	5.10	5.70	42.9	35.1	5.6	3.9	2.45

53	1160	465	0	5.10	5.10	6.65	47.9	32.3	5.2	3.1	3.19
54	1160	484	0	5.10	11.13	16.63	29.3	33.3	6.3	2.6	4.88
55	1160	472	0	5.63	8.99	12.64	45.0	32.4	4.4	2.2	5.69
56	1020	432	250	0.28	6.04	8.82	65.8	22.1	4.1	4.4	5.80
57	1020	432	250	1.10	6.04	8.35	62.0	25.6	4.6	3.5	5.18
58	1020	432	250	2.00	6.04	7.77	56.8	29.4	4.8	2.9	4.41
59	1020	432	250	2.00	4.83	7.41	72.3	18.2	2.8	2.6	5.36
60	1020	432	250	2.00	7.91	9.85	52.3	31.8	8.2	2.4	5.15
61	1020	454	250	10.15	5.35	7.11	61.0	22.6	6.4	3.2	4.34
62	1020	454	250	10.15	3.00	4.69	72.8	15.6	2.3	2.7	3.42
63	950	409	250	0.45	3.92	6.93	77.8	10.4	2.9	3.9	5.39
64	960	403	250	0.48	6.83	9.80	60.4	21.5	6.3	6.2	5.92
65	960	403	250	3.00	6.83	9.10	47.6	31.8	6.0	2.3	4.33
66*	920	395	250	3.33	9.57	11.82	21.8	46.3	8.4	1.3	2.57
67*	920	379	250	3.33	9.57	11.09	35.8	43.6	8.6	1.4	3.97
68*	920	367	250	3.33	9.57	11.84	48.9	34.7	8.4	1.2	5.79
69*	920	364	250	0.88	9.57	12.63	59.0	24.4	10.8	2.8	7.45
70*	920	357	260	0.32	9.57	13.17	64.3	20.6	9.3	3.1	8.47
71*	920	358	260	0.32	11.00	15.20	62.2	22.3	9.9	2.2	9.45
72*	920	358	260	6.39	9.57	14.49	31.0	30.6	5.1	1.9	4.49

Table 1. *Continued*

Empty Reactor Vol = 5 mL, Pre-heater Vol = 5 mL   Quartz Lined Reactor   (* Pyrex Liner)   DMM = CH$_2$(OCH$_3$)$_2$

Reaction Number	Pressure Reactor psi	T Reactor °C	T Pre-heater °C	Flow Rate mL/s	[O$_2$] Conc. %	CH$_4$ Conv. %	Select. CH$_3$OH %	Select. CO %	Select. CO$_2$ %	Select. DMM %	Yield CH$_3$OH %
73*	920	352	275	6.39	9.57	13.75	52.5	28.0	5.0	2.7	7.22
74*	1,160	344	275	1.00	4.20	7.20	75.1	11.1	1.6	8.3	5.42
75*	1,160	347	275	1.00	5.67	8.55	69.3	15.2	3.8	8.2	5.93
76*	1,160	355	275	1.00	8.62	12.37	63.8	22.1	6.3	5.6	7.89
77*	1,160	360	275	0.05	9.57	14.22	59.0	20.8	6.3	1.9	8.39
78	1020	356	285	0.95	11.71	15.81	57.3	22.2	10.7	5.7	9.06
79	950	372	260	0.13	13.75	16.31	45.8	32.7	14.1	2.8	7.47
80	930	362	275	1.16	11.70	15.22	59.0	25.1	10.8	3.3	8.98
81	930	369	275	2.51	11.70	12.16	29.8	40.8	19.5	3.1	3.62

methylal) and methylformate, decreased methanol selectivity (see Figure 2 and Figure 3). The formation of oxygenated products probably can explain the maximum in methanol selectivity observed with increasing pressure [30]. Interestingly, DMM became an important product at high pressures and low flow rates though DMM formation was not predicted by any model of methane oxidation. DMM formation was confirmed by GC-MS analysis (basic peaks with m/e of 45 and 75). It is difficult to explain methylal formation though it may be a result of some radical reactions or polycondensation reaction of methanol and formaldehyde. Methylal formation in methanol conversion at 200-400°C has been reported previously [31]. The presence of ethanol was detected only at the highest pressures studied. This is consistent with a cage effect [32] involving the intermediate reactions which have been previously postulated

$$CH_3O_2 + CH_4 \rightleftarrows CH_3O_2H + CH_3 \longrightarrow CH_2O_2H + CH_4$$

$$CH_3O_2H \rightleftarrows CH_3O + OH \longrightarrow CH_2O + H_2O$$

$$CH_3O + CH_4 \rightleftarrows CH_3OH + CH_3 \longrightarrow CH_2OH + CH_4$$

where the reactions indicated by --> occur in a cage.

Rotation of the reaction products in the cage can thus result in alternate products which should include ethylene glycol as well as the ethanol which was detected. Ethylene glycol was, however, not detected at the pressures studied.

The results indicate that optimal pressures for methanol formation are 60-80 atm, where higher pressures lead to higher selectivity of secondary products formation. At lower pressures, e.g. 50 atm, higher temperatures or very low flow rates are necessary to obtain complete oxygen conversion.

Effect of Residence Time (Feed Flow Rate)

It is not easy to elucidate an effect of residence time alone since it may also depend on other reaction conditions, namely pressure, temperature and oxygen concentration. Most experiments were performed at complete oxygen conversions and it was interesting to study the effect of residence time under such conditions. As can be seen in Figure 2, methanol selectivity decreases slightly at lower flow rates (longer residence time), at high pressure (10 atm), and near 100% oxygen conversion. This effect obviously results from DMM and methylformate formation (Figure 2 and Figure 3) in spite of a decrease in selectivity of CO_2 at the lower flow rates. DMM and methylformate selectivity increases with increasing residence time indicating that they are probably secondary products. In contrast, CO_x, C_2H_6, CH_2O and H_2 production increases with increasing flow rates. For lower pressures (60-80 atm), when oxygenated product formation is not significant, the methanol selectivity usually drops at higher flow rates. This flow rate effect is especially noticeable at high oxygen concentrations; i.e., when the flow rate has doubled CH_3OH selectivity falls to half its value (exp. 80-81). For lower oxygen concentrations this effect was not as pronounced (see exp. 56-58). The CO selectivity always increases with increase in flow rate, but CO_2 selectivity sometimes decreases (see exp. 64-65).

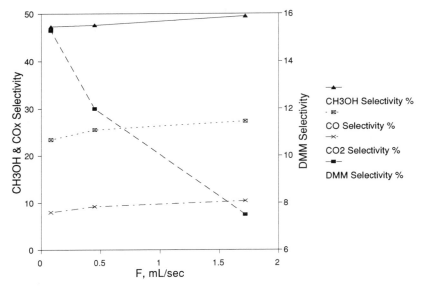

Figure 2. Products selectivity versus flow rate at complete oxygen conversion under high pressure. P = 1480 psi, T = 405 °C, O_2 concentration = 10.09 % (runs 36-38)

It is important to note, that we observed a significant amount of hydrogen, which is several times lower in comparison with CO_x concentration, although hydrogen easily undergoes oxidation at the reaction conditions.

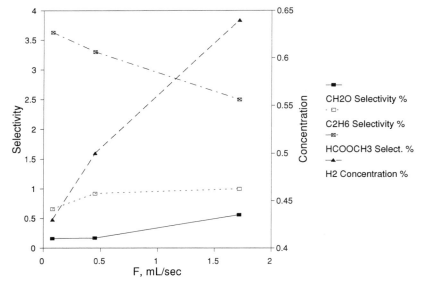

Figure 3. Products selectivity and hydrogen concentration versus flow rate at complete oxygen conversion under high pressure. P = 1480 psi, T = 405 °C, O_2 concentration = 10.09 % (runs 36-38)

As was found for flow rate effect and temperature effect (see below), the increase in H_2 formation was always accompanied by higher CO selectivity. Therefore it is reasonable to suggest that there is a process leading to H_2 and CO formation; most probably the decomposition of formaldehyde, as previously proposed [33], and from methanol.

$$CH_2O \rightarrow CO + H_2$$

$$CH_3OH \rightarrow CO + 2H_2$$

It has been reported [13] that methanol is stable in the absence of oxygen till 550^0 C. But methane oxidation produces "hot" methanol molecules by radical recombination or disproportionation as follows:

$$CH_3 + OH \rightarrow CH_3OH$$

$$2CH_3O_2 \rightarrow CH_3OH + CH_2O + O_2$$

These "hot" molecules may partially decompose to H_2 and CO. Higher pressure can stabilize such molecules, and a third body effect has been demonstrated [29].

At high flow rates the amount of reaction heat released increases at the same oxygen conversion level while heat transfer is limited owing to high pressure and low diffusion rates. This may create a local overheating with enhanced methanol and formaldehyde decomposition. Thus high flow rates, like high temperature, should increase CO and H_2 formation as observed experimentally. Part of the H_2 is probably formed by H atom recombination or abstraction. Another possible reaction forming H_2 may be the following, as proposed in [34]:

$$CO + H_2O \rightarrow CO_2 + H_2$$

This reaction looks less probable, because H_2 formation is related more with CO than CO_2 formation (see also temperature effect) and in this case H_2 production should increase at low flow rates as a result of reaction between products, namely CO and water. On the contrary, the higher flow rate is associated with higher H_2 production (at near 100% O_2 conversion). The C_2H_6 selectivity, which increases at higher flow rate, indicates a higher CH_3 radical generation rate (fast oxidation). The cause of higher CH_2O selectivity is not clear, though perhaps higher flow rates (shorter residence time) decrease secondary formaldehyde reactions or result in an increase in methanol decomposition:

$$CH_3OH \rightarrow CH_2O + H_2$$

For higher process productivity the high feed flow rates are desirable. This will be difficult because methanol selectivity drops at high flow rates, though mixing of reactants and heat transfer should improve.

Effect of Temperature

The temperature effect also depends on other reaction variables. At low (3-6%) and medium (7-8%) oxygen concentrations and low flow rates there is a wide range of temperatures where CH_3OH selectivity depends only slightly on temperature. One can see in Figure 4, the decrease is methanol selectivity is not significant at 405-448 °C. Only at T=465°C was methanol selectivity significantly reduced. The concentration of H_2 in the product stream increased with increase in temperature whereas the selectivity of CO and CO_2 remained essentially constant. The selectivity of C_2H_6 monotonically increased with increasing temperature indicating higher chain reaction rates.

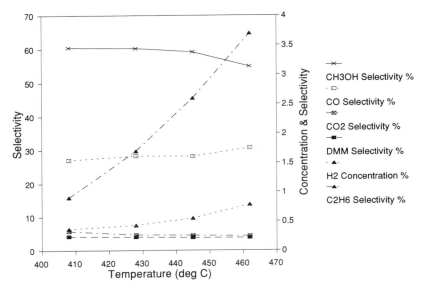

Figure 4. The effect of temperature on products selectivity at low flow rate and middle oxygen concentration. P = 950 psi, F = 0.41 mL/s, O_2 concentration = 7.67 %, complete oxygen conversion (except T = 408 °C) (runs 45-48)

Other peculiarities were observed at high O_2 concentrations (10-13%) and higher flow rates. In this case the methanol selectivity drastically drops with increase in temperature, Figure 5. A few degrees temperature difference (see exp. 72-73) causes a big difference in methanol selectivity. The decrease in methanol selectivity is accompanied by an increase in H_2, CO and C_2H_6 selectivity. The increase in ethane selectivity is especially marked (see Figure 5). At higher temperatures, ethylene and C_3 products were found, indicating that a transition to the oxidative coupling of methane was occurring. In this case the heat released in addition to the high temperature undoubtedly facilitates the rapid oxidation and decomposition of methanol. This temperature is a crucial parameter at high flow rates and high oxygen concentrations. The increase in pre-heater temperature from 250 to 275-285°C allowed the oven temperature to be lowered and enhances the methanol selectivity (see exp. 72-73 and 76-77). This opportunity of improvement, however, is limited and is drastically reduced when using high flow rates and high oxygen concentrations. The optimal temperature is defined as the minimal temperature for full oxygen consumption.

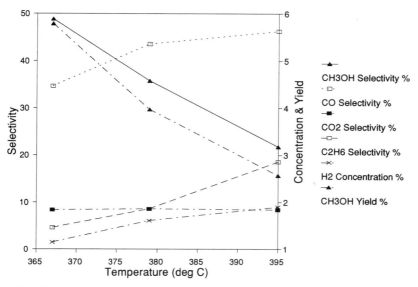

Figure 5. The effect of temperature on products selectivity and methanol yield at high flow rate and high oxygen concentration. P = 920 psi, F = 3.33 mL/s, O_2 concentration = 9.57%, $T_{preheater}$ = 250 °C (runs 66-68)

Effect of Oxygen Concentration

A higher oxygen concentration always leads to lower methanol selectivity (exp. 49-51, 58-60, 74-77, Figure 6 and Figure 7). As shown in Figure 6, the decrease in methanol selectivity is followed by an increase in CO, CO_2 and C_2H_6 selectivity and H_2 concentration. This effect is not very pronounced at optimal reaction conditions, such as minimal temperature at full oxygen conversion and low feed flow rates (see Figure 7). Therefore it is possible to increase methanol yield to over 9%. A higher oxygen concentration also requires higher temperature to reach full oxygen conversion (see Figure 8). The same effect has been reported by Dooley [29]. Unfortunately, the higher oxygen concentration, the lower range of reaction conditions with relatively high methanol selectivity, e.g., at O_2 concentrations of 3-5%, methanol selectivity higher than 60% was observed at flow rates near 10 mL/sec (exp. 61-62), while at 11.7% O_2 concentration the methanol selectivity near 60% was observed only at flow rate of 1 mL/sec (exp. 80-81). At oxygen concentration higher than 13% the methanol selectivity higher 50% was not found at any reaction conditions - the best selectivity is shown in exp. 79. Figure 9 shows the range of flow rates (residence time) that are suitable for high methanol selectivity (70 and 60%) for different oxygen concentrations. This process looks preferable for low oxygen concentrations, where there is a wide range of reaction conditions with high methanol selectivity.

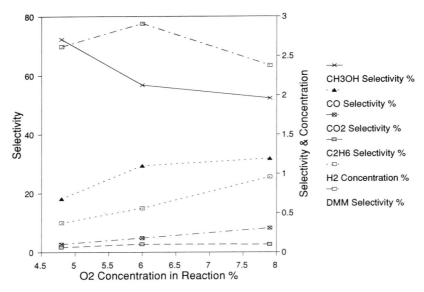

Figure 6. The dependence of product selectivity on oxygen concentration in the reaction mixture. $T_{oven} = 432\ ^oC$, $T_{preheater} = 250\ ^oC$, $P = 1020$ psi, $F = 2.0$ mL/s (runs 58-60)

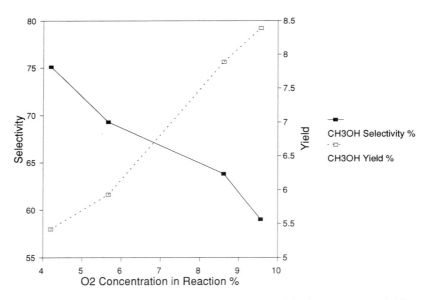

Figure 7. Methanol selectivity and yield versus oxygen concentration at minimal temperature needed for complete oxygen conversion. $P = 1160$ psi, $T_{preheater} = 275\ ^oC$, $F = 1.0$ mL/s (runs 74-77)

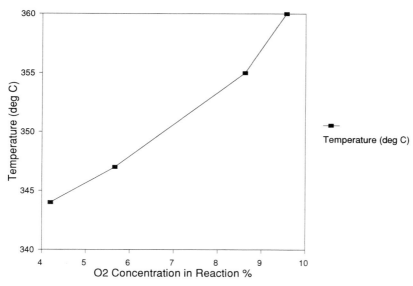

Figure 8. The increasing of minimal temperature needed for complete oxygen conversion with increasing oxygen concentration. $P = 1160$ psi, $T_{preheater} = 275$ °C, $F = 1.0$ mL/s (runs 74-77)

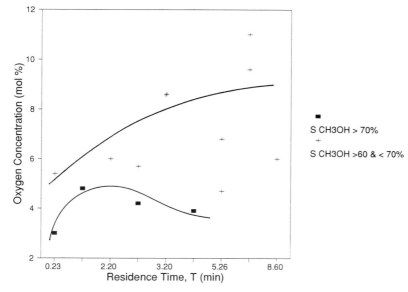

Figure 9. The range of residence times, τ, for conditions where the selectivity of CH_3OH (S_{CH_3OH}) is $S_{CH_3OH} \geq 70\%$, ■ and where $70\% > S_{CH_3OH} \geq 60\%$, O for different $[O_2]$ which is completely consumed. No pre-heater conditions.

283

Effect of Pre-Reactor Heating

The results indicate that pre-reactor heating has no any noticeable influence on product selectivity at near 100% oxygen conversion. Pre-heater temperature was maintained at 250 - 285°C where methane conversion was not observed in pre-reactor zone. The use of a pre-heater leads to a decrease in reaction temperature needed for full oxygen consumption and an increase in methanol selectivity.

This effect is especially significant at high flow rates when it is difficult to heat the reactants quickly to the desired temperature. The use of the pre-heater appears to be crucial to obtain relatively high methanol selectivity at high oxygen concentrations and flow rates (see exp. 72-73).

Effect of Reactor Material

Pyrex and quartz liners were used. There was no significant difference in selectivity for both types of liners at 63-80 atm. However, at the lower flow rates more formic acid was usually obtained with the Pyrex liner.

CONCLUSION

The effect of different reaction variables in methane oxidation to methanol at elevated pressures has been studied. The optimal reaction conditions involve a minimal temperature at complete oxygen conversion and low flow rates (slow oxidation conditions). Reaction pressures of 60-80 atm are preferable. Methanol selectivity of 50-75%, depending on oxygen concentration, was obtained. Higher oxygen concentration leads to a narrower range of reaction conditions that are suitable for high methanol selectivity. The effect of variables on other reaction products (products of total oxidation, recombination product, oxygenated products) was studied. Use of a pre-heater allowed for a decrease in reaction temperature and increase in methanol selectivity.

Concerning the discrepancies between different researchers dealing with methanol selectivity in methane conversion it is necessary to note the following. Analysis has shown that those [21, 29] who used a short reaction zone (10-20 cm) and low flow rates found high methanol selectivity (70-80%). And those [8, 20] who have used longer reaction zones (40-50 cm) and higher flow rates found low methanol selectivity. Present results indicate that the transition to higher feed flow rates reduced methanol selectivity, and it is difficult to obtain high methanol selectivity under these conditions, especially at high oxygen concentrations. Long reaction zones may favour further oxidation and methanol decomposition. For short reaction zones the opportunities to obtain a high productive process are very limited. At low oxygen concentrations the restrictions are not as severe. At pressures of 100 atm or more a cage effect may result in a decrease in methanol selectivity due to the creation of alternate reaction channels.

ACKNOWLEDGEMENT

Grateful acknowledgment is made to the Department of Energy, Mines and Resources, the Natural Sciences and Engineering Research Council of Canada, and Imperial Oil Canada for financial support.

REFERENCES

1. H.D. Gesser, N.R. Hunter, and C.B. Prakash, The direct conversion of methane to methanol by controlled oxidation, *Chem. Rev.* 85:235 (1985).

2. H.D. Gesser, and N.R. Hunter, The direct conversion of methane to methanol (DMTM), in "Direct Methane Conversion by Oxidative Processes. Fundamental and Engineering Aspects," E.E. Wolf, ed., Reinhold-Van Nostrand (1992).

3. M. Yu. Sinev, V.N. Korshak, and O.V. Krylov, The mechanism of the partial oxidation of methane, *Russ. Chem. Rev.* 58:38 (1989).

4. M.J. Brown, and N. D. Parkyns, Progress in the partial oxidation of methane to methanol and formaldehyde, *Catal. Today*, 8:305 (1991).

5. P.S. Casey, T. McAllister, and K. Foger, Selective oxidation of methane to methanol at high pressures, *Ind. Eng. Chem. Res.* 33(5):1120 (1994).

6. J-W. Chun, and R.G. Anthony, Partial oxidation of methane, methanol, and mixtures of methane and methanol, methane and ethane, and methane, carbon dioxide, and carbon monoxide, *Ind. Eng. Chem. Res.* 32, 788 (1993).

7. V.S. Arutyunov, V.I Vedeenev, V.E. Leonov, N. Yu. Krymov, and A.D. Sedykh, The influence of pressure and some gas additives on the high pressure oxidation of methane to methanol, *Proc. Xth International Symposium on Alcohol Fuels*, Colorado Springs, USA, 1993.

8. A.M.A. Bennett, G.A. Foulds, B.F. Gray, S.A. Miller, and G.S. Walker, Homogeneous gas-phase oxidation of methane using oxygen as oxidant in an annular reactor, *Ind. Eng. Chem. Res.* 32:780 (1993).

9. S.Y. Chen, and D. Wilcox, Effect of vanadium oxide loading on the selective oxidation of methane over V_2O_5/SiO_2, *Ind. Eng. Chem. Res.* 32:584 (1993).

10. J-W. Chun, and R.G. Anthony, Catalytic oxidations of methane to methanol, *Ind. Eng. Chem. Res.* 32:259 (1993).

11. J.H. Dygos, R.A. Periara, D.J. Taube, E.R. Evitt, D.G. Loffler, P.R. Wentrcek, G. Voss, and T. Masuda, A mercury-catalyzed, high-yield system for the oxidation of methane to methanol, *Science*, 259: 340 (1993).

12. K.J. Thomas, R. Willi, and A. Baiker, Partial oxidation of methane-the role of surface reactions, *Ind. Eng. Chem. Res.* 31:2272 (1992).

13. D.E. Walsh, D.J. Martinik, S. Han, and R.E. Palermo, Direct oxidative methane conversion at elevated pressure and moderate temperatures, *Ind. Eng. Chem. Res.* 31:1259 (1992).

14. V.I. Vedeneev, V.S. Arutyunov, N. Yu. Krymov, P.M. Cherbakov, and A.D. Sedykh, Some features of methane oxidation at high pressures, *Catal. Today*, 13: 613 (1992).

15. K. Omata, N. Fukuoka, and K. Fujimoto, Methane partial oxidation to methanol-solid initiated homogeneous methane oxidation, *Catal. Lett.* 12:227 (1992).

16. J.E. Lyons, et al., *Active iron oxo centers for the selective catalytic oxidation of alkanes* in "Studies in Surface Science and Catalysis Vol. 68: Structure-Activity and Selectivity Relationships in Heterogeneous Catalysis," R.K. Grasselli and A.W. Sleight, eds. Elsevier, Amsterdam, Neth. (1991).

17. D.W. Rytz, and A. Baiker, Partial oxidation of methane to methanol in a flow reactor at elevated pressure, *Ind. Eng. Chem. Res.* 30:2287 (1991).

18. V.A. Durante, D.W. Walker, S.M. Gussow, J.E. Lyons, and R.C. Hayes, Catalytic oxidation of alkanes with

improved yield, U.S. Patent No. 5,132,472 issued July 21, 1992.

19. V.A. Durante, D.W. Walker, S.M. Gussow, and J.E. Lyons, Silicometallate molecular sieves and their use as catalysts in oxidation of alkanes, U.S. Patent No. 4,918,249 Issued April 17, 1988.

20. R. Burch, G.D. Squire, and S. Chi Tsang, Direct conversion of methane to methanol., *J. Chem. Soc. Faraday Trans. I*, 85:3561 (1989).

21. P.S. Yarlagadda, L.A. Morton, N.R. Hunter, and H.D. Gesser, Direct conversion of methane to methanol in a flow reactor, *Ind. Eng. Chem. Res.* 27:252 (1988).

22. L.A. Morton, H.D. Gesser, and N.R. Hunter, The partial oxidation of CH_4 to CH_3OH at high pressure in a packed reactor, *Fuel Sci. Technol. Int'l.* 9:913 (1991).

23. P.S. Yarlagadda, L.A. Morton, N.R. Hunter, and H.D. Gesser, Temperature oscillations during the high pressure partial oxidation of methane in a tubular flow reactor, *Combustion and Flame.* 79:216 (1990).

24. H.D. Gesser, N.R. Hunter, and P.C. Das, The ozone sensitized oxidative conversion of methane to methanol and ethane to ethanol, *Catal. Lett.* 16:217 (1992).

25. S.S. Shepelev, H.D. Gesser, and N.R. Hunter, Light paraffin oxidative conversion in a silent electric discharge, *Plasma Chem. Plasma Process.* 13:479 (1993).

26. P.S. Yarlagadda, L.A. Morton, N.R. Hunter, and H.D. Gesser, Direct catalytic conversion of methane to higher hydrocarbons, *Fuel Sci. Technol. Int'l.* 5:162 (1987).

27. H.D. Gesser, *et al.*, Proceeding Natural Gas Conversion Symposium, Sydney, Australia July (1993).

28. B.G. Charlton, B.T. Le, J.C. Jones, B.F. Gray, and G.A. Foulds, The use of jet-stirred CSTR to study the homogeneous gas phase partial oxidation of methane to methanol, Proceeding Natural Gas Conversion Symposium, Sydney, Australia July (1993).

29. W. Feng, F.C. Knopf, and K.M. Dooley, The effects of pressure, third bodies, and temperature profiling on the noncatalytic partial oxidation of methane, *Energy and Fuels.* 8(4):815 (1994).

30. (a) E.H. Boomer, and V. Thomas, The oxidation of methane at high pressures II. Experiments with various mixtures of Viking natural gas and air, *Can. J. Res.* 15(B):401 (1937).
    (b) E.H. Boomer, and V. Thomas, The oxidation of methane at high pressures III. Experiments using pure methane and primarily copper as catalyst, *Can. J. Res.* 15(B):414 (1937).

31. L.F. Marek, and D.A. Hahn, "The Catalytic Oxidation of Organic Compounds in the Vapour Phase," Chem. Catalog Co. Inc., New York (1932). Russian Edition, p. 147

32. (a) J.M. Zellweger, and H. Vanden Bergh, The photolytic cage effect of iodine in the gas phase, *J. Chem. Phys.* 72:5405 (1980).
    (b) J.C. Dutoit, J.M. Zellweger, and H. Vanden Bergh, The photolytic cage effect of iodine in gases and liquids, *J. Chem. Phys.* 78:1825 (1983).

33. P.E. Oberdorfer, and R.F. Winch, Chemicals from methane in a high-compression engine, *Ind. Eng. Chem.* 53: 41 (1961).

34. M.M. Karavaev, V.E. Leonov, I.G. Popov, and E.T. Shepelev, "Technology of Synthetic Methanol" Khimiya, Moscow (1984) (in Russian).

SOME FEATURES OF HIGH PRESSURE METHANE OXIDATION TO METHANOL AND FORMALDEHYDE

V.S.Arutyunov, V.Ya.Basevich, V.I.Vedeneev

Semenov Institute of Chemical Physics, Russian Acad. Sci.,
Kosygin st., 4, Moscow, 117334, Russia
Phone: (7-095) 939 7287; Fax: (7-095) 938 2156; E-mail: KINET@glas.apc.org

INTRODUCTION

The partial oxidation of methane to methanol is one of the most promising routes of natural gas conversion into liquid hydrocarbons. The process may be accomplished both catalytically and at homogeneous conditions. In the last case the high pressures exceeding 50 atm necessary for appropriate product yield.[1] Due to the most catalytic works the process has heterogeneous - homogeneous nature[2] and includes steps of both heterogeneous and homogeneous radical generation and their subsequent transformation. It was demonstrated that at pressures above 30 atm the process is practically independent of the nature of surface and introduction of catalyst do not rise up efficiency of the process.[3,4]

Although high pressures favour methanol formation resulting from methane oxidation, the cost of gas compression is one of the most prominent factors in the cost of production.[5] Therefore, it is important to know the actual pressure dependence to determine the optimum operating pressure. The aim of this paper is to reveal the role of the pressure in this process and to obtain its optimum region.

KINETIC SIMULATIONS

The kinetic simulation of homogeneous partial methane oxidation was carried out on the base of the model of methane oxidation at ambient temperatures and elevated pressures.[6] To make it possible to conduct calculations at lower pressures the pressure dependence of some rate constants in model was taken into account.

The simulation revealed that the main feature of high pressure oxidation of methane is the existence of two phases of the reaction distinctly differing by their time scales. The very beginning stage of the process proceeds by chain-branched mechanism very like to the mechanism of hydrogen oxidation. Any reactions of radicals with intermediate products in this phase are unimportant. The very fast initial autoacceleration in this phase completes by a subsequent stationary state which is characterized by approximate equality of the rates of branching and radical recombination. The mechanism of the main phase of the reaction may be considered as a degenerated chain-branching where branching is connected with intermediate products.

The calculations conducted for methane-oxygen (9:1) mixture at temperatures 600-750 K have shown the existence of critical phenomena at pressure change (Figure 1).

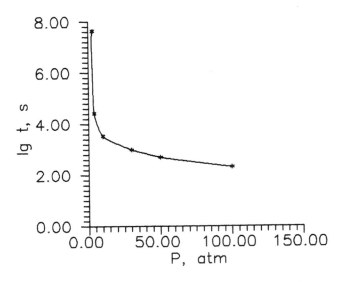

Figure 1. Calculated dependence of the reaction time (t) from pressure. Methane/oxygen = 9, T=650 K.

The transition via some critical pressure leads to dramatic (more than 1000-fold) change of the reaction time (the time of 95% oxygen consumption) (Figure 1). The explanation is that the quasistationary concentration of rate determining methylperoxide radicals in short initial stage below the critical pressure is determine by their heterogeneous or homogeneous generation and above the critical pressure by chain-branching. But the velocity of chain-branching exceeds the velocity of radical generation by several orders of magnitude so the quasistationary concentration of methylperoxide radicals in short initial stage approximately 4 orders of magnitude differ above and below critical pressure. It explains the negligible influence of catalyst on process at sufficiently high pressures because catalytic generation of radicals is not able to compete with a chain-branched mechanism of their generation. The value of this critical pressure depends on temperature. For 600 K the calculated critical pressure lies between 6 and 7 atm where as at 700 K between 1 and 2 atm.

EXPERIMENTAL RESULTS

To reveal the real influence of pressure the experiments were carried out in the range of 30-230 atm on the constantly working flow pilot set with preliminary prepared NG-air mixtures containing approximately 2.8% of oxygen. The most prominent changes in yield and composition of liquid phase take place at pressures below 100 atm (Figure 2).

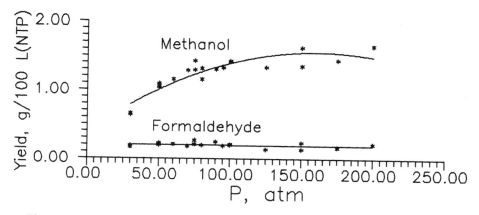

Figure 2. Experimental methanol and formaldehyde yield *vs.* pressure. T = 673K, $[O_2]$=2.8%.

First of all the pressure rise leads to increase of the total yield of liquid oxidation products which reaches maximal values at P=100-200 atm. The methanol and ethanol concentrations in liquid product reach maximum at approximately 150 atm, and that of formaldehyde monotonously fall down with pressure. It leads to fast change in the methanol to formaldehyde ratio with pressure. But due to increase of the total yield of liquid phase the formaldehyde yield is approximately constant in this pressure range.

The composition of the gas phase changes mainly due to rise of carbon dioxide concentration with pressure, while carbon monoxide concentration remains practically constant. The rise of carbon dioxide concentration with pressure is, probably, at any rate partly, connected with formic acid formation,[7] followed by its decay to carbon dioxide.

The estimated selectivities of products relative to consumed methane (approximately 40% for methanol and 7% for formaldehyde) are very similar to those obtained in[4,8] and followed from kinetic modelling of the process.

The main conclusion from these experiments that there is no any reason to use in this process pressures exceeded 100 atm.

On the base of this investigations a new organization of the process has been suggested, including changes in reactor construction. The tests were carried out on the pilot plant with productivity 100 ton of methanol per year and showed that the new organization permits to get approximately 30% increase in productivity. Now in Russia is worked out the project of a pilot plant with productivity 10.000 ton of methanol per year.

REFERENCES

1. H.D.Gesser, N.R.Hunter, and C.B.Prakash, The direct conversion of methane to methanol by controlled oxidation, Chem.Rev. 85:235 (1985).
2. M.Yu.Sinev, V.N.Korchak, and O.V.Krylov, The mechanism of the partial oxidation of methane, Russian Chem.Rev. 58:22 (1989).
3. N.R.Hunter, H.D.Gesser, L.A.Morton, and D.P.C.Fung, Methanol formation at high

pressure by the catalyzed oxidation of natural gas and by the sensitized oxidation of methane, Appl.Catal. 57:45 (1990).

4. R.Burch, G.D.Squir, and S.C.Tsang, Direct conversion of methane into methanol, J.Chem.Soc.Faraday Trans.1. 85:3561 (1989).

5. J.H.Edwards, N.R Foster, The potential for methanol production from natural gas by direct catalytic partial oxidation, Fuel Sci.Techn.Int. 4:365 (1986).

6. V.I.Vedeneev, M.Ya.Gol`denberg, N.I.Gorban`, and M.A.Teitel`boim, Quantitative model of the oxidation of methane at high pressures. 1.Description of model, Kinetics and Catalysis. 29:1 (1988).

7. V.I.Vedeneev, V.S.Arutyunov, N.Yu.Krymov, P.M.Cherbakov, and A.D.Sedykh, Some features of methane oxidation at high pressures, Catal.Today. 13:613 (1992).

8. O.T.Onsager, P.Soraker, and R.Lodeng, Preprints, Experimental investigation and computer simulation of the homogeneous gas phase oxidation of methane to methanol, Methane Activation Symp. Pasifichem.89., Honolulu, Hawaii, 113 (1989).

STRATEGIES IN METHANE CONVERSION

Rameshwar D. Srivastava,[1] Sai V. Gollakota,[1] Gary J. Stiegel,[2] and
Arun C. Bose[2]

[1]Burns and Roe Services Corporation
[2]U.S. Department of Energy
Pittsburgh Energy Technology Center
Pittsburgh, PA 15236

ABSTRACT

Because of its economic potential, the development of energy-efficient
technologies for the conversion of low-value, light alkane gases, such as
methane, to high-value liquid fuels and chemicals is a key element of the
Department of Energy's (DOE) Gas Conversion Program. The current program,
which focuses on methane conversion to liquid fuels and chemicals, comprises
four major research areas: partial oxidation, oxidative coupling, pyrolysis, and
derivatization. This paper outlines the various scientific and engineering
strategies involved in the development of viable methane conversion
technologies under the current program.

INTRODUCTION

The domestic supplies of petroleum in the United States are limited, and
exploration and production of petroleum are declining because of a rapidly
decreasing reserve base and because of decreasing success in finding
economically recoverable new resources. The demand for liquid transportation
fuels, however, is persistent and expected to rise significantly in the coming
decades. Techniques such as enhanced oil recovery have limited potential in
recovering additional petroleum. On the other hand, domestic supplies of natural
gas are significant. A considerable portion of these natural gas sources,
however, are in remote locations. Current commercial methods to transport gas
over long distances cannot effectively utilize these resources. Thus, the
development of novel processes to convert natural gas to high-value liquid fuels

Methane and Alkane Conversion Chemistry
Edited by M. M. Bhasin and D. W. Slocum, Plenum Press, New York, 1995

facilitates the utilization of large domestic resources, reduces our dependence on imported crude oil, and extends the life of our petroleum reserves.

In view of these strategic objectives, the Pittsburgh Energy Technology Center (PETC) of the Department of Energy (DOE) has embarked on an aggressive Gas Conversion Program to develop new, cost-effective technologies to utilize the nation's natural gas resources. A comprehensive review of the state-of-the-art processes in direct conversion of methane to liquid fuels and chemicals is available in the literature.[1]

The current program comprises the development of four primary advanced direct conversion technologies: partial oxidation, oxidative coupling, pyrolysis, and derivatization including other novel technologies (Figure 1). This paper outlines DOE's research activities designed to achieve the strategic objective of developing economically viable and environmentally sound technologies for methane conversion to high-value products.

PARTIAL OXIDATION

Partial oxidation of methane refers to the selective oxidation of methane to methanol or the conversion of methane to synthesis gas. In both technologies, controlling the extent of the oxidation reaction is extremely important.

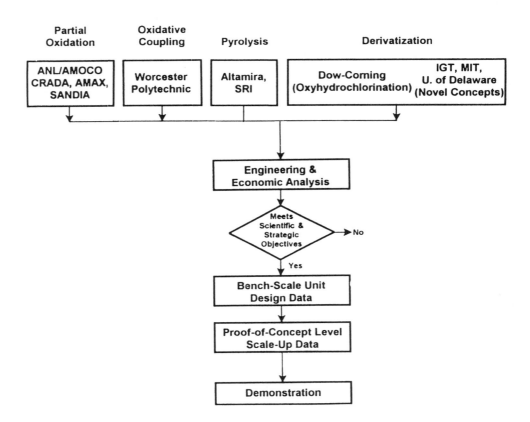

Figure 1. Technology development algorithm for methane conversion

Oxygen-permeable, dense-phase ceramic membranes are being developed for the conversion of methane to synthesis gas or methanol. These oxygen-specific membranes permit oxygen (from air) to pass through the membrane while totally excluding nitrogen. The ceramic membranes are formed by using ceramic powders in the La-Sr-Co-Fe-O system. These powders, produced by a solid-state reaction of the constituent carbonates and nitrates, are used to fabricate the membrane tubes by a plastic extrusion technique. The extruded tubes are heated and sintered before they are installed in methane conversion reactors, where high methane conversion efficiencies are expected.[2] Ceramic membranes are anticipated to control the oxygen flux in the reactor to prevent overoxidation. This will substantially improve the economics both of existing synthesis gas production technologies and direct conversion technologies for methanol. Preliminary tests exhibited high methane conversion efficiencies (> 98%) and CO selectivities (90%) for synthesis gas production.

In the partial oxidation research area, two projects are focusing on the development of catalysts for the conversion of methane to methanol. These catalysts are halogenated iron molecular catalysts and vanadium phosphate catalysts. The iron molecular catalysts facilitate mimicking the high activity and selectivity of enzymatic reactions. In another project, a process is being developed in which a palladium acetate catalyst is used to convert methane to methyl trifluoroacetate, which can be readily hydrolyzed to produce methanol.

OXIDATIVE COUPLING

In oxidative coupling, methane and oxygen are catalytically reacted to yield ethylene and/or other intermediates that may be converted further to petrochemicals or other "end use" fuels via existing commercial technologies. Despite numerous investigations, commercially acceptable yields have not yet been achieved.[1]

Inorganic membrane reactors are being developed for the production of C_2 hydrocarbons by oxidative coupling of methane. Membrane reactors provide optimal quantities of oxygen to provide high selectivities and high yields of C_2 hydrocarbons. Significant work has been performed in developing and testing perovskites, which are selective to oxygen permeation inside the pore structure of alpha-alumina membranes. Selective catalysts are being synthesized and tested for applications in the membrane reactor.[3]

PYROLYSIS

In pyrolysis, methane is converted directly to olefins, aromatics, or other higher hydrocarbons by decomposition of methane under controlled thermal and catalytic conditions. In one project, fullerene-based catalysts are being developed to convert methane to higher hydrocarbons.[4] In another project, a process is being developed to produce aromatics by methane pyrolysis using a novel reactor system that allows rapid quenching of products to minimize solid carbon formation.[5]

DERIVATIZATION AND OTHER NOVEL TECHNOLOGIES

The derivatization approach for the conversion of methane to fuels and chemicals is characterized by a two-stage process. In the first stage, methane is converted to a reactive intermediate, which, in the second stage, can be processed to yield the desired product.

The oxyhydrochlorination (OHC) research is aimed at developing a process for converting methane directly to methyl chloride for subsequent use in the production of silicone polymers.[6,7] When methane is reacted with oxygen and hydrogen chloride gas in the presence of a catalyst, it is converted to methyl chloride in high yield. Considerable work has been performed in the development of copper-based oxyhydrochlorination catalysts. This direct route is economically attractive since it will replace methanol (produced from methane via synthesis gas), which is used for the commercial production of methyl chloride. Commercialization of the OHC technology would reduce dependency on methanol, the price of which fluctuated considerably in recent years.

Another process is being developed in which low-cost alkaline earth metal oxides are used in a plasma reactor to convert methane to solid metal carbides that can be readily hydrolyzed to acetylenes. Hydrogen is a by-product of this reaction.

In other projects, advanced catalysts are being developed for converting methane to carbon disulfide, which can then be converted to higher hydrocarbons, and superacid catalysts are being developed for converting light hydrocarbons to higher hydrocarbons.

SUMMARY

In DOE's current Gas Conversion Program, various technologies are being developed for the conversion of methane to liquid fuels and chemicals. Of the various approaches being investigated, the oxyhydrochlorination route is more advanced than any other direct conversion technology. If sufficient economic incentives warrant further development, OHC could be demonstrated at a proof-of-concept level. Technology development for generation of partial oxidation products from methane using ceramic membranes may result in significant savings in synthesis gas generation costs due to the elimination of the oxygen separation plant and replacement of the conventional synthesis gas generation loop. In the near-term, one or more of the other technologies may also lead to a proof-of-concept demonstration.

REFERENCES

1. R.D. Srivastava, P. Zhou, G.J. Stiegel, V.U.S. Rao, G. Cinquegrane, Catalysis 9:183 (1992).

2. U. Balachandran, S.L. Morissette, J.T. Dusek, and R.B. Poeppel, Argonne National Laboratory; M.S. Kleefisch, T.P. Kobylinski, and C.A. Udovich, Amoco Research Center.*

3.   Y.H. Ma, W.R. Moser, and A.G. Dixon, Worcester Polytechnic Institute.[*]

4.   A.S. Hirschon, R. Malhotra, R.B. Wilson, SRI International.[*]

5.   G. Marcelin, R. Oukaci, Altamira Instruments, Inc.[*]

6.   B.M. Naasz, A.I. Smith, A.I. Toupadakis, and B.R. Crum, Dow Corning
     Corporation, Proceedings of the Liquefaction Contractors' Review
     Conference, Eds. G.J. Stiegel, and R.D. Srivastava, U.S. Department of
     Energy, Pittsburgh, Pennsylvania, September 22-24, 1992 (pp. 581).

7.   B.M. Naasz, J.S. Smith, S.P. Ferguson, and C.G. Knutson, Dow Corning
     Corporation.[*]

---

[*]   Proceedings of the Liquefaction and Gas Conversion
      Contractors' Review Conference, Eds. S. Rogers and P.
      Zhou, U.S. Department of Energy, Pittsburgh, Pennsylvania,
      September 27-29, 1993.

DIRECT OXIDATION OF METHANE WITH ORGANOMETALLIC REACTIONS

Charles E. Taylor, Richard R. Anderson, Curt M. White,
and Richard P. Noceti

U.S. Department of Energy
Pittsburgh Energy Technology Center
P.O. Box 10940
Pittsburgh, PA 15236-0940

Selective, direct oxidation of methane to methanol is a process of scientific interest and industrial importance. Reports have appeared in the literature describing the use of organometallic complexes to effect this transformation.[1,2] Investigation of one of these reaction schemes in our laboratory has produced interesting results. Our research effort was an extension of work reported by Sen and co-workers.[3-4] The purported reaction occurs between methane (at 800 psig 5.52 MPa) and palladium(II) acetate in trifluoroacetic acid at 80°C (Equation 1). The product, methyl trifluoroacetate, is readily hydrolyzed to produce methanol and trifluoroacetic acid.

$$CH_4 + Pd(O_2CCH_3)_2 \xrightarrow[80°C]{CF_3COOH} CF_3CO_2CH_3 + Pd \tag{1}$$

INTRODUCTION

Methane is produced as a by-product of coal gasification. Depending upon reactor design and operating conditions, up to 18% of total gasifier product may be methane. In addition, there are vast proven reserves of geologic methane in the world. Unfortunately, a large fraction of these reserves are in regions where there is little local demand for methane and it is not economically feasible to transport it to a market. There is a global research effort under way in academia, industry, and government to find methods to convert methane to useful, more readily transportable and storable materials. Methanol, the initial product of methane oxidation, is a desirable product of conversion because it retains much of the original energy of the methane while satisfying transportation and storage requirements. A liquid at room temperature, methanol could be transported to market utilizing the existing petroleum pipeline and tanker network and distribution infrastructure. Methanol may be used directly as a fuel or be converted to other valuable products (*i.e.* other transportation fuels, fuel additives, or chemicals). Currently, the technology for direct oxidation of methane to methanol suffers from low methane conversion and poor methanol

Methane and Alkane Conversion Chemistry
Edited by M. M. Bhasin and D. W. Slocum, Plenum Press, New York, 1995

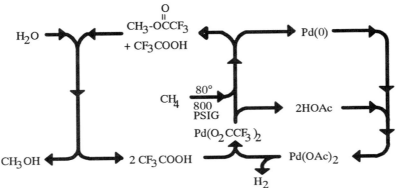

FIGURE 1.   REACTION PATHWAY FOR THE DIRECT METHANE TO METHANOL REACTION

selectivity. A process for the direct oxidation of methane to methanol, in high yield and with high selectivity, is highly desirable.

Formation of methanol (as an ester) directly from methane using Pd(II) reagents has been demonstrated. Sen and co-workers at The Pennsylvania State University report that they have produced methyl trifluoroacetate with reported conversions, calculated on palladium (0) recovered, of ~60 %.

As written, this reaction scheme depicted in Equation 1 requires stoichiometric amounts of the palladium(II)acetate. In addition, a second step, hydrolysis, is required to produce free methanol (Figure 1). Gretz et al.[3] had speculated that it might be possible to make the reaction catalytic by including a co-oxidant, and, in fact, reported a catalytic reaction cycle using potassium peroxydisulfate, $K_2S_2O_8$, as a co-oxidant with the Pd(II) in the monotrifluoroacetoxylation of 1,4-dimethoxybenzene. Further, it was reported that the hydrolysis step can usually be carried out quantitatively (100% yield of 2,5-dimethoxyphenol).[4]

Our objective was to investigate this reaction scheme, reproduce the results reported in the literature and explore the possibility of using this reaction pathway in a commercial process. On initial examination, this reaction scheme appears not to have much commercial application because of long reaction times (~150 hours) and batch mode operation. The reported conversions of 60% were intriguing, but were calculated on the amount of Pd metal recovered from the reactor system. When conversions are calculated on the basis of methane consumed, they are on the order of about 3 percent, of which only 33 percent of the converted methane is methyl trifluoroacetate, a net conversion of methane to product of 1 percent. It spite of this and in light of the reported catalytic reaction, it was decided to examine this reaction in some detail. A reaction scheme similar to that shown in Figure 1 could be of commercial interest if all steps could be realized at reasonable conversions and with minimal side reactions. For purposes of discussion, the scheme may be separated into three major parts: (1) conversion of methane to methyl trifluoroacetate, (2) hydrolysis of methyl trifluoroacetate to methanol and trifluoroacetic acid, and (3) conversion of palladium(0) to palladium acetate. The focus of this paper is the conversion of methane to methyl trifluoroacetate.

EXPERIMENTAL

All reactions were conducted in a sealed, 0.5-in (1.27-cm) o.d. x 12-in (30.5-cm) silica-lined stainless-steel batch reactor (Figure 2). Total volume of the reactor system was

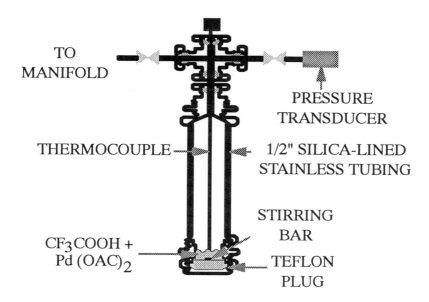

TO MANIFOLD

PRESSURE TRANSDUCER

THERMOCOUPLE

1/2" SILICA-LINED STAINLESS TUBING

STIRRING BAR

CF_3COOH + $Pd(OAC)_2$

TEFLON PLUG

FIGURE 2. METHANE OXIDATION REACTOR

<35 mL. The reactor was lined by Restek, Inc. In order to reproduce the earlier work, experimental conditions were the same as those reported by Sen and co-workers.[3]

The reactor was rinsed with 5.0 g (3.3 mL) of dry trifluoroacetic acid, purged several times with helium at 1000 psig, (6.89 MPa) and charged with reactants, typically 0.15 g of palladium(II) acetate dissolved in 5.00 g of trifluoroacetic acid. After being connected to the gas manifold, the reactor was purged several times with helium at 1000 psig (6.89 MPa), followed by several purges of methane with a final methane pressure of 800 psig (5.52 MPa), and isolated from the gas manifold.

When $^{13}CH_4$ was used, the reactor was immersed in a liquid nitrogen bath. When the temperature had reached -190°C, the reactor was connected to a mechanical vacuum pump and evacuated. The reactor was then isolated from the vacuum pump and, while still immersed in liquid nitrogen, was connected to the cylinders of $^{13}CH_4$ to allow transfer of the cylinder's contents to the reactor. After transfer of the cylinder's contents, the reactor was sealed and placed in the silicone oil bath.

The $^{13}CH_4$, rated by the supplier at 99+% isotopic purity, was supplied in 1-L cylinders at 20 psig (0.14 MPa) pressure. The contents of two cylinders were required to obtain the necessary >800 psig (5.52 MPa) pressure for reaction.

When necessary, the gases were dried prior to entering the reactor by passing through an 8-ft (2.44-m) x 1/4-in (0.64-cm) coil of stainless-steel tubing immersed in an acetone/dry ice bath.

The reactor was heated in a bath of silicon oil to 80°C and the reaction was allowed to proceed for 5 days. The reactor pressure was monitored by a pressure transducer and recorded during the run. Blank runs followed the same procedure with the exception of helium replacing methane.

After removal of the reactor's liquid contents, the reactor was filled with deionized water, capped and placed in an 720-W ultrasonic cleaner for one hour. The solution was then passed through a preweighed .50-μm Teflon® filter and air dried. The residue on the

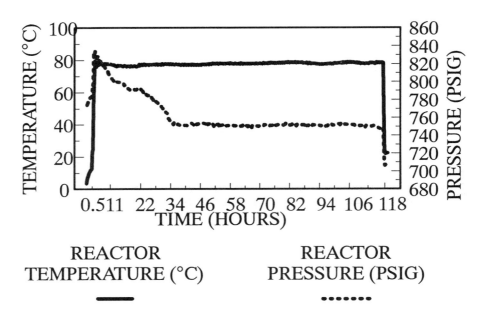

REACTOR
TEMPERATURE (°C)

REACTOR
PRESSURE (PSIG)

FIGURE 3. TYPICAL PRESSURE-TEMPERATURE
PROFILE OF OXIDATION RUN

filter was removed and identified by scanning electron microscopy (SEM) and energy dispersion spectroscopy (EDS). Gaseous components were analyzed on a Hewlett-Packard 5730 gas chromatograph. Liquid samples were analyzed on a Hewlett-Packard 5988A GC/MS system.

RESULTS

The results reported by Sen and co-workers were reproduced; in Run 241 production of methyl trifluoroacetate was observed and a fine metallic powder was recovered. SEM and EDS analysis of the powder confirmed it to be palladium metal with crystallites of the order of 1 micron in size. Quantitative analysis of the palladium metal residue indicated >80% of the palladium acetate recovered as palladium metal. Methane conversion, calculated by the difference in pressure from the beginning to the end of the run, was ~3 mol% (Table I). Figure 3 shows a typical plot of temperature and pressure as a function of time during the course of the run. As shown in Figure 3, the majority (~93.5%) of the methane was consumed during the first 30 hours of the reaction.

Analysis of the reaction mixture from Run 241 (Table II) identified several other oxygenated compounds and water. To determine the source of these compounds, the palladium(II) acetate dissolved in trifluoroacetic acid for Run 242 was analyzed prior to introduction of methane. This revealed the presence of methyl trifluoroacetate, the product of methane oxidation, prior to introduction of methane, and the same components identified before.

The trifluoroacetic acid was analyzed to determine if the methyl trifluoroacetate and other compounds found in Runs 241 and 242 were present. All the unexpected compounds except methyl trifluoroacetate were detected including a significant quantity of water. A blank run (243) was conducted wherein the methane was replaced with helium at 800 psig, 5.52 Mpa. All reaction conditions and operations were identical to previous runs. This

TABLE I.   COMPARISON OF Pd RECOVERED WITH CH$_4$ CONSUMED

RUN	% Pd (0) RECOVERED	% CH4 CONSUMED
241	83.50	3.00
242	77.00	3.13
243	68.30	N/A
244	0.00	N/A
245	83.20	N/A
246	39.20	6.42
248	7.88	4.08
250	4.21	4.01
252	7.46	10.30

TABLE II. GC/MS ANALYSIS OF METHANE OXIDATION REACTION MIXTURES

COMPOUND IDENTIFIED	241	242 START	242 FINISH	242 FILTERED	243 (NO CH4)	CF3COOH
AIR/CO2	X	X	X	X	X	X
CH3-O-CH3		X	X	X		
CF3C(O)OCH3	X	X	X	X	X	
CH3C(O)H	X	X	X	X	X	X
C2H5-O-C(O)H	X	X	X	X		
H2O	X	X	X	X	X	X
CH3COOH	X	X	X	X	X	X
C2H5COOH	X	X	X	X		X
CH3(CH2)2COOH	X					X
CF3COOH	X	X	X	X	X	X

TABLE III. GAS COMPOSITION (PERCENT)

SAMPLE	CH$_4$	He	N$_2$	C$_2$H$_6$	O$_2$	CO$_2$
CH$_4$	99.32		0.66	0.02		
He		100.00				
241	99.70	0.25				TRACE
242	93.19	0.11	5.28	0.02	1.39	0.02
243		100.00				TRACE
244		100.00				TRACE
^{13}CH$_4$	93.5% ^{13}CH$_4$ 6.5% ^{12}CH$_4$					

experiment resulted in the production of methyl trifluoroacetate and a 68.30% recovery of palladium metal. The only logical origin of the ester's methyl group is via decomposition of the starting material's acetate ligand.

After completion of the above experiments, the gas in the reactor was sampled prior to venting and recovery of liquid products. Analysis of the gas samples by GC (Table III) showed only the components present in the feed gas.

For use in the remaining experiments, dry, high purity trifluoroacetic acid was obtained in sealed ampules containing enough acid for a single use. Analysis of this trifluoroacetic acid revealed no detectable quantities of water or the other impurities previously detected. The blank run was repeated. After 150 hours at 80°C and 800 psig helium, the reactor was opened and the solution removed for analysis and comparison with the starting material. No difference in composition was detected between the two samples. Water (8×10^{-3} moles, a 10 fold excess) was then added to the mixture of Run 244, the reactor was charged with helium at 800 psig, and held at 80°C for 150 hours. Analysis of the products of reaction (Run 245) revealed the presence of methyl trifluoroacetate and methyl acetate. Since no methane was present in the system, the only source of the methyl group in the products is from the displaced acetate. This observation is inconsistent with that of Sen[5] in that he did not observe any deuterium incorporation into the methyl trifluoroacetate when $Pd(O_2CCD_3)_2$ was used.

The first experiment (Run 246) to use both methane and the dry, high purity trifluoroacetic acid resulted in products similar to previous experiments with the exception that the amount of palladium metal recovered was only 39.20 mol%, a reduction of >50%. We attribute this decrease to the absence of side reactions caused by the water in the trifluoroacetic acid.

An experiment (Run 248) was conducted using methane that was isotopically enriched in carbon-13. Oxidation products arising solely from the labelled methane, determined by GC-MS, would eliminate the possibility of products arising from the acetate ligand on the palladium(II) acetate. The reactor was filled as described above. Operating under conditions similar to previous experiments resulted in similar methane conversions but a recovered palladium metal amount of only 7.88 mol%. Analysis of the product mixture revealed both $CF_3C(O)O^{13}CH_3$ and $CF_3C(O)OCH_3$. Single Ion Monitoring (SIM) analysis of the isotopic ratio of the labeled products gave a $^{13}C/^{12}C$ ratio of 4.98. The composition of the labeled methane was determined by mass spectroscopy to be 93.5% $^{13}CH_4$ and 6.5% $^{12}CH_4$; a ratio of 14.38. This means that ~11% of the methyl carbon in the methyl trifluoroacetate comes from a source other than the labeled methane. This confirms our postulate that not all of the product arises from the methane introduced as a reactant.

To test the postulate that the presence of water in the reactor system was responsible for the observed decrease in palladium metal recovery, three experiments were conducted (Runs 250, 252, and 254) where the gases were dried prior to entering the reactor as described above. The reactor was prepared as in the $^{13}CH_4$ experiment. After warming to room temperature, 4.8×10^{-3} grams (2.7×10^{-4} moles) of water was recovered from the drying trap. The result of these experiments is that the methane consumption and product distribution remained the same as previously observed, but that recovery of palladium metal was only 4.21, 7.46, and 8.29 mol% respectively. This suggests that the presence of water in either the reactants or in the reactor system is responsible for the greater quantities of palladium metal reported in the literature. Water was not detected in the product mixture by GC/MS for these experiments. Quantitation of reaction products for Run 254 revealed that 0.13 weight percent of the product is methyl trifluoroacetate. This corresponds to a conversion of reactants to the desired product of 0.8 %.

Sen and co-workers postulated that the mechanism of this reaction is electrophilic attack on the methane by Pd(II), followed by reductive elimination to give Pd metal and

TABLE IV. GC/MS ANALYSIS OF METHANE OXIDATION REACTION MIXTURES

COMPOUND IDENTIFIED	243 (NO CH4)	244 START	244 FINISH (NO CH4)	245 (244 + H2O)	246	248	250	252
					13CH4			
AIR/CO2	X	X	X	X	X	X	X	X
CF3C(O)OCH3	X			X	X	X	X	X
CH3C(O)H	X	X	X	X	X			X
CH3C(O)OCH3				X	X	X	X	X
H2O	X				X			
CH3COOH	X	X	X	X	X			X
CF3C(O)CF3				X	X	X	X	X
CF3COOH	X	X	X	X	X	X	X	X

TABLE V. CH$_4$ AND PALLADIUM MOLAR BALLANCE

RUN	Pd(2+) START (x 10^4)	Pd(0) RECOVERED (x 10^4)	CH4 START (x 10^2)	CH4 END (x 10^2)	CH4 CONSUMED (CALC x 10^3)	CH$_4$ CON. / Pd(0)
241	6.69	4.70	7.05	6.59	4.69	9.98
242	6.69	5.14	7.07	6.72	3.59	6.99
243	6.69	4.30	0.00	0.00	0.00	N/A
245	6.92	5.76	0.00	0.00	0.00	N/A
246	6.92	2.72	7.07	6.62	4.54	16.72
248	6.71	0.55	6.83	6.52	3.07	56.32
250	6.99	0.29	8.14	7.55	5.94	203.84
252	6.54	0.49	6.79	6.10	6.90	141.15

the alcohol derivative. This mechanism is supported only by the fact that palladium is a strong electrophile and a good two electron oxidant. Our observations do not support the original assumption that the reaction, as stated in Equation 1, is a 1:1 stoichiometric reaction between methane and palladium(II) trifluoroacetate. Table V lists the molar balance for the experiments. The last column of the table shows the ratio of methane consumed to palladium metal recovered. The data from the early experiments, when water was present, show that the molar ratio of methane consumed to palladium metal recovered is of the order of 10. In later experiments, when water was removed from the reactants, this ratio is an order of magnitude larger. These inconsistencies leave open the important questions about the reaction mechanism which is key to proper evaluation of this reaction as a method for direct oxidation of methane. If the methane is being consumed by some other reaction not involving palladium, is the palladium(0) being reoxidized to a palladium(II) complex, or are other impurities present in the system?

CONCLUSION

This study has shown that the reaction expressed in Equation 1 does occur, as confirmed by the production of $CF_3CO_2{}^{13}CH_3$ from $^{13}CH_4$, but that this reaction is not responsible for all product methyl trifluoroacetate. When a blank experiment was performed using the highest purity starting materials and replacing methane with helium, methyl trifluoroacetate was detected in the product if water was not excluded from the system. The presence of water in the reaction mixture appears to cause the palladium acetate/trifluoroacetate complex to decompose and produce methyl trifluoroacetate and account for the high yields of palladium metal reported in the literature.

ACKNOWLEDGMENT

We would like to acknowledge the technical assistance of Joseph R. D'Este with construction and operation of the reactor unit and Donald V. Martello and Joseph P. Tamilia with the SEM/EDS analyses.

DISCLAIMER

Reference in this report to any specific commercial product, process, or service is to facilitate understanding and does not necessarily imply its endorsement or favoring by the United States Department of Energy.

REFERENCES

1. Crabtree, R.H. *Chem. Rev.* **1985**, *85*, 245-269.
2. Schwartz, J. *Acc. Chem. Res.* **1985**, *18*, 302-308.
3. Gretz, E.; Oliver, T.F.; Sen, A. *J. Am. Chem. Soc.* **1987**, *109*, 8109-8111.
4. Kao, L.-C.; Hutson, A.C.; Sen, A. *J. Am. Chem. Soc.* **1991**, *113*, 700-701.
5. Sen, A. *Platinum Metals Review* **1991**, *3*, 126-132.

COUPLING OF CATALYTIC PARTIAL OXIDATION AND STEAM REFORMING OF METHANE TO SYNGAS

V.R.Choudhary[*], A.M.Rajput and B.Prabhakar

Chemical Engineering Division,
National Chemical Laboratory, Pune 411 008 (India)

ABSTRACT

Methane-to-syngas (i.e.CO and H_2) conversion reactions involving exothermic oxidative conversion of methane and endothermic steam reforming of methane have been carried simultaneously NiO-CaO (Ni/Ca = 3.0) catalyst at different temperatures (700- 850°C), CH_4/O_2 (1.8 - 6.0) and CH_4/H_2O (1.6- 10.0) ratios and space velocities (16 x 10^3 - 1090 x $10^3 h^{-1}$). By the coupling of the exothermic and endothermic reactions, it is possible to make, the CMOCSR (coupled methane oxidative conversion and steam reforming) process mildly endothermic, near thermoneutral or mildly exothermic by manipulating the process conditions (viz. temperature and feed ratios) and also to operate the process in a most energy efficient and safe manner, with requirement of little or no external energy. Under the present energy crisis, the conversion of methane to syngas (with H_2/CO ratio close to 2.0 which is desirable for methanol and Fischer-Tropsch synthesis) with high conversion, selectivity and productivity in a most energy efficient and safe manner with requirement of little or no external energy is of great practical importance for the effective utilization of natural gas by its conversion to value added and/or easily transportable products via syngas routes.

INTRODUCTION

Syngas (a mixture of CO and H_2) is a versatile feed stock mainly used for methanol, ammonia and Fischer- Tropsch synthesis processes and also for carbonylation, hydroformylation, hydrogenation and reduction processes. Since the last decade, world-wide efforts are being made for the conversion of methane (which is a main constituent of natural gas) to value added and easily transportable products like methanol, liquid hydrocarbon fuels and petrochemical feed stocks (e.g. ethylene and other lower olefins) for the effective utilization of natural gas and the syngas conversion routes are considered as important means for achieving the above goal[1]. Syngas is produced from methane or natural gas mostly by the conventional catalytic steam reforming process[2], which is a highly endothermic and hence highly energy intensive process and also involves energy inefficient operation for obtaining H_2/CO ratio (≤ 2) required for methanol and Fischer-Tropsch synthesis processes.

Recently a number of studies[3-8] have been reported on the catalytic oxidative conversion of methane to syngas (which is an exothermic reaction) with H_2/CO ratio of about 2.0 and with high conversion, selectivity and productivity. However, in this process, a small decrease in the selectivity causes a large increase in the exothermic reaction heat, making the process hazardous and/or difficult to design and operate or control.

The present energy crisis and high energy cost and also the environmental problems have created a great need for developing catalytic processes that require much less external energy, operate in a most energy efficient manner and have no hazards. This goal could best be achieved for the methane-to-syngas conversion by coupling of the above endothermic and exothermic processes by carrying out them simultaneously over the same catalyst.

In our earlier studies[5], NiO-CaO catalyst showed high activity, selectivity and productivity in the oxidative conversion of methane to CO and H_2. In this paper, we report our highly promising results on the coupling of the exothermic oxidative conversion of methane to syngas with the endothermic steam reforming of methane over NiO-CaO catalyst at different process conditions.

EXPERIMENTAL

The catalyst used in this work is NiO-CaO (Ni/Ca = 3.0) prepared by mixing thoroughly finely ground high purity nickel nitrate and calcium hydroxide (with Ni/Ca mole ratio of 3.0) along with deionized water just sufficient to form a thick paste, drying and decomposing the mass at 600°C for 4h, powdering, pressing and crushing to 22-30 mesh size particles and calcining in air at 930°C for 4h. The surface area of the catalyst is found to be 2.5 $m^2.g^{-1}$. The XRD analysis showed the presence of NiO and CaO phases in the catalyst.

The oxidative methane-to-syngas conversion and methane steam reforming reactions were carried out simultaneously over the NiO-CaO catalyst at atmospheric pressure by passing continuously a mixture of pure methane, oxygen and steam over the catalyst in a tubular reactor (i.d. : 10 mm) made up of quartz at different temperatures, CH_4/O_2 and CH_4/H_2O mole ratios in feed and gas hourly space velocities (measured at 0°C and 1.0 atm pressure). The reaction temperature was controlled by Chromel-Alumel thermocouple kept in the catalyst bed. Before carrying out the reaction, the catalyst was heated insitu at 930°C for 1 h in a flow of moisture free N_2 (50 $cm^3.min^{-1}$). The water in the product stream was condensed at 0°C. The product gases (after removing water) were analyzed by an on-line gas chromatograph with thermal conductivity detector, using a spherocarb column. The C, H and O mass balances were $100 \pm 4 \%$.

RESULTS AND DISCUSSION

When the catalytic process involving the coupling of steam reforming with oxidative conversion of methane to syngas over the NiO-CaO catalyst at 800°C (CH_4/O_2 = 2.0 and CH_4/H_2O = 4.6 and GHSV = 58,500 h^{-1}) was carried out continuously for about 20 h, no significant change in the activity and selectivity of the catalyst was observed (Fig.1). The net heat of reaction (ΔH_r) and adiabatic temperature rise (ATR) for the above have been estimated to be about -3.0 kcal per mole of methane converted and 90°C, respectively.

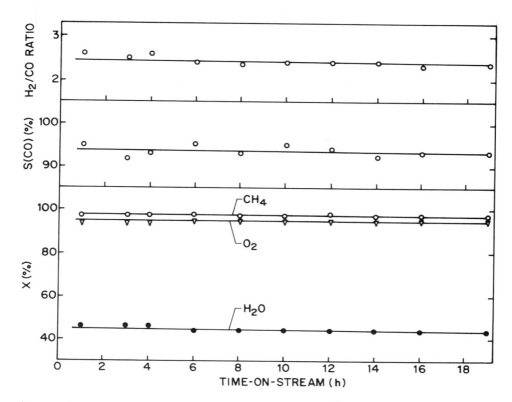

Figure 1. Time-on-stream activity/selectivity of the catalyst (at 800°C, CH_4/O_2 = 2.0, CH_4/steam = 4.6 and GHSV = 58,500 h^{-1}). X = conversion and S = selectivity.

Typical results of the catalytic process, showing high conversion and selectivity are given in Table-1. Following exothermic and endothermic reactions are expected to occur simultaneously over the same catalyst in the catalytic process.

$$CH_4 + 1/2\ O_2 \rightarrow CO + 2H_2 \quad + \quad 5.2\ kcal \quad\quad (1)$$

$$CH_4 + 2\ O_2 \rightarrow CO_2 + 2H_2O \quad + \quad 191.5\ kcal \quad\quad (2)$$

$$CO + H_2O \rightleftharpoons CO_2 + H_2 \quad + \quad 8.0\ kcal \quad\quad (3)$$

$$CH_4 + H_2O \rightleftharpoons CO + 3H_2 \quad - \quad 54.2\ kcal \quad\quad (4)$$

(The endothermic CO_2 reforming reaction is also expected to occur but only to a small extent.)

The results reveal that by carrying out the endothermic steam reforming simultaneously with the exothermic oxidative conversion of methane over the same catalyst, it is possible to make the overall process mildly exothermic, near thermoneutral or mildly endothermic with a small adiabatic temperature rise or fall (Table-1).

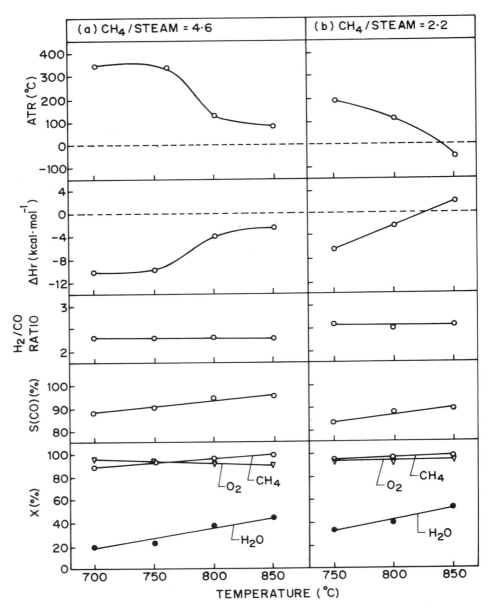

Figure 2. Effect of reaction temperature (CH_4/O_2 = 2.0 and GHSV = 74,000 h^{-1}).

The influence of different process variables (viz. temperature, CH_4/O_2 and CH_4/H_2O ratios in feed and space velocity) on the conversion, CO selectivity and H_2/CO ratio in products and also the estimated net heat of reaction (ΔH_r) and adiabatic temperature rise (ATR) in the catalytic processes has been shown in Figs.2-5.

Table 1. Typical results on the coupling of steam reforming with oxidative methane-to-syngas conversion over NiO-CaO (Ni/Ca = 3.0) (GHSV = 70($\pm$ 10) x 10^3 h^{-1})

Temp (^{0}C)	CH_4/O_2 ratio	CH_4/H_2O ratio	Conversion (%) CH$_4$	O$_2$	H$_2$O	S(CO) (%)	H_2/CO ratio	ΔH_r (kcal mol^{-1})	ATR (^{0}C)
800	2.3	2.3	95.3	90.6	53.6	90.0	2.6	+0.17	-5
800	2.3	4.9	93.6	95.6	51.3	95.1	2.0	-0.35	+10
800	2.0	2.2	97.3	96.5	40.0	87.9	2.5	-2.16	+60
800	2.0	4.6	96.5	95.4	39.3	94.1	2.3	-3.92	+120
800	2.0	2.2	99.3	96.0	54.0	89.0	2.6	+2.12	-55
850	2.0	4.6	98.6	91.1	44.4	95.2	2.3	-2.66	+80

From these results, following general observations have been made.

1. The methane conversion is increased with increasing the reaction temperature but it is decreased with increasing the CH_4/O_2 and CH_4/H_2O ratios in the feed and also the space velocity, as expected.

2. The conversion of water is increased with increasing the reaction temperature and CH_4/O_2 ratio but it is decreased with increasing the space velocity. The influence of CH_4/H_2O ratio on the water conversion is, however, complex (Fig.4): it depends upon the other process conditions (viz. CH_4/O_2 ratio).

3. The influence of process variables on oxygen conversion is quite complex.

4. The CO selectivity is increased with increasing the reaction temperature and the CH_4/H_2O ratio. Whereas, the influence of other process variables on the selectivity is complex.

5. The exothermicity of the overall process is decreased (or the the endothermicity of the process is increased) with increasing in the reaction temperature and CH_4/O_2 ratio but, it is increased with increasing the CH_4/H_2O ratio and the space velocity.

The coupled methane oxidative conversion and steam reforming (CMOCSR) process has following outstanding features:

* The CMOCSR process can be made mildly exothermic, near thermoneutral or mildly endothermic simply by manipulating the process conditions (viz. CH_4/O_2 and CH_4/H_2O ratios in feed and reaction temperature) and hence it can be operated with requirement of little or no external energy.

* The process is operated in a most energy efficient manner because the heat produced in the exothermic reactions is used instantly by the endothermic reaction(s). Because of this, the process operates also in the most safe manner as the possibility of development of hot spots in the catalyst bed and/or reaction run-away conditions are eliminated or drastically reduced due to a buffer action on the temperature, resulting from the coupling of the exothermic and endothermic reactions.

* In the CMOCSR process, high methane conversion (above 90%) with high CO selectivity (above 90 %) at high space velocity (≈ 80,000 h⁻¹ or even higher) and hence high CO productivity can be obtained.

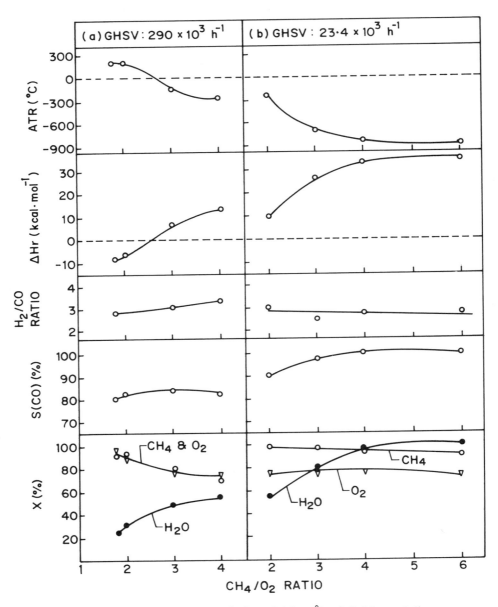

Figure 3. Influence of CH_4/O_2 ratio in feed (at 800°C and CH_4/steam = 1.6).

* Because of a very low net heat of reactions (and consequently a low adiabatic temperature rise or fall), it would be possible to operate the CMOCSR process in an adiabatic reactor (i.e., a very simple reactor with no arrangement to supply or remove heat) requiring much lower capital and process operation costs.

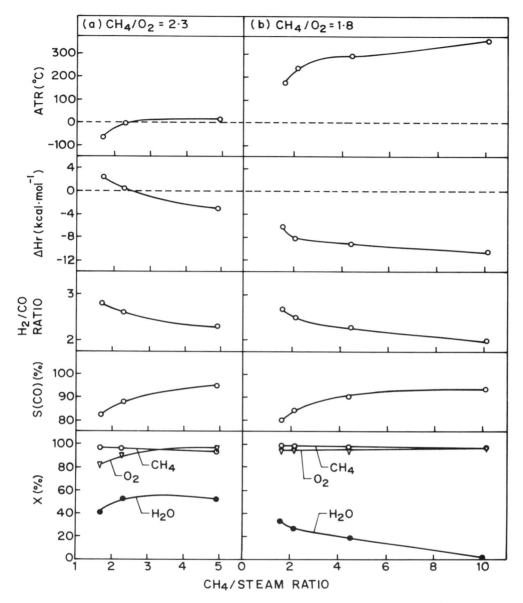

Figure 4. Influence of CH_4/steam ratio in feed (at 800 °C and GHSV = 77,000 h^{-1}).

CONCLUSIONS

By coupling of the exothermic oxidative conversion and the endothermic steam reforming of methane over the same catalyst, almost all the limitations of these two individual processes are eliminated. The coupled methane oxidative conversion and steam

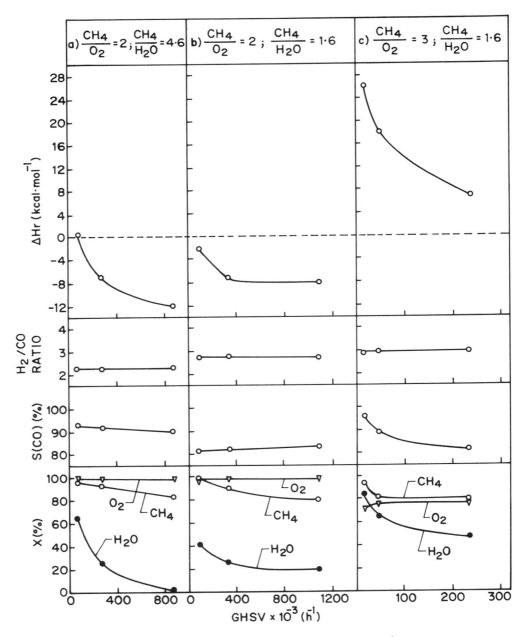

Figure 5. Influence of gas hourly space velocity (at 800°C).

reforming (CMOCSR) process operates in a most energy efficient and safe manner with a requirement of little or no external energy. Also, there is a high possibility of developing CMOCSR process operating adiabatically in a very simple reactor with much favorable process economics.

REFERENCE

1. J.R.Anderson, Methane to higher hydrocarbons, Appl. Catal. 47:177(1989).
2. (a) R.E.Kirk and D.F. Othmer (Eds.), Encyclopedia Of Chemical Technology (3rd edition), Wiley Interscience, N.Y., Vol 12 (1990).
   (b) Ulmann's Encyclopedia of Industrial Chemistry (5th revised edition), VCH Verlagsgesellschaft mbH, Weinheim, Vol 12 (1989).
3. A.T.Ashcroft, A.K.Cheetham, J.S.Foord, M.L.H.Green,C.P.Grey, A.J.Mureell and P.D.F.Vernon, Selective oxidation of methane to synthesis gas using transition metal catalysts, Nature, 344:319 (1990).
4. D.Dissanayake, M.P.Rosynek, K.C.C. Kharas and J.H.Lunsford, Partial oxidation of methane to carbon momoxide and hydrogen over a Ni/Al_2O_3 catalyst, J.Catal. 132:117 (1991).
5. V.R.Choudhary, A.M.Rajput and B.Prabhakar, Low temperature oxidative conversion of methane to syngas over NiO-CaO catalyst, Catal. Lett. 15: 363 (1992).
6. V.R.Choudhary, A.S.Mamman and S.D.Sansare, Selective oxidation of methane to CO and H_2 over Ni/Mgo at low temperature, Angew. Chem. Int. Ed. Engl. 31:1189 (1992).
7. J.A.Lapszewicz and X - Z Jiang, Preprints - Am. Chem. Soc. Dir. Pet. Chem., Investigation of the mechanism of partial oxidation of methane to synthesis gas, 37(1):252 (1992) ; Characteristics and performance of catalysts for partial oxidation of natural gas to syngas, 38(4):815 (1993).
8. D.A.Hickman and L.D.Schmidt, Production of syngas by direct catalytic oxidation of methane ,Science, 259:343 (1993); Synthesis gas formation by direct oxidation of methane over Pt monoliths, J.Catal. 138:267 (1992).

ACETALDEHYDE FROM CHLOROMETHANE AND CARBON MONOXIDE

Y. Soong, R. R. Schehl, and R. P. Noceti

Pittsburgh Energy Technology Center
U.S. Department of Energy
P. O. Box 10940, Pittsburgh, PA 15236

Acetaldehyde (CH_3CHO) is the major oxygenated product of the thermal reaction of carbon monoxide and chloromethane under reaction conditions, i.e., between 600 °C and 700 °C, and between 0.1 MPa and 1 MPa. The reaction conditions have a significant effect on the formation of the products.

INTRODUCTION

Conversion of methane, the principal component of natural gas, to higher molecular weight compounds is of great interest to industry. Methane is the most stable saturated hydrocarbon and is a molecule of low chemical reactivity. Considerable work has been directed toward development of processes for conversion of methane to higher hydrocarbons and oxygenates. Known processes for the direct conversion of methane to higher hydrocarbons are difficult to control or are energetically restrictive. Stepwise processes for conversion of methane to higher hydrocarbons via an intermediate, such as chloromethane or methanol, offer advantages in these areas.

Methane can be converted directly to C_2 and higher hydrocarbons, including benzene, by pyrolysis, either neat or in the presence of chlorine as a reactant or free radical initiator[1-4]. The principal products of the pyrolysis of methane in the presence of chlorine are acetylene, ethylene, hydrogen chloride, and benzene. Chloromethane is thought to be an intermediate in this reaction[5,6]. The products may be explained by a pathway that includes decomposition of chloromethane followed by recombination of the fragments. Oxygenate production through direct oxidation of methane to methanol is also possible but is plagued by low conversions and low selectivity. Well known, multistep routes from methane to higher hydrocarbons include steam reforming of the methane or its direct partial oxidation to synthesis gas, followed by either a Fischer-Tropsch synthesis to produce hydrocarbons or methanol synthesis with subsequent

Methane and Alkane Conversion Chemistry
Edited by M. M. Bhasin and D. W. Slocum, Plenum Press, New York, 1995

conversion by the Mobil Methanol-to-Gasoline (MTG) process[7]. A more recently studied pathway consists of conversion to chloromethane followed by a zeolite-based chloromethane-to-liquid fuel reaction[8]. Only limited work has been directed toward the conversion of methane to higher oxygenates via chloromethane and carbon monoxide. Our previous study reported that C_2 oxygenates could be formed in the carbon monoxide/chloromethane system[9]. The oxygenates may be readily utilized for production of gasoline octane enhancers or chemical feedstocks. The objective of this work is to explore the chemistry leading to formation of C_2 oxygenates in uncatalyzed pressurized carbon monoxide-chloromethane reaction systems.

EXPERIMENTAL

High purity He, Ar, CO, and CH_3Cl (Matheson) were used throughout this study. The details of the high-pressure fixed-bed reactor system are shown elsewhere[10]. The reactor is a 0.79-cm i.d., 20-cm-long empty stainless steel tube. Scoping experiments were conducted in this reactor system in a temperature range of 600 to 900°C, in a pressure range of 0.1 to 1 MPa with gas flows of Ar, CH_3Cl, CO and He of 28, 5, 18 and 52 ml/min, respectively. The space velocity is around 630 h^{-1} in this reactor. The reaction products were analyzed, after three hours of operation, using a Hewlett-Packard 5730A GC equipped with a Porapak PS and a Carbosieve S-II column. No efforts were made to measure the concentrations of HCl and Cl_2 in the effluent or the amounts of carbon deposited in the reactor.

RESULTS AND DISCUSSIONS

The results, showing the effects of temperature and pressure on the reaction, are summarized in Table 1. Methane, C_2, C_3, C_4, and C_5 hydrocarbons (saturated and unsaturated) were detected in the products under the reaction conditions studied. Similar product distributions have been reported previously for the methane and chlorine reaction system but only at ambient reaction pressures[3-6,9]. We postulate that the methyl radical, arising from chloromethane pyrolysis, was responsible for the methane, C_2, C_3, C_4, and C_5 hydrocarbon products. The methyl radical forms more readily than the chloromethyl radical from the decomposition of CH_3Cl because the bond dissociation energy of Cl-CH_3 is 19.3 Kcal/mole less than that of H-CH_2Cl[11]. A detailed mechanism for the formation of hydrocarbons from the pyrolysis of chloromethane has been discussed by Weissman and Benson[5]. Chloromethane conversion trends in our system are not informative whenever reaction temperatures are above 750 °C. Nearly 100 % conversion was observed at the higher reaction temperatures and pressures (those shown in Table 1 with less than 5 % CH_3Cl in the product). It can be seen from Table 1 that the production of methane above 750 °C is independent of reaction temperature or pressure. This observation can be attributed to the limiting, nearly complete conversion of chloromethane or that the deposited carbon, formed from chloromethane, served as a reducing agent for chloromethane. At a lower temperature, 600 °C, the conversion of chloromethane could be enhanced by increasing the reaction pressure. At this temperature, it increases from 30% conversion at 0.1 MPa to 63% conversion at 1.0

Table 1. CH_3Cl and CO Reaction

Product Distribution 10^{-9} g-mole/ml

T (°C)	P (MPa)	CH_4	C_2	C_3	C_4	C_5	⬡	⬡·	⬡-CH_3	CH_3Cl	CH_2Cl_2	$CHCl_3$	C_2H_5Cl	CH_3COCl	CH_3OH	CH_3CHO	C_2H_5OH	CO_2
600	0.1	1248	78	4.9	0.2	0.2	0.3	0.1	–	5837	367	–	2.2	0.35	2.5	369	1.7	–
700	0.1	2027	50	4.1	0.3	0.3	18.6	7.1	–	845	13.3	–	0.3	1.23	9.5	3124	12.3	–
750	0.1	2048	627	8.8	0.3	0.2	53.4	–	–	1.3	0.9	4.3	–	0.7	0.8	223	23	–
800	0.1	1987	547	8.3	0.5	0.1	132	–	–	0.5	2.7	10.8	–	–	0.4	3.4	26	–
850	0.1	1932	380	5.4	0.3	0.2	159	50	–	0.3	1.9	6	–	0.9	2.9	4.2	10.8	–
900	0.1	1855	176	1.4	0.2	0.4	193	11	–	0.4	1.3	5	–	4	1.6	–	0.8	–
600	0.5	1803	171	6.8	0.2	0.8	1.4	15	–	4344	118	7	1.7	0.6	7.5	822	3.5	–
650	0.5	1986	345	6.4	0.2	–	24	6	–	1239	12	21	0.8	1.8	4.3	2457	5.2	–
700	0.5	2071	465	14	0.6	–	69	6	–	2.8	0.6	9	0.2	1.9	1.1	39	11	–
750	0.5	2040	415	13	0.8	–	130	7	–	2	–	7	0.1	–	–	–	11	19
800	0.5	1986	280	4	–	0.3	178	9	–	0.5	–	2	–	–	–	–	1	20
850	0.5	2000	119	1	–	–	125	–	–	–	–	1	–	–	–	–	–	29
900	0.5	1951	12	–	–	–	105	3	–	–	–	3	–	–	–	–	–	60
600	1	1874	190	12	0.2	0.4	10	1	–	3370	7.6	15	1	0.5	5.2	1748	2.3	–
700	1	2079	352	15	0.9	0.2	66	3.6	–	4.3	–	0.2	0.2	–	0.5	3	4.3	–
750	1	3875	297	9	0.3	–	80	3.3	–	2.2	–	5	0.1	–	–	–	1.4	59
800	1	2054	167	2	0.6	–	87	25	–	1.5	1.7	–	–	–	–	–	–	34
850	1	2103	19	0.1	–	–	57	26	–	0.8	–	3	–	–	–	–	–	54
900	1	1947	3	–	–	–	46	44	–	1.7	–	–	–	–	–	–	–	82

C_2, C_3, C_4, and C_5 include saturated and unsaturated hydrocarbons

317

MPa. The presence of CH_2Cl_2, CH_3Cl and trace amounts of C_2H_5Cl were also observed in the products. Production of CH_2Cl_2 and $CHCl_3$ may be explained by sequential reactions of chloromethane and its products as in the following:

$2CH_3Cl \rightarrow CH_4 + CH_2Cl_2$

$CH_3Cl \rightarrow CH_3\cdot + Cl\cdot$

$CH_3Cl + Cl\cdot \rightarrow CH_2Cl\cdot + HCl$

$CH_2Cl\cdot + Cl\cdot \rightarrow CH_2Cl_2$

$CH_2Cl_2 + Cl\cdot \rightarrow CHCl_2\cdot + HCl$

$CHCl_2\cdot + Cl\cdot \rightarrow CHCl_3$

$CH_3\cdot + HCl \rightarrow CH_4 + Cl\cdot$

$CH_2Cl_2 \rightarrow CH_2\text{:} + 2\ Cl\cdot\ (\text{or } Cl_2)$

$CH_3\cdot \rightarrow CH_2\text{:} + H\cdot$

$CH_2\text{:} + Cl\cdot \rightarrow CH_2Cl\cdot$

$CH_3Cl + CH_2Cl\cdot \rightarrow CH_3\cdot + CH_2Cl_2$

$CH_2\text{:} \rightarrow CH\ \vdots\ + H\cdot$

$CH\ \vdots\ + 3\ Cl\cdot \rightarrow CHCl_3$

The major chlorine containing species, HCl and Cl_2, were identified in the products by independent GC/MS analysis.

Not surprisingly, reaction conditions do have a profound impact on product distribution. Table 1 shows that, at 600 °C, the production of C_2-C_5 hydrocarbons increased with increasing reaction pressure but, at 750 °C or higher, the production of C_2, C_3, C_4 and C_5 decreased with increasing reaction pressure. At any of the pressures studied, hydrocarbon production increased with increasing reaction temperature until an optimum, pressure-dependent temperature for hydrocarbon production was reached.

Table 1 also shows that benzene (C_6H_6) and toluene ($C_6H_5CH_3$) were detected in the products. The production of benzene and toluene in a methane-chlorine system has been reported previously[3-6,9]. At lower pressure, 0.1 MPa, the formation of benzene increased as the temperature increased and had not peaked at 900 °C. At the higher pressures, 0.5 and 1 MPa, benzene production was at a maximum around 800 °C.

The production of oxygenates, acetaldehyde (CH_3CHO), and ethanol (C_2H_5OH), is also strongly dependent on the reaction conditions. At a lower pressure, 0.1 MPa, the production of CH_3CHO peaked at 700 °C. Notice at these conditions that more gmole/ml of CH_3CHO were made than that of CH_4. At a higher pressure, 0.5 MPa, the maximum yield was shifted to a lower temperature, 650 °C. Finally, at 1 MPa, the maximum production of CH_3CHO was observed at 600 °C. Increasing the reaction pressure although reducing the optimum temperature for oxygenate production also decreases the overall production of CH_3CHO. A similar trend was found for the formation of ethanol (C_2H_5OH). Methanol (CH_3OH), also detected in the products, may be explained by the reaction of CH_3Cl and H_2O ($CH_3Cl + H_2O \rightarrow CH_3OH + HCl$). Acetaldehyde ($CH_3CHO$) likely arises from acetyl chloride (CH_3COCl); trace amounts of acetyl chloride were observed whenever acetaldehyde was produced. Acetyl chloride is readily synthesized by passing carbon monoxide and chloromethane through a vessel having a smooth, nonmetallic surface at a temperature of 600°C and above (CO + $CH_3Cl \rightarrow CH_3COCl$)[12-14]. Several possible pathways are possible that lead to acetaldehyde. Catalytic reduction of CH_3COCl with hyrogen to produce CH_3CHO and HCl is the well-known Rosenmund reduction, unlikely in this case because of the lack of catalyst. It is postulated that acetaldehyde formation may follow, by analogy, the mechanism suggested by Benson and coworkers in the Br/CH_3CHO system at 300 K in a very low pressure reactor[15].

$CH_3COCl \rightarrow Cl\cdot + CH_3CO\cdot$

$CH_3CO\cdot + HCl \rightarrow Cl\cdot + CH_3CHO$

Alternatively, the production of CH_3CHO may be explained by the following reactions:

$CH_3Cl \longrightarrow CH_3\cdot + Cl\cdot$

$CH_3\cdot + CO \longrightarrow CH_3CO\cdot$

$CH_3CO\cdot + CH_3Cl \longrightarrow CH_3CHO + CH_2Cl\cdot$

The C_2H_5OH could be formed by the hydrogenation of CH_3CHO. The observed C_2H_5OH might also be a result of reaction between C_2H_5Cl and H_2O. The pathways leading to the ethanol/chloroethane product couple are not clear with regard to the relationship of the product and intermediate. Oxygenate formation is in accord with our previous finding that C_2 oxygenates can be formed as major products from the reaction of CO and methylchloride[9]. It is interesting to note that no carbon dioxide was detected at the low reaction pressure, 0.1 MPa. In contrast to the lower pressure results, carbon dioxide was detected at the higher reaction pressures (0.5 and 1 MPa). The production of carbon dioxide may be attributed to the disproportionation reaction ($2\ CO \rightarrow C + CO_2$) or the water gas shift reaction. Carbon deposits were observed in the reactor but the amounts were not determined. A significant result was the almost complete absence of CH_3CHO, C_2H_5OH, and CH_3OH, whenever carbon dioxide was produced or nearly complete conversion of chloromethane was achieved. Loss of CO through disproportionation would naturally result in decreased formation of oxygenates. The nearly complete conversion of chloromethane and the absence of oxygenated products suggests that almost all the chloromethane is converted to polymerized carbon. When conditions favor disproportionation of CO to CO_2 and C, production of oxgyenated products is minimized and, in the absence of other possible pathways, chloromethane decomposes to form an active, reducing carbon. Higher pressure also diminishes the formation of oxygenates from the carbon monoxide-chloromethane system.

SUMMARY

Pressure has significant effects on the carbon monoxide-chloromethane reaction system. Higher pressures increase the carbon dioxide formation at the expense of C_2 oxygenate formation, shift the optimum temperature for the maximum yields of C_2 oxygenates to a lower temperature, and affect CH_3Cl conversion, benzene formation, and the formation of C_2 hydrocarbons. The complex product slate argues for a series of simultaneous parallel reactions with widely divergent mechanisms. Computer simulations of the possible reaction network have been undertaken and will be discussed in the future.

ACKNOWLEDGEMENTS

The authors are much indebted to Mr. J. R. D'Este and Mr. A. G. Blackwell for the careful collection of data.

DISCLAIMER

Reference in this report to any specific commercial product, process, or service is to facilitate understanding and does not necessarily imply its endorsement or favoring by the United States Department of Energy.

REFERENCES

1. W. Bartok and Y. H. Song., *U.S. Pat.* 4,683,419 (1985)
2. E. Bartholome, H. Friz, F. Neumayr., M. Reichect, and U. Wanger., *U.S. Pat.* 3,542,894 (1970)
3. S.W. Benson and P. Verdes., *U.S. Pat.* 4,199,533 (1980)
4. R. G. Minet and S.C. Che., *U.S. Pat.* 4,804,797 (1989)
5. M. Weissman and S. W. Benson, Inter. *J. Chem. Kinet.* 16, 307, (1984)
6. R. G. Minet and J Kim., *Chemical Economy & Engineering Review* Vol 15, No 10, 35, (1983)
7. C. D. Chang., A. J. Silvestri., and R. L. Smith., *U. S. Pat.* 3,928,483 (1975)
8. C. E. Taylor and R. P. Noceti, *U.S. Pat.* 4,769,504 (1988)
9. Y. Soong., R. R. Schehl, and R. P. Noceti., *React. Kinet. Catal. Lett.* 49, 21 (1993)
10. Y. Soong., A. G. Blackwell., R. R. Schehl, and R. P. Noceti., *Fuel Sci. & Tech. Int.* 11, 937 (1993)
11. M. Weissman and S. W. Benson, *J. Phy. Chem.* 87, 243, (1983)
12. A. Wacker., *Brit. Pat.* 773,775 (1957)
13. I. Kuriyama, *Japan Pat.* 76,137,018 (1976)
14. H. Erpenbach, K. Gehrmann, W. Lork, and P. Prinz., *Germany Pat.* 3,016,900 (1981)
15. T. S. A. Islam, R. M. Marshall, and S. W. Benson., *Int. J. of Chem. Kinet.* 16, 1161, (1984)

PARA-SELECTIVE GAS PHASE O$_2$ OXIDATION OF XYLENES OVER Fe/Mo/BOROSILICATE MOLECULAR SIEVE, CVD Fe/Mo/DBH

Jin S. Yoo, Chin Choi-Feng and Gerry W. Zajac
Amoco Research Center, P.O. Box 3011, Naperville, IL 60566

INTRODUCTION

Terephthaldehyde is emerging as a promising reactive bifunctional monomer for a variety of commercial applications including liquid crystals, electron conductive polymers, thermal engineering polymers, optical brighteners, and other novel specialty polymers.

Conventionally, terephthaldehyde has been synthesized from the tetrachlorinated side chain derivatives of p-xylene in the liquid phase process [1]. The cost of terephthaldehyde, $12-13 per pound, is prohibitive even for the specialty product application. Recently, a new commercial process has been developed with the modified ZrO$_2$ catalyst to produce aromatic, aliphatic and alicyclic aldehydes via hydrogenation of the corresponding acids [2]. Nonetheless, because of a concern for the sustainable availability of cost effective aldehydes, the chemical industry has sought direct synthesis of terephthaldehyde by the selective partial oxidation of p-xylene.

The selective oxidation of alkylaromatics to the corresponding aldehydes, especially dialdehydes in the para-position, is a formidable challenge since the formation of acid is much more favored, in particular, in the liquid phase process such as the MC oxidation process [3] In the liquid process, the aldehydes are typically present as undesirable impurities in the main product, terephthalic acid. Despite these difficult challenges facing the direct oxidation approach, the pursuit to achieve a one-step process by the gas phase oxidation route has been continuing in industry.

Recently, we found that the CVD Fe/Mo/DBH catalyst was effective for selective formation of terephthaldehyde and p-tolualdehyde in the gas phase O$_2$ oxidation under mild conditions. Despite the large surface areas, 150-280 m^2/g, of the catalyst, burning was limited to less than 10 mole % under controlled oxidation conditions [4,5].

Besides the activity for para-selective oxidation of xylenes and other methylaromatics to aldehydes, the CVD Fe/Mo/DBH catalyst also capable of catalyzing various reactions such as

Methane and Alkane Conversion Chemistry
Edited by M. M. Bhasin and D. W. Slocum, Plenum Press, New York, 1995

oxydehydrogenation of alkylaromatics to styrene derivatives [6], one-step hydroxylation of benzene to phenol with N_2O [7], and activation of CO_2 [5]. This paper presents the chemistry involved in these catalytic reactions.

EXPERIMENTAL

The detailed procedures for preparing the chemical vapor deposition (CVD) catalysts, characterizing the CVD catalysts using electron microscope, NH_3 TPD, XRD, and Raman spectroscopy, and analyzing the product using an on-line GC were described in detail previously [4].

RESULTS AND DISCUSSIONS

Oxidation of p-xylene and its isomers

The Fe/Mo/DBH catalysts having different metal compositions were prepared by depositing $FeCl_3$ and MoO_2Cl_2, $MoCl_5$, or $MoOCl_4$ on the borosilicate molecular sieve, HAMS-1B-3, according to the chemical vapor deposition (CVD) technique described in a previous paper [4]. The partial removal of boron from the HAMS matrix occurs during the CVD catalyst preparation, particularly in the iron vapor deposition and washing step. The catalyst compositions of these catalysts are summarized in Table 1.

These catalysts were evaluated in a premixed gas blend (O_2/p-xylene = 10/1) in a micro-reactor loaded with 0.50 g of the catalyst. The results compared in Table 2 show that all the catalysts prepared from $MoCl_5$, MoO_2Cl_2 and $MoOCl_4$ are very active and selective for the aldehyde synthesis.

Reactivity of Xylene Isomers

Ortho- and m-xylene were evaluated for the gas phase O_2 oxidation in a micro-reactor loaded with 0.5 g of the fresh catalyst 2 at 350-400°C under the O_2/xylene ratio of 43:1, and the results

Table 1. Composition of CVD Fe/Mo/DBH catalysts 1-5.

ID	Metal compound	Metal composition					Surface area
		Mo %	Fe %	Cl ppm	B ppm	Mo/Fe	m²/g
1	$MoCl_5$ $FeCl_3$	5.2	1.5			2.05	
2	$MoOCl_4$ $FeCl_3$	4.8	1.2	0.0	272	2.38	
3	MoO_2Cl_2 $FeCl_3$	7.0	2.2	700	540	1.80	283
4	" "	9.4	2.7			2.03	263
5	" "	6.8	1.9		540	2.10	

322

Table 2. Gas phase oxidation of p-xylene over catalysts 1, 2 and 3
Catalyst weight: 0.501 g (1 mL), O_2/pX = 10/1, premixed gas: 0.10% pX, 1.0% O_2, and 1.0% N_2 in He.

Run no.	1		2		3	
Mo-compound	$MoCl_5$		$MoOCl_4$		MoO_2Cl_2	
Temperature °C	325	350	350	375	325	350
Gas flow rate (sccm)	400	400	400	400	400	400
Contact time (sec)	0.15	0.15	0.15	0.15	0.23	0.23
WHSV (h^{-1})	0.22	0.22	0.22	0.20	0.22	0.22
O_2 conversion mol%	11.2	19.4	12.6	24.8	12.0	25.7
pX conversion mol%	29.7	57.9	42.8	69.2	43.3	74.7
Product selectivity %						
p-Tolualdehyde	25.7	21.7	29.2	25.2	33.7	24.9
Terephthaldehyde	54.8	47.6	48.6	49.2	48.3	39.4
Benzaldehyde	1.5	1.8	1.9	2.2	2.0	2.5
p-Toluic acid	0.0	0.0	0.0	0.0	0.4	1.3
Co	4.1	5.0	4.3	5.8	2.7	5.3
CO_2	8.3	12.3	10.6	16.1	6.6	14.0
Byproducts	1.5	1.7	1.5	1.8	2.3	2.6

Table 3. Reactivity of xylene isomers
Catalyst 1: 5.2 wt% Mo, 1.5 wt% Fe, and Mo/Fe=2.05, O_2:xylene:43:1, contact time:0.15 sec. WHSV:0.32 h^{-1}.

Run no.	1		2		3		
Xylene isomer	o-xylene		m-xylene		p-xylene		
Temp. °C	350	400	350	400	350	325^1	350^1
Conv. mole %	6	14	3	9	78	32	58
Product selectivity mo%							
Monoaldehyde	33	14	67	33	0	26	22
Dialdehyde	trace	trace	0	0	44	55	48
PTAN	0	29					
TMBPM	0	21					
MA	0	0	0	0	12	0	0
CO		29				4	5
CO_2		36				8	12
$CO+CO_2$	66			33	67	18	

$^1O_2$:xylene=10:1, PTAN:phthalic anhydride,
TMBPM:trimethylbiphenylmethane.

are compared with that of p-xylene in Table 3. The conversion of xylene at 350°C was 78 mol%, 6 mol% and 3 mol% for p-, o-, and m-xylene, respectively under identical conditions. Selectivity to terephthaldehyde was 44 mol%, trace and zero for p-, o-, and m-xylene, respectively. Extensive burning (65-67 mol%) was accompanied in the oxidation of both o- and m-xylene. The catalyst 2 clearly exhibits the para-selective oxidation function for p-xylene, and the reactivity of xylene isomers on the CVD Fe/Mo/DBH catalyst is in the order of p-xylene >> o-xylene > m-xylene.

A synthetic xylene mixture feed containing ethylbenzene, p-xylene, m-xylene and o-xylene in a molar ratio of 2:5:10:5 was blended to mimic the commercially available xylene feed. The blend was subjected to the gas phase O_2 oxidation over catalyst 5 in-situ coated with $Si(OCH_3)_4$ under standard conditions. The results plotted in Figure 1 show that p-xylene can be preferentially oxidized from the mixed blend by keeping conversions of other components at minimum levels.

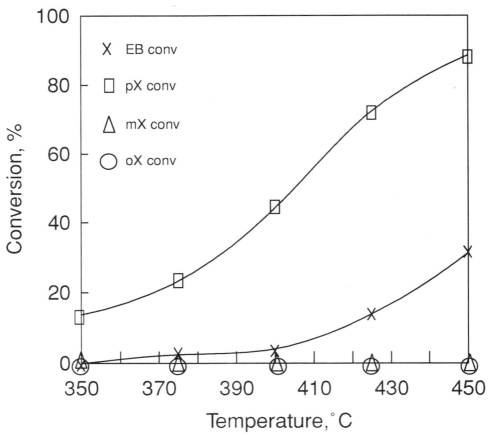

Figure 1. Oxidation of xylene isomers mixed with ethylbenzene
Catalyst 5: CVD Fe/Mo/DBH in-situ coated with tetramethylorthosilicate.

Effect of CO₂

In an effort to minimize burning of the substrates in the gas phase O_2 oxidation over the CVD Fe/Mo/DBH catalyst by using CO_2, oxidation of p-xylene was carried out in the two feed streams having different compositions, namely p-xylene/O_2/N_2/He, p-xylene/O_2/CO_2/N_2. Nitrogen was, in general, used as an internal standard for the GC. The CVD Fe/Mo/DBH catalyst (catalyst 4) was evaluated for the gas phase oxidation of p-xylene in a micro-reactor loaded with 0.5 g (1.4 mL) catalyst at temperatures in a range of 250-375°C. The results are summarized in Table 4, and plotted in Figure 2.

As shown in Fig. 2, a dramatic increase in the conversion of p-xylene to terephthaldehyde (TPAL) and p-tolualdehyde (TOAL) was observed in the presence of the commercial grade (99.5 % minimum purity) CO_2 for the gas phase O_2 oxidation at all temperatures studied. For example, the conversions of p-xylene were 22.1 mol% vs 40.2 mol% at 300°C and 57.0 mol% vs 82.5 mol% at 350°C. Selectivities to terephthaldehyde and p-tolualdehyde remained approximately the same in these two feed streams, although the conversion levels were significantly different at 300-325°C.

Table 4. Oxidation of p-Xylene with CVD Fe/Mo/DBH
Catalyst 5: 6.8 wt% Mo, 1.88 wt% Fe, 540 ppm B, Mo/Fe=2.1,
       0.508 g (1.4 mL) in a quartz micro-reactor
Oxidation conditions: WHSV: 0.22 h⁻¹, Contact time: 0.21 sec.,
Gas flow rate: 400 sccm
Feed gas blends: I: 0.1% p-xylene, 1.0% O_2, 1.0% N_2 in He
               II: 0.1% p-xylene, 1.0% O_2, 1.0% N_2 in CO2
               III: 0.1% p-xylene, 1.0% N_2 in CO_2

Temp. °C	250			300			350			375		
Feed	I	II	III	I	II	III	I	II	III	I	II	III
Conv.%	8.6	23.4	0.6	17.6	33.3	5.1	41.2	65.5	4.1	60.7	84.1	8.1

Product selectivity %

TOAL	32.4	54.5	44.9	57.9	57.2	69.4	50.2	40.9	60.5	40.6	40.6	62.9
TPAL	3.3	4.7	0.0	16.4	27.5	13.8	23.5	33.6	28.4	32.6	30.2	26.4
BZAL	0.0	0.0	0.0	1.3	1.5	0.0	2.4	2.7	1.2	2.7	3.1	1.6
MA	0.0	0.0	0.0	0.0	0.0	0.0	2.4	6.0	0.0	5.8	13.7	5.7
TOL	20.0	21.4	29.2	6.2	6.9	8.1	3.5	4.8	5.4	3.1	4.8	0.0
PCUMENE	0.0	0.5	0.0	0.0	0.0	0.0	0.0	0.0	0.0	0.0	0.0	0.0
TMBPM	23.8	17.8	25.9	7.5	6.8	8.7	1.7	0.6	4.5	0.4	0.0	3.3
CO	0.0	0.0	0.0	0.0	0.0	0.0	0.6	3.4	0.0	4.7	7.5	0.0
CO₂	0.0	-	-	10.7	-	-	15.6	-	-	20.2	-	-

The lower metal loading CVD Fe/Mo/DBH catalyst (catalyst 5) was studied in the same feed streams, and the results are shown in Table 5 and Fig. 3. The conversions of p-xylene were 17.6 mol% vs 33.3 mol% at 300°C and 41.2 mol% vs 65.5 mol% at 350°C, and selectivities to terephthaldehyde were 16.4 mol% vs 27.5 mol% at 300°C and 23.5 mol% vs 33.6 mol% at 350°C in feed 1 vs 2.

Selectivities to terephthaldehyde are calculated by excluding CO_2 produced by substrate burning based on an assumption that burning of p-xylene is negligible in the presence of CO_2 under the conditions employed for these runs. This assumption is valid for lower temperatures, in particular over catalyst 4. The basis of the assumption was verified experimentally. In comparison to catalyst 5, catalyst 4 seems to be more active, exhibits higher p-xylene conversion, and shows better selectivity toward aldehydes at 300-375°C. Practically no CO_2 produced by burning was detected with catalyst 4 until the reaction temperature reached 375°C, while, over catalyst 5, burning (10.7%) started at 300°C and it became significant (20.2 %) at 375°C . Also the higher molar ratio of terephthaldehyde to p-tolualdehyde was attained over catalyst 4 than catalyst 5. Although the catalyst 4 behaved somewhat differently from the catalyst 5, the results over these two catalysts led to a conclusion that CO_2 indeed promotes the gas phase O_2 oxidation of p-xylene.

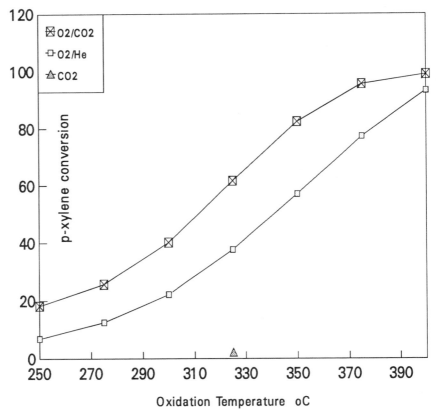

Figure 2. CO_2 effect on p-xylene conversion
Catalyst 4: CVD Fe/Mo/DBH, 9.4% Mo, 2.7% Fe, Mo/Fe=2.03.

As described briefly above, the most significant difference in catalysts 4 and 5 lies in the burning of the substrate. With catalyst 5, burning started at 300°C while it started at a much higher temperature, 375°C, over catalyst 4. This different catalytic behavior may be attributed to the difference in the loading level of the metal components. Over catalyst 4 at lower temperatures, a net increase in the p-xylene conversion between feed stream 1 (O_2) and feed stream 2 (O_2 in CO_2) becomes more significant, and at the same time, the side reactions also become more noticeable in the presence of CO_2. An increase in side reactions may partly be due to the enhanced acidity generated by the CO_2 molecules adsorbed on the surface of the DBH matrix. Above 375°C, the promoting effect of CO_2 becomes less significant.

The yields of aldehydes (terephthaldehyde plus p-tolualdehyde) in different feed streams over catalysts 4 and 5 were calculated based on an assumption that burning is negligible in the presence of CO_2, and plotted in Figures 4 and 5, respectively. Again these clearly confirm the beneficial effect of CO_2 in the gas phase O_2 oxidation of p-xylene over the Fe/Mo/DBH catalyst at a certain range of reaction temperatures. It also shows that the range of temperature effective for the promoting function of CO_2 shifts depending on the loading level of metal components on the DBH matrix.

Table 5. Oxidation of p-Xylene with CVD Fe/Mo/DBH
Catalyst 4: 9.4 wt% Mo, 2.7 wt% Fe, Mo/Fe=2.03, surface area 263 m^2/g
     0.508 g (1.4 mL) in a quartz micro-reactor
Oxidation conditions: WHSV: 0.22 h^{-1}, contact time: 0.21 sec.,
     gas flow rate: 400 sccm
Feed gas blends: I: 0.1% p-xylene, 1.0% O_2, 1.0% N_2 in He
                 II: 0.1% p-xylene, 1.0% O_2, 1.0% N_2 in CO_2
                 III: 0.1% p-xylene, 1.0% N_2 in CO_2

Feed	I	II	I	II	I	II	III	I	II
Temp.°C	250		300		325			350	
pX conv. %	6.9	18.2	22.1	40.2	37.7	61.6	2.2	57.0	82.5
Product selectivity %									
TOAL	51.7	48.3	51.8	48.3	47.5	41.8	55	40.1	33.1
TPAL	13.9	16.1	35.8	38.8	42.0	40.0	37	40.8	34.6
BZAL	0.0	0.0	2.3	2.3	3.4	3.0		4.2	3.7
MA	0.0	0.0	0.0	0.0	0.0	0.5		3.2	11.5
TOL	26.1	24.9	7.7	8.1	6.1	7.4	6	5.7	8.6
PCUMENE	4.1	3.7	0.4	0.4	0.1	0.1		0.0	0.0
TMBPM	7.0	7.0	3.2	2.2	1.7	0.9	2	0.9	0.0
CO	0.0	0.0	0.0	0.0	0.0	3.3		3.3	17.0
CO_2	0.0	-	0.0	-	0.0	-		0.0	-

Near Edge X-ray Absorption Fine Structure Study

In order to gain some insight into the site for activating CO_2, a synchrotron radiation study was conducted using the Amoco beamline at The Synchrotron Radiation Center, Stoughton, Wisconsin. The total electron yield near edge x-ray absorption fine structure spectroscopy of the dosed ferric molybdate are summarized in Figures 6-8. The results of the CVD Fe/Mo/DBH catalyst were of poorer signal to noise but reflected the same trends as those found in the pure ferric molybdate material.

The carbon 1s spectra of the CO_2 dosed ferric molybdate at 330°C is shown in Figure 6 after the background carbon signal is subtracted. The presence of carbon is ubiquitous throughout the optics of the 6m TGM so that the subtraction of the background carbon absorption is important. This data is compared to a spectrum of condensed multilayer benzene at 125°K on gold. Comparison of the C 1s absorption data to published data for condensed CO_2 indicates

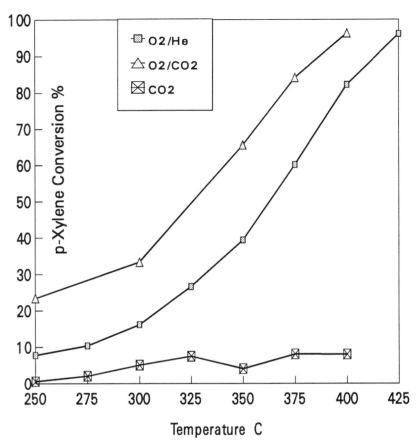

Figure 3. CO₂ effect on p-xylene conversion
Catalyst 5: CVD Fe/Mo/DBH, 6.8% Mo, 1.88% Fe, 0.05% B, Mo/Fe=2.10.

a close correspondence [8,9]. The different spectra of the x-ray absorption total electron yield data indicates the presence of adsorbed CO_2 at reaction temperature of 330°C.

The total electron yield data at the Mo $3d_{5/2,3/2}$ region is shown in Figure 7. No noticeable differences in the Mo $3d_{5/2,3/2}$ regions are observed between the fresh ferric molybdate and after the 2.5 KL CO_2 at either O°C or 300°C. At the dosing temperature (-125°C) the total yield is suppressed and a poorer signal to noise ratio is observed. The total electron yield absorption data of the Fe $2p_{3/2,1/2}$ is shown in Figure 8. The observed spin-orbit splitting of 13.5 eV is consistent with a Fe^{3+} ion. The changes in the distribution of states before and after CO_2 dosing persist to reaction temperature (330°C).

A plausible explanation of the observed redistribution of states could stem from the convolution of the Fe $2p_{3/2,1/2}$ with 3d unoccupied orbitals. The observed decrease in density of unoccupied states might reflect a charge transfer interaction with the CO_2.

The CO_2 and O_2 molecules may be simultaneously activated on the surface of the CVD Fe/Mo/DBH catalyst. These activated CO_2 and O_2 molecules interact with each other to create a synergy and form an active intermediate species, possibly peroxocarbonate, on the Fe-site rather than the Mo-site shown previously [4], which could function to abstract two hydrogen atoms from the methyl groups in the p-xylene molecule in a concerted manner to generate the diradical species, $\cdot CH_2\text{-}C_6H_4\text{-}CH_2\cdot$. The peroxocarbonate intermediate may also serve to abstract one hydrogen atom from one methyl group to form the monoradical, $\cdot CH_2\text{-}C_6H_4\text{-}CH_3$. The mono- and di-radical resulting in this manner lead to form p-tolualdehyde and terephthaldehyde, respectively, in the subsequent step. The following peroxocarbonate species is proposed by following an analogy of the Rh-peroxocarbonate intermediate postulated by Dubois et al., [10], and proposed for the liquid phase oxidations by Aresta, et al. [11].

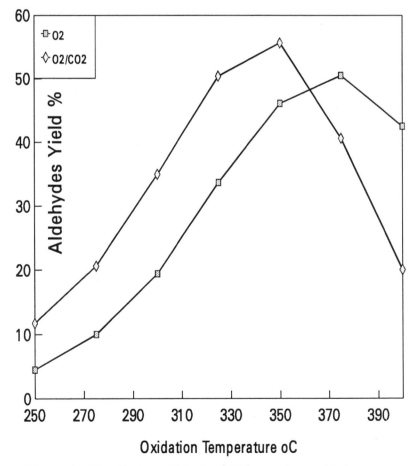

Figure 4. CO_2 effect on aldehyde yield in p-xylene oxidation
Catalyst 4: CVD Fe/Mo/DBH, 9.5% Mo, 2.7% Fe, Mo/Fe=2.03.

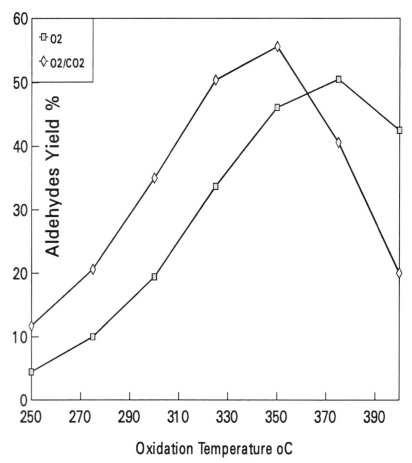

Figure 5. CO_2 effect on aldehyde yield in p-xylene oxidation
Catalyst 5: CVD Fe/Mo/DBH, 6.8% Mo, 1.9% Fe, 0.05% B, Mo/Fe=2.10.

The role of supporting matrix

CVD Fe/Mo/ZSM-5. The CVD Fe/Mo/ZSM-5 catalyst (referred to as catalyst A) was prepared from $FeCl_3$, MoO_2Cl_2 and ZSM-5(Si/Al=30) containing 1.07 wt% Al, and its composition was 5.3 wt% Mo, 2.21 wt% Fe, Mo/Fe=1.4. In order to increase the Mo/Fe ratio above 1.5/1, an aqueous solution of ammonium paramolybdate was impregnated onto a portion of catalyst A by the incipient wetness technique. The resulting catalysts is referred to as catalyst B. The compositions of the resulting two catalysts, A and B, are shown below.

These catalysts were subjected to the gas phase O_2 oxidation of para-xylene in a micro-reactor loaded with 0.5 g catalyst in a feed containing 0.1 vol% para-xylene, 1.0 vol% O_2, and 2.0 vol% N_2 in He under the standard oxidation conditions.

The results in Table 6 show that the main reaction over the ZSM-5 counterparts, catalysts A and B, is disproportionation of para-xylene to toluene and pseudocumene, and dehydrocoupling to trimethylbiphenylmethane with substantial burning. Although catalyst B produced more aldehydes than catalyst A, and selectivities to terephthaldehyde and para-tolualdehyde were respectively 21% and 6% at 24-43% conversion of p-xylene, these were much lower than those over the DBH counterpart, catalyst 1.

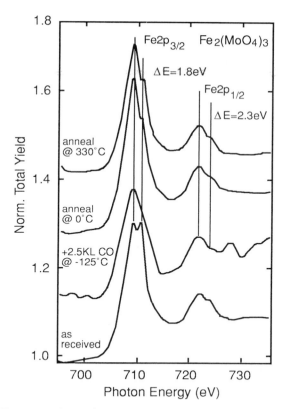

Figure 6. C_{1s} x-ray absorption near edge spectroscopy (XANES) taken at Synchrotron Radiation Center showing presence of CO_2 at 330°C after dosing at liquid nitrogen temperature.

Catalyst composition of Fe/Mo/ZSM-5
ZSM-5: Si/Al=30, 1.7% Al

ID	Metal composition		
	Mo%	Fe%	Mo/Fe
A	5.3	2.1	1.4
B	8.5	2.5	2.5

CVD Fe/Mo/silicalite. A portion (100 g) of Silicalite S-115 (Union Carbide) was steamed and heated at a rate of 5°C per min to 650°C and then held at 650°C for 1 day. The temperature was then raised while steaming at 5°C to 800°C, and held at 800°C for 6.9 days. After calcining 97.2 g were recovered. The sample was then heated with $SOCl_2$ vapor to remove aluminum. The resulting solid was subjected to the standard CVD technique. Small amounts of iron and molybdenum were deposited probably because there must be a limited number of anchoring sites of Si-OH on the resulting silicalite surface. ICP analysis showed 0.58 wt% Mo and 0.13 wt% (Mo/Fe=2.6). A premixed feed of 0.1 vol% p-xylene, 1.0 vol% O_2, 2.0 vol% N_2 in He was oxidized over the catalyst. and the results are summarized in Table 7. Both terephthaldehyde and p-tolualdehyde were formed along with some side products with extensive burning.

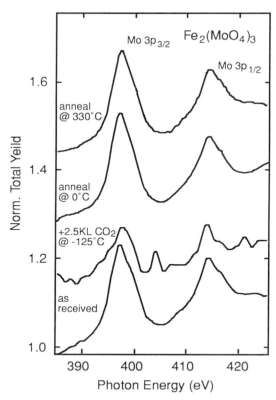

Figure 7. Mo^{6+} $3P_{3/2, 1/2}$ XANES showing minimal effect of CO_2 on Mo^{6+} $4d^*$ orbitals.

Bulk Ferric Molybdate, $Fe_2(MoO_4)_3$. Ferric molybdate, $Fe_2(MoO_4)_3$, was purchased from Atomergic Chemetal Corp., and phase purity of the sample was determined by the X-ray diffraction method. Ferric molybdate with monoclinic symmetry was mixed with approximately 15% of molybdenum trioxide with orthorhombic symmetry in the sample. A physical mixture of the ferric molybdate as received and borosilicate molecular sieve, HAMS-1B-3, in a weight ratio of 1:10 was ground together, and pressed to prepare 20-40 mesh material. This material was evaluated in the micro-reactor loaded with approximately 0.5 g for the p-xylene oxidation with the premixed gas blend, 0.1 vol% p-xylene, 1.0 vol% O_2, and 1.0 vol% N_2 in He under the standard conditions.

The main objective of studying the physical mixture was to gain some insight on the nature of interaction between iron-molybdenum moieties with the molecular sieve matrix in the CVD catalyst. The results shown in Table 8 indicate that the bulk ferric molybdate requires much higher temperature than the CVD Fe/Mo/DBH to obtain comparable p-xylene conversion. The physical mixture catalyzed mainly disproportionation reaction of p-xylene to produce toluene and pseudocumene, and exhibited poor selectivity to aldehyde with extensive burning. These results suggest that the para-selective activity of the Fe/Mo/DBH catalyst for the gas phase O_2 oxidation of p-xylene stems from the intimate interaction of the metal components with the molecular sieve matrix.

Table 6. Gas Phase O_2 Oxidation of p-Xylene over Fe/Mo/ZSM-5
Catalyst A: Fe/Mo/ZSM-5, 5.3 wt% Mo, 2.21 wt% Fe, 1.07 wt% Al, Mo/Fe=1.4
Catalyst B: modified catalyst with an aq. solution of ammonium para-
molybdate, 8.5 wt% Mo, 2.47 wt% Fe, 1.07 wt% Al, Mo/Fe=2.0
Gas stream: 0.1% p-xylene, 1.0% O_2, 2.0% N_2 in He

Catalyst	A			B				
Temp °C	300	350	400	325	350	375	400	425
Conv.%	26.6	58.0	94.9	24.0	42.5	61.0	77.4	95.0

Product selectivity %

	A-300	A-350	A-400	B-325	B-350	B-375	B-400	B-425
TPAL				20.0	20.9	16.8	13.8	8.1
TOAL	0.0	2.5	0.9	6.4	6.4	5.9	7.0	6.1
BZAL	3.9	0.8	1.3	2.1	2.8	4.3		
BZAC	6.4	0.5	0.4	0.3	0.1	0.1		
TOL	53.1	62.0	33.9	39.4	35.4	33.5	24.3	17.6
PCUMENE		15.1	3.7	0.0	3.6	1.9	1.0	0.3 0.0
TMBPM				8.2	4.1	1.4	0.3	0.0
CO				0.0	8.1	11.9	14.0	18.2
CO_2				21.0	21.5	26.1	34.0	38.9
$CO+CO_2$	26.5	28.4	47.2					

Table 7. Gas phase O_2 oxidation of para-xylene over CVD Fe/Mo/silicalite

Catalyst	Fe/Mo/silicalite			Catalyst 1
Temp °C	325	350	375	350
pX conversion %	35.7	49.7	72.4	67.0

Product selectivity %

	325	350	375	350
Toluene	15.8	12.8	11.8	3.0
Pseudocumene	1.2	0.5	0.3	0.0
p-Tolualdeyde	27.8	22.6	15.2	28.0
Terephthaldehyde	9.9	13.0	9.2	50.2
Benzaldehyde	0.8	1.2	1.6	1.5
Maleic anhydride	0.0	0.00.0		7.4
p-Toluic acid	0.0	0.0	0.0	7.9
TMBPM	7.2	3.2	0.8	0.0

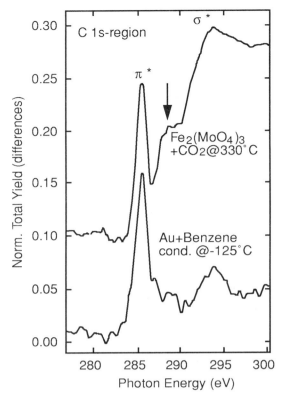

Figure 8. Fe^{+3} $2P_{3/2,\ 1/2}$ XANES indicating major effects on CO_2 on Fe^{+3} $3d^*$ orbitals.

Table 8. Oxidation of p-Xylene with Bulk $Fe_2(MoO_4)_3$
Catalyst: 0.51 g (0.25 mL), WHSV:0.05, contact time: 0.15 sec
Premixed gas: 0.10% p-xylene, 1.0 % O_2, 1.0 % N_2 in He

Catalyst	$Fe_2(MoO_4)_3$[1]		10% $Fe_2(MoO_4)_3$ in HAMS-1B-3 (physical mixture)			
Temp. °C	450	400	450	500	550	600
pX conversion %	8.7	5.6	16.1	39.9	66.2	87.4
Produce selectivity %						
TPAL	23.6	2.2	3.0	4.0	4.7	4.2
TOAL	58.5	23.0	18.1	12.1	7.4	3.5
Toluene		51.9	39.1	30.6	26.1	19.9
BZAC		3.2	4.4	6.7	9.7	11.4
CO		17.6	22.0	24.1	26.5	28.7
CO_2	18.2	7.7	19.8	20.2	15.9	

[1]purchased from Atomergic Corp. (XRD analysis show 15% MoO_3 in $Fe_2(MoO_4)_3$

Figure 9a. TEM of CVD Fe/Mo/DBH.

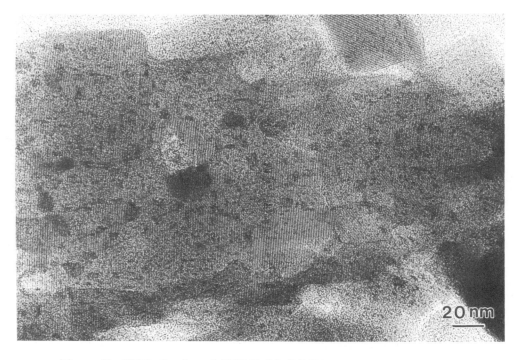

Figure 9b. TEM of activated CVD Fe/Mo/DBH by prolonged calcination.

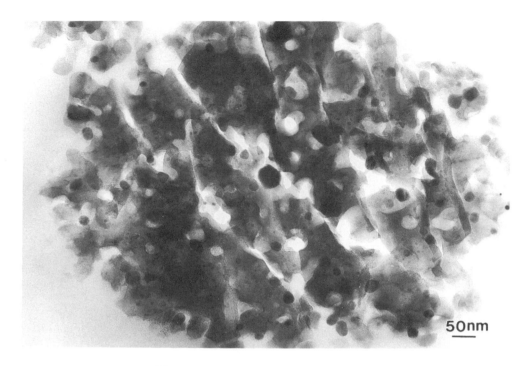

Figure 9c. TEM of CVD Fe/Mo/ZSM-5.

Figure 9d. TEM of CVD Fe/Mo/silicalite.

Electron Microscopy Study

The difference in catalytic behavior among three supporting matrices can, at least partially be explained by their structural differences. The transmission electron micrographs (TEM) of CVD Fe/Mo/ZSM-5 and CVD Fe/Mo/silicalite are compared against that of the CVD DBH counterpart in Figure 9. The TEM studies revealed that in the CVD catalyst, Fe and Mo are finely and uniformly deposited onto DBH primarily along the micropore channels (Figure 9a). After calcination at 650°C for a prolonged period, most of these cylindrical domains have changed into nearly spherical domains via diffusional processes. However, nearly all Mo/Fe still reside inside the channels of the DBH sieve, as shown in Figure 9b. The composition of the Mo/Fe rich phases were analyzed by energy dispersive x-ray spectrometry (EDXS), which indicated that the Mo/Fe domains are predominantly $Fe_2(MoO_4)_3$ and $FeMoO_4$. Very little DBH destruction was observed in this catalyst.

Electron microscopy of the ZSM-5 counterpart shows a very different microstructure (Figure 9c). Unlike the CVD Fe/Mo DBH, significant sieve destruction has occurred during the preparation of the catalyst. Besides the large number of pores developed inside ZSM-5, the surface regions of the ZSM-5 particles have broken into small crystallites which are clustered together. In this catalyst, most of Mo/Fe rich particles are either at the exterior or inside the large pores of the ZSM-5 crystallites. The composition of these Mo/Fe rich particles varies from MoO_3, $FeMOO_4$, $Fe_2(MoO_4)_3$, to Fe_2O_3.

The electron micrograph of catalyst CVD Fe/Mo/silicalite shown in Figure 9d revealed some sieve destruction of the silicalite. However, most of the voids are concentrated near the center of the sieve particles. Most voids have a size of approximately 10- 40 nm, which are smaller than the voids observed in the CVD ZSM-5 counterparts. Many of them are angularly shaped suggesting the preferential fracturing of the sieve along certain crystallographic planes. In some regions, several voids are channeled together. Only a few small Fe and Mo rich particles (less than 10 nm) are seen. Overall, a low level of Fe and Mo is detected by EDXS.

CONCLUSIONS

The Fe/Mo/borosilicate molecular sieve catalyst was prepared from $FeCl_3$ and MoO_2Cl_2 and partially deboronated borosilicate molecular sieve (DBH) by the chemical vapor deposition technique. The resulting CVD Fe/Mo/DBH catalysts exhibited para-selective oxidation activity by preferentially oxidizing p-xylene over its isomers and ethylbenzene in the xylene mixture feed. p-Xylene was oxidized to terephthaldehyde and p-tolualdehyde in high selectivities under mild gas phase O_2 oxidation conditions. The aldehyde yield was dramatically increased in the presence of CO_2, and the resulting activity enhancement was addressed by in-situ synchrotron radiation studies. Near edge X-ray absorption fine structure(NEXAFS) evidence is provided for a strongly chemisorbed CO_2 species on the Fe-site on the catalyst at the reaction temperature.

The partially deboronated borosilicate matrix (DBH) resulting from the CVD procedure for preparing CVD Fe/Mo/DBH played a unique role for the selective synthesis of terephthaldehyde by the gas phase O_2 oxidation of para-xylene. Although the CVD Fe/Mo/ZSM-5 catalyst exhibited some catalytic activity for the synthesis of terephthaldehyde and para-tolualdehyde, it mainly catalyzed side reactions.

The CVD Fe/Mo/silicalite containing lower levels of iron and molybdenum was active for aldehyde synthesis with the accompanying side reactions, but extensive burning occurred. The difference in their catalytic behavior can at least partially be caused by their structural differences, in addition to the inherent acidity differences of these matrices.

REFERENCES

1 E. Pajda, Chemik (Gliwice) 23, (8) 291 (1970).

2. T. Yokoyama. T. Setoyaman, N. Fujita, M. Nakajima, T. Maki, and K. Fuji, Appl. catal., A: 88(2), 149 (1992)

3. W. Partenheimer, J. Mol. Catal., 67, 35 (1991); W. Partenheimer and J. Kaduk, Stud. Surf. Sci. Catal., 66, 613 (1991).

4. J.S. Yoo, J.A. Donohue, M.S. Kleefisch, P.S. Lin and S.D. Elfline, Appl. Catal., A: 105 (1993) 83.

5. J.S. Yoo, P.S. Lin and S.D. Elfline, Appl. Catal. A: 106 (1993) 259.

6. J.S. Yoo, J.A. Donohue and M.S. Kleefisch, Appl. Catal. A: (1994) 80.

7. J.S. Yoo, R.A. Sohail, S.S. Grimmer, and C.Choi-Feng, Catal. Lett., 29 (1994) 299.

8. M. Bader, B. Hillbert, A. Puschmann, J. Hasse and A.M. Bradshow, Europhysics Lett., 5, (1988) 443.

9. G. Illing, D. Hoekett, E.W. Plummer, H.J. Feund, J. Sommers, Th. Lindner, A.M. Bradshaw, U. Buskotte, M. Neuman, U. Starke, K. Heinz, P.O. DeAndres, D. Saldin and J.B. Pendry, Surf. Sci. 206 (1988) 1.

10. J. L. Dubois, M. Mimoun, and C.J. Cameron, Catal. Lett. 6 (1990) 967.

11. M. Aresta, C. Fragale, E. Quaranta, and I. Tommasi, Chem. Soc., Chem. Commun. 1992, 315.

METHANOL OXIDATION OVER OXIDE CATALYSTS

Stuart H. Taylor, Justin S.J. Hargreaves, Graham J. Hutchings, and Richard W. Joyner

Leverhulme Centre for Innovative Catalysis
Department of Chemistry
University of Liverpool
PO Box 147
Liverpool L69 3BX

ABSTRACT

The catalytic partial oxidation of methane to methanol has proved to be an extremely demanding reaction. It is clearly important that methanol should be stable over any potential catalysts, so a systematic study has been undertaken to determine the types of oxides over which methanol might have the required stability. The results should be valuable for the design of future catalysts. Methanol stability by catalytic oxidation has been investigated in the temperature range 150-500°C and 1 atm pressure. Over the majority of materials a substantial proportion of methanol was converted to carbon oxides below 350°C. Several oxides exhibited improved performance, these were MoO_3, Nb_2O_5, Ta_2O_5 and WO_3. Methanol conversion over these materials was high, but the major products were formaldehyde and dimethylether. Sb_2O_3 showed the best performance, producing a methanol conversion of only 3% at 500°C. The results have been compared with both physical and chemical properties of the oxides in an attempt to establish trends. In particular we have looked at relationships with oxygen exchange data. A weak but significant correlation between methanol decomposition to carbon oxides and the rate of oxygen exchange is discussed.

INTRODUCTION

In recent years much research endeavour has focussed on the direct partial oxidation of methane to oxygenates, in particular methanol. A simple one step process would have considerable advantages over the existing technology for methanol manufacture. The direct conversion has proved to be an extremely demanding reaction, due to the harsh conditions required to activate the methane molecule, and the thermodynamic instability of the desired products, with respect to those of combustion. Numerous catalysts have been tested for this reaction but, to date, none have exhibited any outstanding performance [1-4]. The majority of catalysts which have shown some success for this reaction have been oxides. Clearly it is important that methanol should be stable over any potential catalysts, under appropriate conditions.

This study was undertaken to investigate methanol stability over a wide range of oxide materials. By studying the interaction of methanol with single oxides, we aim to identify types of materials over which methanol possesses the required degree of stability, and hence are suitable components on which to base the design of improved catalysts.

EXPERIMENTAL

Methanol oxidation studies were carried out in a conventional laboratory microreactor. Helium (99.995%) and oxygen (99.5%) were used as diluent and reactant respectively. Methanol (BDH, Analar grade) was introduced to the reactor via a syringe pump into a Pyrex vaporiser maintained at 100°C, where it was mixed with oxygen and helium in the required ratios. The reactor was a fused silica tube (8 mm internal diameter), heating was by a furnace capable of maintaining temperatures in the range 100-520°C, with a 50 mm uniform heated zone.

Catalysts were pelleted to a 0.6-1.0 mm uniform particle size range before testing, the catalyst bed length was typically 10 mm. The reactant feed composed of methanol/oxygen/helium in the ratio 1/4/12, with gas hourly space velocities in the region of 12000 h^{-1}, at 1 atm pressure. Methanol oxidation experiments were carried out in the temperature range 150-500°C.

Product analysis was by on-line GC-MS, fitted with a thermal conductivity detector, in series with a Varian Saturn Mass Spectrometer.

The oxides for these studies were used as supplied, the purity was greater than 99% in all cases, generally exceeding 99.9%. The oxides were characterised by powder X-ray diffraction and nitrogen surface area measurements in accordance with the BET method.

RESULTS

In total 33 different oxides (Table 1) were tested and characterised. These materials were selected by potential interest as catalysts, based on previous studies of this oxidation reaction [1-4]. In addition it was our intention to examine a wide range of the periodic table.

Surface area measurements showed a large variation, ranging from 278 m^{2}g^{-1} to less than 3 m^{2}g^{-1} for several of the materials. In many cases the values obtained were below that of the of the limit of measurement, in these cases the areas were recorded as less than a specific value, which vary due to the differing masses of catalyst which were characterised.

Results from powder X-ray diffraction showed that the majority of materials were present as a single phase. The exceptions were Al_2O_3, Ga_2O_3, Nb_2O_5, Pr_6O_{11} and TiO_2 which were a mixture of two or more phases. All the oxides were highly crystalline except SiO_2, which was a fumed silica, and Al_2O_3 which exhibited transitional phases [5].

Methanol decomposition studies showed a wide range of stabilities, but in many cases the conversion and selectivity to products followed similar patterns. On this basis it was possible to group oxides together, table 2 shows the resultant groups.

Table 3 shows example results for methanol oxidation over various oxides in our classification.

The oxides in groups 1 and 2 produced mainly carbon oxides from methanol. Group 1 oxides showed high selectivity to CO_2 throughout the range of conversion, in some cases trace quantities of CO were also observed. Materials in group 2 exhibited similar behaviour, but CO selectivities were considerably higher. Below 300°C in the region low conversion some selectivity to oxygenated products was exhibited.

Oxides in group 3 showed appreciable selectivity to formaldehyde and/or dimethylether up to 400°C. Above this temperature only carbon oxides were produced.

Oxides belonging to group 4 showed selectivity towards formaldehyde and dimethylether at all temperatures investigated. Increasing temperature increased the production of formaldehyde and CO_x at the expense of dimethylether.

MoO_3 was the only material in group 5, this showed high selectivity to formaldehyde across the temperature range, with low levels of CO_x at 500°C.

Group 6 contains only Sb_2O_3, which showed low methanol conversion, a maximum of 3% was observed at 500°C. Up to 400°C dimethylether was the exclusive product, at higher temperatures this selectivity fell steadily as formaldehyde and CO were produced. The wide range of results obtained from these oxides make comparisons complex. To rationalise this information we have devised a ranking order to compare performance, which has been based on the temperature at which 30% of the reactant methanol feed was converted to carbon oxides (T_{30}). This ranking is shown in figure 1.

Table 1. BET surface areas and phases identified by powder X-ray diffraction for oxides used in this study.

Oxide	BET Surface Area/m^2g^{-1}	Identified phase	Crystal system
Al_2O_3	78	transitional	N/A
Bi_2O_3	<0.5	Bismite	Monoclinic
CaO	3.1	Lime	Cubic
CdO	1.3	Monteponite	Cubic
CeO_2	11.0	Cerianite-(Ce)	Cubic
Co_3O_4	3.3	Co_3O_4 (Spinel)	Cubic
Cr_2O_3	4.2	Eskolaite	Rhombohedra
CuO	1.4	Tenorite	Monoclinic
Fe_2O_3	5.2	Hematite	Rhombohedral
Ga_2O_3	22	Alpha-Ga_2O_3	Hexagonal
		Beta-Ga_2O_3	Monoclinic
		Gamma-Ga_2O_3	Cubic
Gd_2O_3	1.3	Gd_2O_3	Cubic
La_2O_3	0.5	La_2O_3	Hexagonal
MgO	51	Periclase	Cubic
Mn_2O_3	3.8	Bixibite-C	Cubic
MoO_3	0.5	Molybdite	Orthorhombic
Nb_2O_5	2.1	T-form	Orthorhombic
		M-form	Monoclinic
Nd_2O_3	1.6	Nd_2O_3	Hexagonal
NiO	1.9	Bunsenite	Cubic
Pr_6O_{11}	2.3	$PrO_{1.83}$	Cubic
		PrO_2	Cubic
Sb_2O_3	1.8	Senarmontite	Cubic
SiO_2	278	amorphous	N/A
Sm_2O_3	4.3	Sm_2O_3	Cubic
SnO_2	4.6	Cassiterite	Tetragonal
Ta_2O_5	<0.5	Ta_2O_5	Monoclinic
Tb_4O_7	0.7	Tb_4O_7	Cubic
TiO_2	48	Rutile	Tetragonal
		Anatase	Tetragonal
V_2O_5	6.7	Shcherbinaite	Orthorhombic
WO_3	0.8	WO_3	Monoclinic
Y_2O_3	3.8	Y_2O_3	Cubic
Yb_2O_3	3.8	Yb_2O_3	Cubic
ZnO	7.4	Zincite	Hexagonal
ZrO_2	6.5	Baddelyite	Monoclinic

Table 2. Oxide groupings based on methanol conversion and product distribution.

Group	Oxide
1	Bi_2O_3, CdO, CuO, Cr_2O_3, PbO
2	CaO, CeO_2, Gd_2O_3, La_2O_3, MgO, Mn_2O_3, Nd_2O_3, NiO, Pr_6O_{11}, Sm_2O_3, SnO_2, Tb_4O_7, Y_2O_3, Yb_2O_3, ZnO
3	Al_2O_3, Co_3O_4, Fe_2O_3, Ga_2O_3, SiO_2, TiO_2, V_2O_5, ZrO_2
4	Nb_2O_5, Ta_2O_5, WO_3
5	MoO_3
6	Sb_2O_3

Table 3. Results from methanol oxidation experiments.

Group	Oxide	Temp/°C	Conv/%	Selectivity/%			
				CO	CO_2	HCHO	DME
1	CuO	200	100	–	100	–	–
2	CaO	300	1	100	–	–	–
		400	19	21	79	–	–
		500	100	2	98	–	–
3	Ga_2O_3	300	97	5	4	–	91
		400	100	38	62	–	–
		500	100	2	98	–	–
4	Nb_2O_5	300	6	–	26	–	74
		400	52	2	8	17	73
		500	97	15	17	40	28
5	MoO_3	400	28	–	–	–	100
		500	91	2	1	97	–
6	Sb_2O_3	300	1	–	–	–	100
		400	1	–	–	–	100
		500	3	14	–	86	–

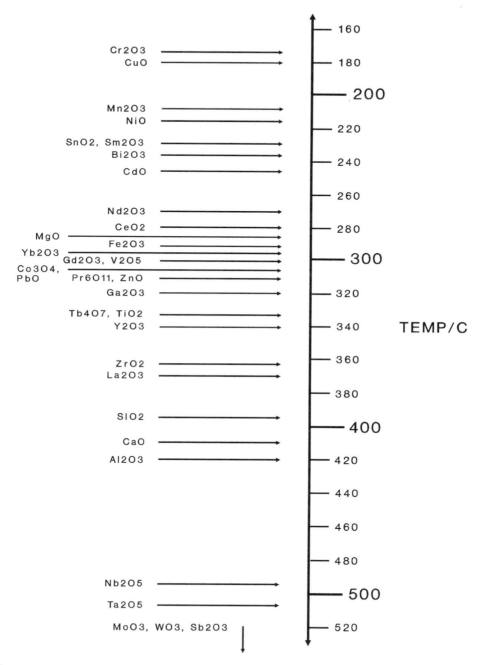

Figure 1. Oxide ranking based on the temperature at which 30% of reactant methanol was converted to carbon oxides.

In many cases the actual methanol conversion at this temperature was considerably higher, since products other than CO_x were also formed. However, we consider these products to be acceptable by-products to methanol in any methane oxidation process, ie. formaldehyde and dimethylether. Although the surface areas of the materials vary considerably (Table 1), no endeavour has been made to normalise for such effects. It is considered that the surface areas of these oxides are typical of those expected in any catalyst formulation. Ranking the performance in this manner produced a wide scale of activity, ranging over 340°C. Similar rankings were produced if data were ranked for other levels of CO_x production. There was a large grouping of materials in the low to mid-range of T_{30} values. The best performance for methanol stability were shown by five oxides; Nb_2O_5, MoO_3, Sb_2O_3, Ta_2O_5 and WO_3, which were clearly superior to the others investigated. In particular it was not possible to rank MoO_3, Sb_2O_3 and WO_3 on this scale, since with these oxides very low selectivities to CO_x were always obtained.

DISCUSSION

The substantial number of oxides employed in this study permits the possibility for common trends to be examined for the methanol decomposition reaction. We have examined the data for relationships between the ability of an oxide to produce CO_x from methanol with a range of physical and chemical properties. In particular data has been compared with nearest metal-oxygen distance, nearest oxygen-oxygen distance, unit cell volume, metal ionic radius and electrical conductivity characteristics, but no correlations were found. This is perhaps not surprising, since the oxides have metal ion valencies ranging from +2 to +5, and belong to different crystallographic systems, therefore exhibiting a broad range of surface structures. Taking into consideration these discrepancies, and grouping together oxides with similar valencies and crystal types did not improve the situation.

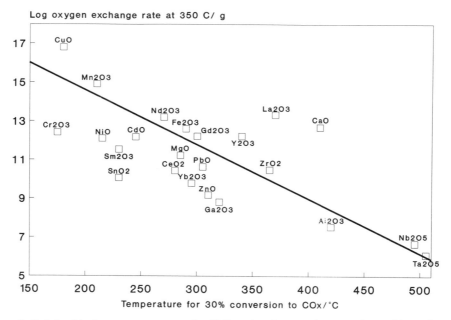

Figure 2. Relationship between temperature for 30% methanol conversion to carbon oxides and oxygen exchange activity.

The oxygen exchange over oxides has been previously studied [6,7]. Three mechanisms of exchange were identified, R_1 and R_2 correspond to exchange of gaseous oxygen with the oxide surface, either by an atomic or molecular reaction. The third mechanism, R_3, operated via surface exchange, both R_1 and R_2, but rapid oxygen diffusion throughout the oxide produced exchange with the whole of the bulk and not merely the surface. A plot of oxygen exchange rate for R_1 and R_2 mechanisms at 350°C, normalised for weight, against T_{30} is shown in figure 2.

The data shows a weak but significant correlation, as $R = 0.472$ is significant at the 98% level for two independent variables. It is interesting to note that Nb_2O_5 and Ta_2O_5, which are two of the best oxides, showed the lowest activity for oxygen exchange. On the other hand CuO, which readily decomposed methanol, was one of the most active materials, hence the weak correlation may be important for the design of improved catalysts.

In conclusion this study emphasises just one of the problems involved with the partial oxidation of methane to methanol, that is the instability of the product. With the exception of Sb_2O_3, methanol was unstable over all of the materials under conditions employed in our experiments. Over some materials formaldehyde and dimethylether are relatively more stable products. Over MoO_3 formaldehyde was a particularly stable molecule. This study presents considerable data covering methanol stability over simple oxide catalysts. These data are relevant to the scientific design of catalysts for direct methane partial oxidation to methanol, since it is crucial that surfaces are identified that do not decompose methanol to carbon oxides. This approach may be used as a first step in the successful design of novel catalysts for this demanding conversion. In addition the data are weakly correlated with the rate of surface oxygen exchange. It can be expected that improved catalyst formulations could be based on MoO_3, Nb_2O_5, Sb_2O_3, Ta_2O_5 and WO_3.

ACKNOWLEDGMENT

Financial support for this work was provided by the Gas Research Institute, Chicago.

REFERENCES

[1] N.R. Foster, Appl. Catal., 19:1 (1985).
[2] R. Pitchai, K. Klier, Rev. Sci. Catal. Eng., 28:13 (1986).
[3] M.J. Brown, N.D. Parkyns, Catal. Today, 8:305 (1991).
[4] N.D Parkyns, C.I. Warburton, J.D. Wilson, Catal. Today, 18:385 (1993).
[5] G. Busca, V. Lorenzelli, G. Ramis, R.J. Willey, Langmuir, 9:1492 (1993).
[6] E.R.S. Winter, J. Chem. Soc. (A), 2289 (1968).
[7] E.R.S. Winter, J. Chem. Soc. (A), 1832 (1969).

INDEX

Pt coated ceramic foam monoliths, 227
Pulsed studies, 3, 143

Radicals, 287, 288
Raman spectra, 25, 219, 322
Raman spectroscopy, 25, 219, 322
Reaction scheme for partial oxidation of methane
    to formaldehyde, 224
Reverse steam reforming, 237
   Of carbon, 229
Rhenium, 95
   On alumina, 95
   On HZSM-5, 95
   On MgO, 95
   On silica, 95
   On supports, 95
   On zeolites, 95
Rhodium, 59, 63
Ruthenium, 59, 63
   Copper-ruthenium, 63

Sb_2O_3, 341, 344
Scanning electron microscopy (SEM), 300
Silica, 340
Silica support, 179, 195, 207, 219, 220, 241
Silicone polymers, 294
Single Ion Monitoring (SIM), 302
SiO_2 support, (see silica support)
SnO_2, 219, 220
Sodium tungstate/silica, 45
Solid carbon C_s, 228
Solid carbon formation, 294
Spectroscopy, 327
Stainless steel, 265
Stannic oxide catalyst, 265, 267
Steam reforming, 305
Strontium fluoride, 22
   Oxide, 22, 74
Strontium/lanthanum oxide, 31
   Sulfated, 32, 34
Styrene derivatives, 322

Supported molybdenum oxide catalyst, 182
Supported vanadium oxide catalyst, 180
Surface reactions in n-butane oxidation, 235
Syngas, 293, 305
Synthesis gas, 292, 293, 305

Ta_2O_5, 344, 345
Tantalum oxide, 339
Temporal analysis of products, 113
Terephthaldehyde, 321, 324, 325, 326, 327, 330
Terminal V=O double bond, 224
Tetramethylorthosilicate, 324
TiO_2 support, 219, 220, 340
Toluene, 318, 330, 332
TPD, 322
Transmission electron micrographs (TEM), 187,
   337
Transmission electron microscopy, 187, 337
Transient Analysis of Products (TAP), 195
Transient studies, 113, 143
Trifluoroacetic acid, 297, 298, 300, 302
Trimethylbiphenylmethane, 330
Tungsten oxide, 339, 344

UV-visible diffuse reflectance spectra, 180

V=O terminal groups, 245
Vanadium oxide catalyst, 195, 196, 219, 241
Vanadium phosphate catalyst, 293

WO_3, 339, 344

X-ray absorption, 328
   Near edge spectroscopy (XANES), 337
X-ray diffraction (XRD), 24, 73, 80, 146, 195, 306,
   322, 332
X-ray photoelectron spectroscopy (XPS), 74, 82,
   195, 215, 242
XPS spectra, 180
XRD, 24, 73, 80, 146, 195, 306, 322, 332
Xylenes, 321, 322, 324